Axel Volmar, Olga Moskatova, Jan Distelmeyer (eds.)
Video Conferencing

Digital Society | Volume 53

Die E-Book-Ausgabe erscheint im Rahmen der »Open Library Medienwissenschaft 2023« im Open Access. Der Titel wurde dafür von deren Fachbeirat ausgewählt und ausgezeichnet. Die Open-Access-Bereitstellung erfolgt mit Mitteln der »Open Library Community Medienwissenschaft 2023«.

Die Formierung des Konsortiums wurde unterstützt durch das BMBF (Förderkennzeichen 16TOA002).

Die Open Library Community Medienwissenschaft 2023 ist ein Netzwerk wissenschaftlicher Bibliotheken zur Förderung von Open Access in den Sozial- und Geisteswissenschaften:

Vollsponsoren: Technische Universität Berlin / Universitätsbibliothek | Universitätsbibliothek der Humboldt-Universität zu Berlin | Staatsbibliothek zu Berlin – Preußischer Kulturbesitz | Universitätsbibliothek Bielefeld | Universitätsbibliothek Bochum | Universitäts- und Landesbibliothek Bonn | Technische Universität Braunschweig | Universitätsbibliothek Chemnitz | Universitäts- und Landesbibliothek Darmstadt | Sächsische Landesbibliothek, Staats- und Universitätsbibliothek Dresden (SLUB Dresden) | Universitätsbibliothek Duisburg-Essen | Universitäts- und Landesbibliothek Düsseldorf | Goethe-Universität Frankfurt am Main / Universitätsbibliothek | Universitätsbibliothek Freiberg | Albert-Ludwigs-Universität Freiburg / Universitätsbibliothek | Niedersächsische Staats- und Universitätsbibliothek Göttingen | Universitätsbibliothek der FernUniversität in Hagen | Staats- und Universitätsbibliothek Hamburg | Gottfried Wilhelm Leibniz Bibliothek - Niedersächsische Landesbibliothek | Technische Informationsbibliothek (TIB) Hannover | Karlsruher Institut für Technologie (KIT) | Universitätsbibliothek Kassel | Universität zu Köln, Universitäts- und Stadtbibliothek | Universitätsbibliothek Leipzig | Universitätsbibliothek Mannheim | Universitätsbibliothek Marburg | Ludwig-Maximilians-Universität München / Universitätsbibliothek | FH Münster | Bibliotheks- und Informationssystem (BIS) der Carl von Ossietzky Universität | Oldenburg | Universitätsbibliothek Siegen | Universitätsbibliothek Vechta | Universitätsbibliothek der Bauhaus-Universität Weimar | Zentralbibliothek Zürich | Zürcher Hochschule der Künste

Sponsoring Light: Universität der Künste Berlin, Universitätsbibliothek | Freie Universität Berlin | Hochschulbibliothek der Fachhochschule Bielefeld | Hochschule für Bildende Künste Braunschweig | Fachhochschule Dortmund, Hochschulbibliothek | Hochschule für Technik und Wirtschaft Dresden - Bibliothek | Hochschule Hannover - Bibliothek | Hochschule für Technik, Wirtschaft und Kultur Leipzig | Hochschule Mittweida, Hochschulbibliothek | Landesbibliothek Oldenburg | Akademie der bildenden Künste Wien, Universitätsbibliothek | Jade Hochschule Wilhelmshaven/Oldenburg/Elsfleth | ZHAW Zürcher Hochschule für Angewandte Wissenschaften, Hochschulbibliothek

Mikrosponsoring: Ostbayerische Technische Hochschule Amberg-Weiden | Deutsches Zentrum für Integrations- und Migrationsforschung (DeZIM) e.V. | Max Weber Stiftung – Deutsche Geisteswissenschaftliche Institute im Ausland | Evangelische Hochschule Dresden | Hochschule für Bildende Künste Dresden | Hochschule für Musik Carl Maria Weber Dresden Bibliothek | Filmmuseum Düsseldorf | Universitätsbibliothek Eichstätt-Ingolstadt | Bibliothek der Pädagogischen Hochschule Freiburg | Berufsakademie Sachsen | Bibliothek der Hochschule für Musik und Theater Hamburg | Hochschule Hamm-Lippstadt | Bibliothek der Hochschule für Musik, Theater und Medien Hannover | HS Fresenius gemGmbH | ZKM Zentrum für Kunst und Medien Karlsruhe | Hochschule für Grafik und Buchkunst Leipzig | Hochschule für Musik und Theater »Felix Mendelssohn Bartholdy« Leipzig, Bibliothek | Filmuniversität Babelsberg KONRAD WOLF - Universitätsbibliothek | Universitätsbibliothek Regensburg | THWS Technische Hochschule Würzburg-Schweinfurt | Hochschule Zittau/Görlitz, Hochschulbibliothek | Westsächsische Hochschule Zwickau | Palucca Hochschule für Tanz Dresden

Axel Volmar, Olga Moskatova, Jan Distelmeyer (eds.)

Video Conferencing

Infrastructures, Practices, Aesthetics

[transcript]

Funded by the Deutsche Forschungsgemeinschaft (DFG, German Research Foundation) – Project-ID 262513311 – SFB 1187 "Media of Cooperation".

Bibliographic information published by the Deutsche Nationalbibliothek
The Deutsche Nationalbibliothek lists this publication in the Deutsche Nationalbibliografie; detailed bibliographic data are available in the Internet at http://dnb.d-nb.de

First published in 2023 by transcript Verlag, Bielefeld

Cover layout: Maria Arndt, Bielefeld
Printed by: Majuskel Medienproduktion GmbH, Wetzlar
https://doi.org/10.14361/9783839462287
Print-ISBN: 978-3-8376-6228-3
PDF-ISBN: 978-3-8394-6228-7
EPUB-ISBN: 978-3-7328-6228-3
ISSN of series: 2702-8852
eISSN of series: 2702-8860

Printed on permanent acid-free text paper.

Contents

Video Conferencing: Infrastructures, Practices, Aesthetics

An Introduction

Axel Volmar, Olga Moskatova, and Jan Distelmeyer

In preparation for a private meeting in late 2022, one of us was invited to install the Discord app, which they had no previous knowledge of. Being used to Zoom as a default mode of video conferencing, they recall, they found interacting with less common applications felt disorienting. Discord certainly confused our colleague and provoked them to search for familiar operations, functions, and aesthetics. After two years of pandemic Zooming, video conferencing meant for them joining a session by clicking on a link or typing a meeting ID; looking at the grid of symmetrical tiles; smoothly switching camera and mic on and off; sharing screens; observing a list of participants on the right side of the screen; and occasionally chatting in a small window. Therefore, our coeditor's first approach to Discord was a comparative one as they quickly started to look for the typical "Zoom experience" on a new platform, but the differences were too apparent. Whereas Zoom's opening interface mainly affords the planning and coordinating of meetings ahead of time and thereby connotes formal modes of interaction (for instance, in professional contexts), Discord's main user interface resembles a blend of an instant messenger, such as Skype, and a social media platform, such as YouTube or Twitch, thus foregrounding more informal social interaction as well as the consumption of content. After starting Discord, the user finds herself in the middle of a chat interface. It is divided into an area dedicated to a phone list of *friends*, an indifferent and all-encompassing term for contacts typical for social media such as Facebook; an oblong bar for managing contacts; and an area for the actual chatting, which invites the user to join "popular communities"—channels dwelling on music, gaming, education, science and technology, or general entertainment—that seem to emulate the recommendation logics and patterns of interaction and valorization prevalent on many social media platforms.[1]

1 Popular content or content to explore is recommended, for example, on the home pages or opening interfaces of YouTube, Twitch, or TikTok ("For You"). Streaming platforms, such as Netflix, also rely on popularity as a means to personalize content. On the tensions between personalization and popularity, see Unternährer (2021).

Figures 1–2: Discord inferface with fixed chat channels on the left and activated video chat feature (upper image); group video feature, added in response to the pandemic as a complement to voice channels (lower image)

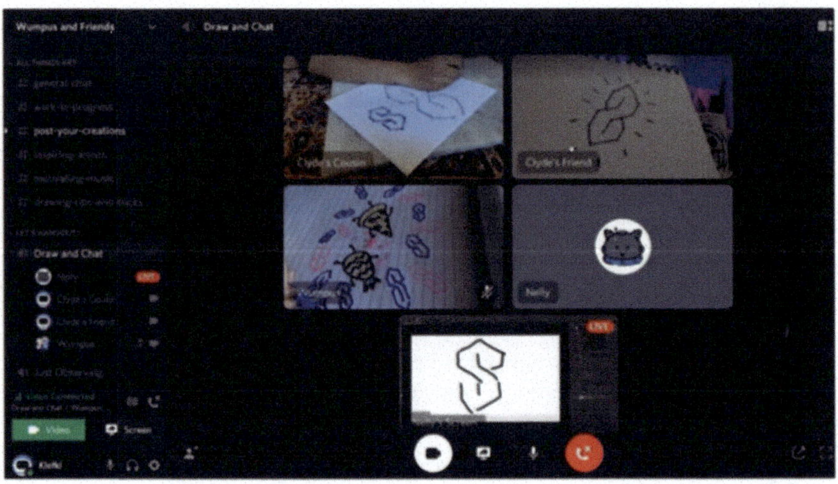

Sources: Engadget, https://www.engadget.com/2017-10-06-discord-video-chat-screen-share-rollout.html; Engadget, https://www.engadget.com/discord-adds-dropin-dropout-video-chats-120545406.html.

To start a video conversation, you do not join a preplanned "room," "meeting," or "session" but rather call a "friend"—spontaneously or after chatting with your contact.[2] After the connection is established, you can switch your camera on and start experimenting with screen sharing, chatting, or transferring documents via chat. Interestingly, the chat still occupies half of the screen; however, the caller can change the size of the video tiles, arranging them hierarchically (according to size) or symmetrically (according to size and spatial arrangement) or switching between full mode, pop-up, or image-in-image view. Unlike in Zoom, it is thus the chat area that is fixed, while the video stream comes on top of it and can be adopted individually, including actively blinding out the chat. The most confusing and defamiliarizing effect, however, certainly comes into play when trying to share the screen: instead of replacing the video tiles with the shared screen, Discord multiplies the video tiles, even resulting in a recursive and disorienting mise en abyme of tile-in-a-tile-in-a-tile, completely subverting the spatial clarity of Zoom.[3] This effect can be worsened when both callers start to share their screens simultaneously: each sharing adds a small video tile and then becomes repeated in the "shared tile."

An obvious cause of this defamiliarizing irritation is related to the equally global and incisive mainstreaming of video conferencing at the onset of the Covid-19 pandemic in early 2020. For sure, a majority of computer and smartphone users had been familiar with video chat applications—such as Skype, FaceTime, and Google Hangouts—for years. It was, however, the global pandemic and, more particularly, the various measures to fight the spread of the disease that significantly contributed to the dissemination of a set of video-based synchronous media practices that were hitherto far less common among the general population and that we subsume and address, in the following, under the term *video conferencing*. But what, then, constitutes video conferencing? In other words, what sets video *conferencing* software—such as Zoom, WebEx, Teams, Jitsi, and BigBlueButton—apart from video *chat* applications like Discord or Skype? While both types of applications share synchronous video link capability as a common mediatechnological denominator, they seem to diverge most strongly on the level of practice and, ultimately, in functionality, for they each cater to different use cases and are furthermore embedded in different practical contexts. Discord, for instance, emerged in 2015 as a messaging system for the global online gaming community and thus addresses

---

2 Discord also affords several other activities reminiscent of social media, such as exchanging emojis, GIFS, and stickers with friends or buying virtual "Nitros" (gifts, emojis, stickers, animation, etc.) for a month or a year—a form of monetarization reminiscent of virtual gifts used in live streaming on Twitch or TikTok.

3 Recursive image feedback can be prevented in Zoom by sharing the content of a particular application or window rather than transmitting the "screen" signal as such. Sharing the entire screen, however, may indeed result in the abovementioned "tile-in-a-tile-in-a-tile" effect.

people who want to get and stay in touch with other players or community members both during and between online sessions. Discord thus generally considers itself a chat or communications application that offers a range of different communicative channels, including video. This self-conception is underlined by their motto "Your place to talk," which clearly puts conversations at the center of the application. Conferencing software, in turn, generally serves to support goal-oriented group activities and hence formats, such as meetings, events, classes, and other scheduled encounters. The fact that conference applications mobilize the video feature for different purposes than instant messenging results in equally diverging configurations of basic functionalities.

First, in video conferencing applications, the video capability is usually embedded in interface arrangements that expect group settings rather than one-on-one (or one-on-some) conversations, which is why they aim to represent both individual speakers and the audience of participants (or some of them) on the screen, most commonly by means of the tile view. For this reason, video conference calls tend to feel more *public* than the comparatively *private* conversations. Second, video conference calls are initiated differently: rather than being spontaneous dial-up chats with "friends," conference calls are usually scheduled ahead of time and announced via invitation links attributed to an individual session or call. Therefore, video conferencing apps usually offer pre-meeting functionality that allow users to coordinate the planning of meetings, particularly the scheduling of the meeting and the invitation of participants by way of meeting links. Another notable difference here is that video chats are usually initiated by calling a person (or account), while video conference calls are tied to a specific (unique or recurring) time slot. Third, the video link between conference participants is usually not primarily used to facilitate communication as such (although communication is, of course, always a major part of video conference calls) but rather serves as a point of departure for subsequent collective activities, such as group coordination, decision making, and learning. Such goal-oriented group practices often involve the use of collaborative tools, both within the app (particularly screensharing, shared whiteboards, breakout rooms, or polls) and beyond (e.g., Google Docs or similar web-based office tools). On a general level, video conferencing can therefore be distinguished from video chat in the sense that it represents not a *communicative* but rather a *cooperative* technology (see also Volmar et al. 2023). The considerations drawn together in this book generally deal with such purposeful gatherings and the contexts in which they unfold.

Although video conferencing has only recently become a generally accepted form of gathering, it is important to note that the pandemic does not at all mark the beginning of video conferencing. As we will outline below, mediated social encounters based on audiovisual communication technologies have in fact been a possibility for decades. For the longest time, however, the use of video conferencing was largely limited to special use cases and remained fairly invisible and insignificant to the larger public. While Skyping was by all means an established alternative to making phone calls since the mid-2000s, multipoint video conferencing with an increasing number of participants became a widespread phenomenon of private and professional interaction only as a result of the unprecedented political situation that put about one-third of the global population under lockdown. The restrictions on physical encounters proved indeed to be decisive for the rapid normalization of video conferencing across different sectors. Although video conferencing software had been widely available already before the pandemic, remote technologies were largely rejected as a valid alternative to physical, on-site meetings. One of the main reasons for this seems to be related to the gravitational force of infrastructural configurations—that is, the particular ways that social practices were linked to physical resources and habituated forms of interaction. Prior to the pandemic, the experience of collective practices had been strongly shaped by the "infrastructural base" (Star and Ruhleder 1996) of face-to-face encounters. This particularly involved physical spaces especially designed to support group activities (such as offices, meeting rooms, classrooms, and gyms) on the one hand and a plephora of auxiliary practices attached to those activities (such as tendencies of superiors toward habitualized practices of social surveillance or informal practices of socializing, like chatting by the water cooler) on the other. Therefore, it is probably not an exaggeration to say that it took a pandemic to render remote forms of meeting and other collaborative group activities a part of everyday life and a new normal for millions of people within just a few weeks; social distancing measures disentangled people from the infrastructural base that had previously shaped the experience of group activities (see Volmar et al. 2023).

In the course of this perceptual and infrastructural shift in early 2020, the previously little-known video conferencing service Zoom gained so much in popularity that the name became anchored in common usage both as a generic term for video conferencing (the verb "to Zoom") and as part of new labels and neologisms for new video conferencing-related phenomena and experiences ("Zoom bombing" and "Zoom fatigue," for instance). The general perception that Zoom almost seemed to have emerged out of nowhere might have contributed to the public misimpression that video conferencing did not exist before 2020 (see Li et al. 2022). Zoom, however, had been in business since 2012 and even become a global leader in cloud-based video conferencing software by 2019. Zoom had long marketed their software toward early adopters in startups, small and midsize businesses, and enterprises who would use video conferencing to organize meetings with remote workforces or among employees based in different locations or to conduct online courses and webinars in continuing education or open universities. Zoom's growing success throughout the 2010s eventually prompted other providers, such as Cisco and Microsoft, to develop their own cloud-based video conferencing solutions.

Caused by such external necessities, then, the rise of video conferencing represents a rather strange mediatechnological shift in that it unfolded quite differently to prior cases of technological change. To a vast majority of people, video conferencing came not as a choice but as a mere necessity, directive, or workaround—in short, as a technological base they had to adapt to in an extremely short amount of time. The pandemic thus turned the everyday lives of billions into a kind of global experiment in the evolution of digital tools for remote interaction. The endless circulation of Zoom "fails," online resources about remote work, and video conferencing guides on social media, all of which accompanied the appropriation of video conferencing in the early days of the pandemic, evidences the fact that the mainstreaming of digital tools for remote interaction and our familiarization to them took place as a collective learning experience on a massive scale (see Volmar et al. 2023). As such, the boom in remote tools in general and video confencing in particular might be emblematic of a larger shift in the self-conceptualization of contemporary societies, especially in the so-called Global North, from societies guided by the promise of progress and projection to societies of mere reaction and adaptation in light of numerous crises—a process that quietly started at the beginning of the twenty-first century with the war on terror; formed more visibly under the imprint of the financial crisis, which gave rise to the infamous political doctrine of TINA ("There is no alternative"); and further intensified over the past few years due to the growing implications of climate change and the imminent effects of the global pandemic. In a similar vein, the discourse on video conferencing during the onset of the pandemic differed markedly in tone from, for instance, the one on the internet in the 1990s, which, for the most part, was rather playful, experimental, and optimistic. While Twitter conversations about Zoom and similar apps—which oscillated between amazement and bewilderment, curiosity and desperation, cries for help and offers of assistance—revealed manifold experiences with the novel medial situation of remote life, most of them nevertheless remained linked to highly practical contexts and quotidian routines that people tried to keep up by means of digital technologies.

People who switched to video conferencing were nevertheless not simply at the mercy of circumstances; after all, the collective learning experience that unfolded produced new knowledge bases with best practices, workarounds, and troubleshooting advice. Moreover, due to the pandemic situation, people creatively (mis)appropriated video conferencing for use practices previously never associated with it, such as remote yoga classes and dinners. Thus, while the pandemic was a crucial factor in the spread of video conferencing as an everyday medium, the emerging media-cultural situation also had an effect on video conferencing technology itself, which changed in the course of this process toward universalization as developers integrated new features based on user innovation. Discord, for instance, which also became an increasingly popular platform during the pandemic, used this spike in user numbers to emancipate itself from the thematic context it originated from by, among other things, changing its motto from "Chat for Gamers" to "Chat for Communities and Friends." Likewise, Zoom shifted its focus from corporate communication practices and the promise of offering "One consistent enterprise experience" (before March 2020) to a more general user base and slogan, claiming that we are all "In this together" (in March 2020).

Taking stock of these developments, the authors in this book understand video conferencing as a media-cultural formation that has been largely shaped by the effects of the global pandemic. More particularly, what video conferencing constitutes today has been largely determined by a collective experience as well as processes of mutual adaptation: in the same way that video conferencing changed quotidian practices of meeting and collaborating, the mass appropriation and misappropriation of the technology—to fit numerous use cases it had not exactly been designed for—changed the functionality and appearance of the software and the providers' descriptions and understanding of their products. In this respect, video conferencing seems to be a particularly good example for the argument put forth by German media scholar Erhard Schüttpelz that all media are in fact "media of cooperation" (Schüttpelz 2017, 24) in that they are being used not merely to consume content but to organize everyday practices and support individual and collective goals. While the pandemic situation thus prompted what Volmar et al. (2023, 99–100) call "a general socio-technical process of *re-infrastructuring* disrupted ecologies of everyday practices" by way of video conferencing tools—a process that proved to be an exhausting experience for many—it nevertheless resulted in a number of fairly consolidated cultural forms and media practices that most people now deem to be video conferencing. This rapid normalization of expectations and usage habits was not least caused by the market dominance of a small number of individual providers, most prominently Zoom, Cisco WebEx, Microsoft Teams, and Google Hangouts. As users became particularly habituated to the functionality, workflows, and forms of interaction conceived for business communication and continuous learning, they quickly got acquainted to the aesthetics—shaped by, for instance, the presence of the notorious image tiles—as well. Put differently, the resulting habitualized media practices, which now form the nucleus of video conferencing culture, seem to have been shaped to no small degree by a particular confluence of video conferencing applications previously tailored to the business world and of everyday practices of joint action, a conjunction that produced its own formats, subversive practices, and cultural forms.

This contingent and rather surprising formation of video conferencing as a set of widely used media practices calls for scholary investigations that take stock of its specificities in greater detail and interrogate its particular media-historical moment. *Video Conferencing: Infrastructures, Practices, Aesthetics* thus takes the current situation as a starting point for assessing the complex mediality of this new form of distributed social interaction. Linking theoretical reflection to material case studies, the contributors to this volume question video conferencing and the specific meanings it acquires in different social, cultural, and historical contexts. Together, the volume's contributions, most of which stem from media studies and neighboring disciplines, expand the scope of examination beyond the contexts and experiences of the global pandemic—for instance, by connecting them to prior forms and deeper histories of audiovisual communication and remote interaction. Before we discuss the structure of the volume, we would therefore like to provide some historical context regarding the media history of video conferencing and its history as an object of scholarly research to situate the positions presented in this volume within a longer media history of *visual (tele)communications* technologies.

A Brief History of Video Conferencing

Figure 3 a–b: Two-way television booth at AT&T 195 Broadway, 1930 (left); scematic of an early television demonstration over radio and telephone circuits, 1927 (right)

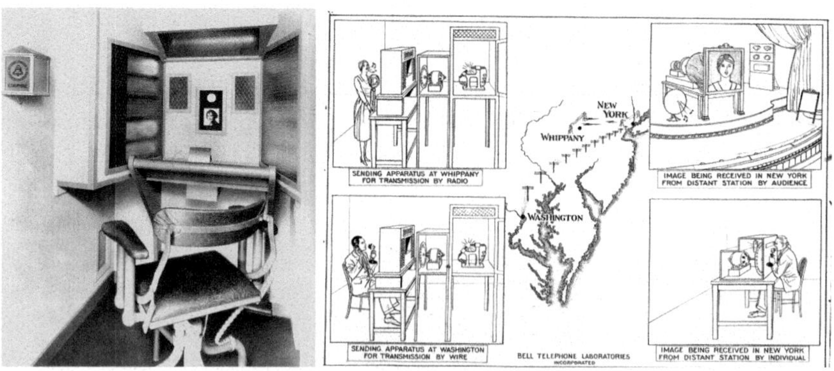

Sources: Courtesy of AT&T Archives and History Center.

As stated above, the Covid-19 pandemic by no means represented the beginning of video conferencing but rather marked the starting point for the formation of what could be termed a global video conferencing culture. The history of video conferencing—as a technology, a practice, and a discourse—is, however, much older than even the memory of Skype conversations from the mid-2000s might suggest. The technological foundation used for video conferencing—the creation and transmission of electronic audio and video signals as a means of telecommunication—is basically as old as television. During its experimental phase in the 1920s and 1930s, the direction in which the technology of television would develop as a public medium remained largely undecided: contemporaries saw both potential for television as a programmed broadcast medium in the form of radio with added moving imagery or as a telecommunication medium modeled after the telephone but enhanced by a televisual channel (the term *tele-vision* not coincidentally reminiscent of the term *telephone*). Unsurprisingly, research on television at Bell Telephone Laboratories also involved experiments in what was termed "two-way television," which consisted of two interconnected camera (recording) and display (reproduction) systems located in different places.

Figure 4 a–b: Camera setup behind the booth (left) and booth of the German Fernsehsprechdienst in the late 1930s (right)

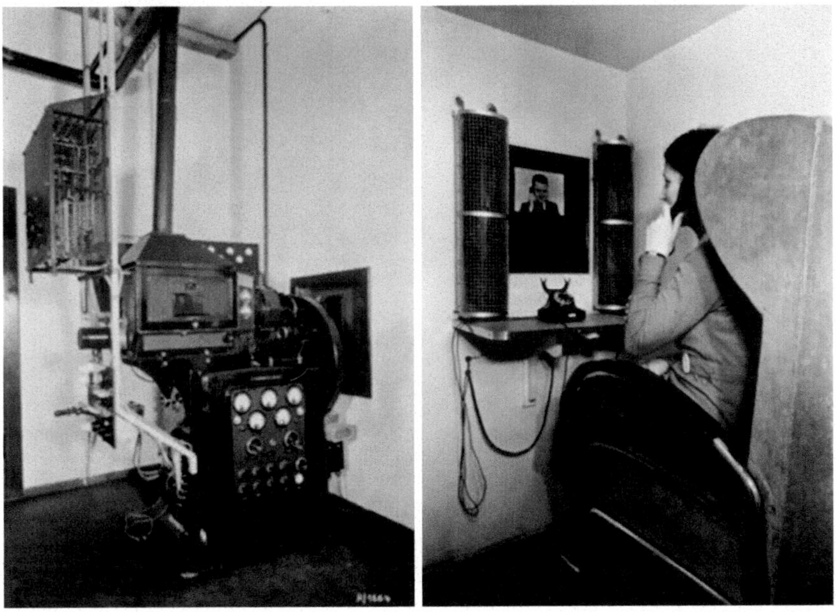

Sources: German Federal Archives.

Between 1936 and 1940, the German Reichspost even developed a national public visual telephone network called *Fernsehsprechdienst* (literally meaning "television-phone"). The service consisted of connection points (*Fernsehsprechstellen*) that were similar in technology and design to the two-way television booths devised at Bell Labs and that were located at central postoffices and public places in a number of larger cities all over Germany (among others, Berlin, Leipzig, Nuremberg, Frankfurt, Munich, and Hamburg) and connected via newly designed and laid broadband cables (see Goebel 1953). While serving as an attraction during the 1936 Olympic

Games and other large-scale events, such as fairs, to boast the German Reich's know-how in electrical and communications engineering, the regular service was rarely used and never proved commercially successful. The further expansion of the network was abandoned in 1940 after the system was considered not essential for the war, and it was never taken up again.

Figure 5 a–b: Images of the AT&T Picturephone Mod-II with accessory pieces, 1970 (left); and as a display device for computer output, 1970 (right)

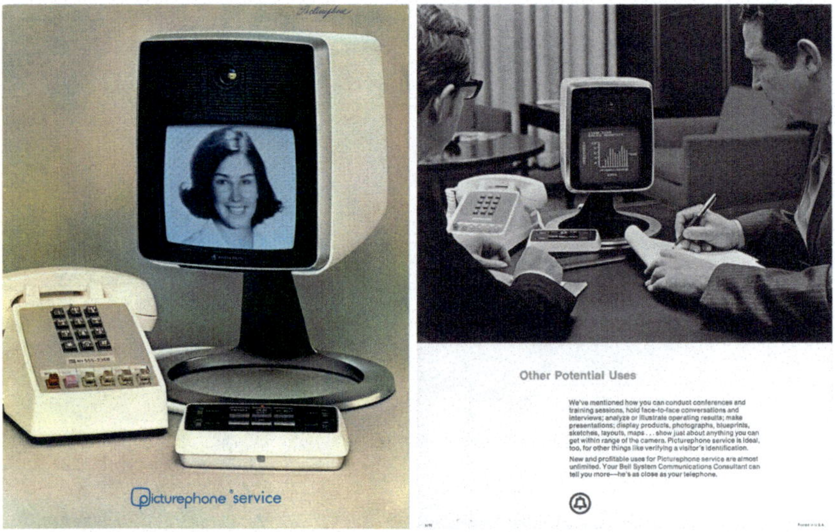

Sources: Courtesy of AT&T Archives and History Center.

In the 1950s, however, communications engineers all over the world picked up the thread again and conceived videophones on the basis of the newly invented transistor technology, which made it possible to squeeze the recording and display systems into single devices. AT&T, for instance, aimed at establishing visual telephony under the trademark Picturephone service as the next big step in the history of telecommunications and invested about half a billion US dollars into research and development of visual communications. After presenting Picturephone to the public in 1964 at the world's fair in New York and in Disneyland as yet another booth-based service, AT&T launched a Picturephone subscriber service using desktop devices in 1970—first in the local network of Pittsburgh and later in Chicago (see Lipartito 2003; Mills 2012; Dietrich 2020). Among other things, the Picturephone "Mod-II" offered to hold video conference calls with more than two participants through automatic image switching by means of voice detection. It also featured "graphics capability" for sharing documents and slides by means of an extractable mirror, which pointed the camera downward by 90 degrees to capture the tabletop. It even offered the possibility to use the device as a computer terminal and display system (see figure; note that the push-button telephone, introduced by AT&T in 1963, served as an input or terminal device to enter commands and alphanumerical information/data).

While the AT&T Picturephone probably represents the most iconic example of analog video telephones, similar services were conceived, tested, or marketed in many countries around the globe, including Great Britain, France, Germany, Sweden, Switzerland, the Soviet Union, Japan, and the Philippines. Technologically feasible yet not publicly available or accepted, audiovisual forms of telecommunication also became part of a broader vision of and discourse on the technological future and the cultural imaginary of the space age—for instance, through the representation of videophones in popular culture, such as in the animated series *The Jetsons*, and not least thanks to AT&T's marketing department, which prominently product-placed their Picturephone in Hollywood motion pictures, such as *2001: A Space Odyssey* (1968) and *Blade Runner* (1982). Despite the hype, however, none of the videophone services proved successful in terms of either revenue or substantial user figures. Judged solely by numbers, Picturephone turned out to be a huge economic failure and did not even come close to becoming the envisioned future of telecommunications (Noll 1992; Lipartito 2003; Schnaars and Wymbs 2004).

In the mid-1970s, due to a lack of interest from the general subscriber (mostly because of the exorbitant cost of the service), AT&T geared their efforts from televised phone conversations toward group-based meeting solutions in business contexts. After several years of testing, the Picturephone Meeting Service, with the rather unfortunate initialism PMS, was first launched in 1982 (*New York Times* 1982; Wright 1983; Menist and Wright 1984). It consisted of a network of specially equipped and interconnected conference rooms in 12 major US cities. Even earlier in the 1970s, the Japanese Electronic Corporation (NEC) and, shortly after, British Telecom (BT) had already introduced group-oriented video conferencing systems based on analog television technology (Wilcox 2000, 4). Several other telecommunications companies, particularly in Europe, would also start to build video conference rooms and experiment with transnational video calls in the 1980s. Due to the equipment costs and high bandwidth requirements, however, video conferencing rooms remained a niche application, the users of which were enterprise customers and, significantly, the telecommunications companies themselves.

From the 1970s onwards, large corporations, among them Procter & Gamble, IBM, and Boeing, also began to establish private video conferencing systems (Johansen and Bullen 1984, 164), thereby giving rise to a new industry dedicated to both analog and, increasingly, digital video conferencing technology. Compression Labs, Incorporated, for instance, introduced the first commercial digital group video conferencing system in 1982. The CLI T1 was designed to enable video communications over leased-line T1 circuits at 1.544 Mbps. Despite the high costs (ca. $250,000 for the codec device alone plus about $1,000 per hour line costs), the prospects of digital video conferencing created incentives for new development in digital image and video compression. PictureTel Corporation, for instance, was founded on data compression research that combined transform coding of digital images with interframe motion compensation.[4] The technology would later become a substantial element of the MPEG video compression standards, which shows that video conferencing represented a major driver of research into digital video coding.

Figure 6: A promotional image for AT&T's Picturephone Meeting Service

Source: Courtesy of AT&T Archives and History Center.

PictureTel's digital codecs did not only result in a lower price range for video conferencing ($80,000 for the codec hardware and $100 per hour line costs) but also led to a generally better compatibility of video communications with dial-up data networks, such as the Integrated Services Digital Network (ISDN). By the end of the

4 Motion compensation means that rather than analyzing and coding each image frame separately, PictureTel's advanced video codecs, such as the C-2000 algorithm from 1986, considered only the alterations—that is, the motion—of successing frames relative to reference frames, a method that proved to be significantly more efficient than existing compression techniques.

1980s, more than 70 percent of the digital video conference systems in use throughout the world were PictureTel systems. In the 1990s and 2000s, companies such as Polycom, Hewlett-Packard, Tandberg, and Cisco would follow. The growing choice of competing systems moved the question of interoperability into focus, ultimately leading to a number of industry-wide standards developed by the ITU-T,[5] first and foremost the H.320 standard (1990), an umbrella recommendation for transmitting multimedia content (i.e., audio, video, and data) over ISDN networks, and the H.261 (1988) and H.263 (1996) video compression standards devised for video coding at low bit rates (see also Wilcox 2000, 119–147). These standards also laid the foundations for the subsequent family of MPEG standards, whose video compression codecs were most prominently used to store video content on DVDs. As the common standards allowed owners of systems from different vendors to dial one another, they were particularly important to increase the chances for a wider adoption of video conferencing (although this never really happened). With increasing computing power and broadband internet connections in the mid-2000s, video conferencing studio systems were gradually supplemented by mobile units—so-called rollabouts—that could be moved between regular meeting rooms, as well as by native video chat applications, such as Skype and later FaceTime, that were based entirely on general purpose technologies, which means that no further hardware was necessary.

To grasp the meaning of video conferencing not only with respect to its technological base (of digital image transmission) but also on the level of user practice, it is, however, necessary to consider a different genealogy of technologically mediated conferencing. It seems noteworthy that for the longest time, this parallel and at times overlapping historical strand of video conferencing did not involve the transmission of moving images at all. Rather, it was centered on the evolution of audio conference calls and the question of how to share documents and other visual content, such as slides and graphs, between conference participants. First attempts to share documents on different computers or display systems connected via local digital networks in real time date back well into the 1960s. Most notably, the intranet-based system PLATO (short for Programmed Logic for Automated Teaching Operations) developed primarily at the University of Illinois's Computer-Based Education Research Laboratory is regarded as an early experiment in distributed instruction and thus a predecessor of what are now known as *webinars*. With the dissemination of internet access and personal computers equipped with operating systems featuring a graphical user interface, the 1990s saw a boom in extended forms of document sharing, first within networked desktop applications, such as Lotus Notes

---

5 The ITU-T is the Telecommunication Standardization Sector within the International Telecommunication Union.

(first released in 1989), and second, through so-called net conferencing or "audio-graphic conferencing" applications (Wilcox 2000, 149–164), such as Microsoft's NetMeeting, Intel's ProShare, PictureTel's LiveShare Plus, and PlaceWare's Auditorium (which came out of Xerox PARC), all of which combined the possibility of hosting audio calls via the software rather than via telephone with application and document sharing capability and other collaborative and communicative tools, such as whiteboard, note pad, and chat functionality. Some of these applications, such as ProShare, later featured video capability too.

Figure 7: Astronaut Marsha Ivins holding a teleconference with student participants in the KIDSAT program using Intel ProShare, the computer screen shows an image of the participants on Earth as well as Ivins's camera image and data of the KIDSAT experiment (photo taken February 27, 1997)

Source: National Aeronautics and Space Administration. Lyndon B. Johnson Space Center.

With the growing popularity of the World Wide Web and graphical browsers in the late 1990s, companies like WebEx moved net conferencing functionality to browser-based conference tools, which were soon termed *web conferencing*. In the early 2000s, WebEx Meeting Center (WebEx), GoToMeeting (Citrix Systems), and Adobe Connect (Adobe) became widely used products for web-based conferencing and webinars. As mentioned above, video support was usually not part of web conferencing before the advent of broadband internet access, when video capabilities were successively added to the existing products. But while video-based conferences and webinars became a technological possibility, the quality of the video transmission turned out to be notoriously unreliable. With its high requirements in terms of transmission bandwidth, connection stability, and signal latency, multipoint video conferencing with high numbers of participants was clearly at odds with the logics of packet-switched networks and the numerous contingencies in terms of hardware, software, and network configurations at the different endpoints. In the early 2010s, Eric Yuan, the founder of Zoom Video Communications, turned to the then-new capabilities of cloud computing to overcome the persistent technical difficulties and thereby gave web conferencing a substantial makeover—with considerable success. Although version 1.0 of the Zoom client, which was released in January 2013, allowed users to host video calls with up to only 25 participants, it was yet able to attract more than a million users within just a few months. As both software and user base grew in the consecutive years, competitors, such as Cisco WebEx, followed suit and developed their own cloud-based video conferencing solutions. By 2019, the year of its initial public offering, Zoom Communications—although largely unnoticed by the general public—had become a global leader of video conferencing software within the business and distance education sectors, which is one of the main reasons, next to its technological edge, that it ended up becoming the go-to application for remote meetings during the global Covid-19 pandemic.

Video Conferencing as a Research Object

In the five decades preceding the pandemic, the diverse manifestations of visual communications and mediated conferencing solutions prompted scientific investigations into video communication as well. It is interesting that, during those years, researchers looked into many of the questions that came up in light of the unfolding pandemic. In the early 1970s, in large part due to the oil crisis and an emerging environmentalist movement, telecommunications research aimed, for instance, to assess the potential impacts of a future adoption of video communication on business travel, commuting, and more generally energy consumption and environmental pollution, such as by contrasting the cost of video conferencing with the cost of travel (Nilles, Carlson, and Gray 1976; Gold 1979). Other studies attempted to estimate the potential range of application of video conferencing, primarily by comparing, differentiating, and rating mediated and nonmediated forms of communication. This line of research actually became the dominant vector of early video conferencing research, as, for instance, a review article from 1984 stated: "Most researchers concentrated their efforts on empirical investigations of the effect of channel type (audio, audio-video or face-to-face) upon meeting outcomes and user attitudes" (Albertson 1984, 394) to determine which types of conferences and tasks might be most effectively shifted to video in the future (see also Williams 1977, 964).

One argument commonly voiced was that the efficacy of a communications technology increased with the amount of "bandwidth"—that is, "communicative channels," such as text, audio, and video—offered to users (Ryan and Craig 1975, 2). Others emphasized the significance of nonverbal communication and concluded that the relative lack thereof in mediated forms of communication would render establishing relationships, treating sensitive topics, and even communication in general more difficult than in face-to-face situations (Kendon 1967; Sacks, Schegloff, and Jefferson 1974; Argyle, Lalljee, and Cook 1968). While not entirely false, these rather general views nevertheless displayed considerable shortcomings as they failed, for instance, to account for task-specific efficiency within a given technology. Moreover, they were unable to explain the high acceptance and efficiency scores of "low-bandwidth," phone-based conference calls. Social psychologists John Short, Ederyn Williams, and Bruce Christie from the University of London therefore sought to find a better explanation. In their 1976 study, *The Social Psychology of Telecommunications*, the researchers proposed the concept of "social presence" to examine and understand technologically mediated communication. According to their theory, the social presence of a given medium is determined first by the *objective* features of that medium—"[qualities] of the communications medium"—and second by *subjective* features resulting from users' perceptions and opinions or attributions regarding that medium—the "perceptual or attitudinal dimension of the user, a 'mental set' towards the medium" (Short, Williams, and Christie 1976, 65). The researchers explained that they thus conceived of social presence "not as an objective quality of the medium, though it must surely be dependent upon the medium's objective qualities, but as a subjective quality of the medium" (66).[6] A follow-up study by Rutter et al. (1981) regarded social presence to be determined by "cuelessness"—that is, the lack of social cues, within a certain conversational setting: "The smaller the aggregate number of available social cues from whatever source—visual communication, physical presence or, indeed, any other—the more task oriented and depersonalized the content, and the less spontaneous the style" (48). According to this understanding, then, "social presence is underpinned by cuelessness. The more cueless a medium, the less its social presence" (49). Another comparative approach, which ventured in a similar direction, proposed the concept of "information richness," or "media richness," as a way to rate different forms of communication and, more concretely, to identify potential fields of application for video conferencing: "Richness is defined as the potential information-carrying capacity of data" (Daft and Lengel 1984, 196). Subsequent research modified and updated the model of media richness (see, for instance, Dennis and Valacich 1999).

---

6 The subjective assessments of media technologies and hence the social presence of different means of interpersonal communication were usually determined by means of questionnaires based on the method of the "semantic differential."

Though offering new terminologies and the consideration of subjective attitudes toward technological settings, the findings of the comparative approaches largely reproduced the results of the earlier 1970s studies focused on communicative bandwidth, which favorably positioned video conferencing closely to face-to-face communication. This proved to be problematic given the fact that none of the developed theoretical concepts—whether social presence, cuelessness, or information richness—provided explanations for the then-notorious rejection of visual communications technologies by users. As an immediate effect of this lack of uptake, research on video conferencing generally declined in the 1980s. A growing number of studies, however, also started to directly address this general disinterest in "teleconferencing" (as it came to be called at the time) and thus took the paradox of teleconferencing as a starting point for conceptualizing mediated conferencing along new lines. Johansen and Bullen (1984), for instance, expressed doubts that video conferencing could in fact replace face-to-face-meetings. Birell and Young (1984) even called into question "the desire to replicate the face-to-face meeting. We should be considering more deeply whether the face-to-face model is really so very valid" (286). The puzzlement over the facts that "teleconferencing expectations in general have failed to realize themselves fully despite consistently brilliant market forecasts" (Egido 1990, 351) and that video conferencing continued to remain a "technology on the fringe" (Mayes and Foubister 1996a, 163; see also 1996b) were echoed in the literature well into the 1990s and 2000s.

As a response to this situation, new approaches came to the fore, which suggested discarding comparative methodologies of assessing different technologies in general in favor of more detailed microanalyses of actual teleconferencing situations. As one researcher put it, "In order to understand the impact of mediated communication on this intersubjective process more fully, research is needed which focuses on the interaction itself rather than on task effectiveness, user attitudes, or simple objective measures of communicative differences" (Hiemstra 1982, 883). Psychologists, for instance, approached this question by measuring the influence of individual parameters—such as image resolution, size, and refresh rate—on speech comprehension and the ability to decode emotional cues (see, for instance, Wallbott 1992; Blokland and Anderson 1998; Barber and Laws 1994). Sociologists, linguists, and computer scientists appropriated conversational analysis (see Sacks, Schegloff, and Jefferson 1974), an approach to studying pragmatic language rooted in Harold Garfinkel's concept of ethnomethdology, to examine conversations via teleconferencing technologies.

On a methological level, researchers made use of new technological possibilities of creating and storing video-based research data. Périn (1983), for instance, proposed using video recordings in conjunction with detailed transcriptions of teleconferencing meetings to study the basic rules and pragmatic strategies of video-based communication, including turn-taking sequences between speakers, the use of the gaze, and other verbal and nonverbal cues. In a similar vein, Cohen (1984) pursued questions regarding how video conferencing results in altered perceptual conditions, which in turn influence the fundamental organization of interpersonal communication (e.g., with respect to turn-taking patterns, turn length, or disruptions of the temporal coordination of communicative activities). By focusing on the specifics of teleconferencing interactions, Cohen and others were able to pinpoint some of the major issues with video-mediated communication. Interestingly, a lot of these issues are still very much part of our video conferencing experience today, most notably transmission delay, which "disrupts the pace of normal conversations, makes the appropriate timing of interruptions more difficult, and impedes the smooth resolution of simultaneous speech events" (Cohen 1984, 292). These approaches were further advanced in the 1990s by, for instance, Abigail Sellen in her work on speech patterns in video-mediated conversations (see Gaver 1992; Sellen 1992; Heath and Luff 1993; O'Conaill, Whittaker, and Wilbur 1993). At the same time, researchers also refined their methodological toolkit, not least by conceiving elaborate transcription methods (see O'Conaill and Whittaker 1995; O'Malley et al. 1996; Ruhleder and Jordan 2001). Apart from this, video conferencing research also branched into studying various fields of application, such as business communication (see Köhler 1993; Schulte 1993; Kydd 1994), education (see Storck and Sproull 1995; Kawalek 1997; Schütze 2000), and medicine (see Guckelberger 1995; Armoni 2000). This line of video conferencing research based on ethnomethodology and conversational analysis is still very much alive today (see Due and Licoppe 2020).

With the gradual advancements of personal computing and digital video compression in the 1980s and early 1990s, video conferencing research too went digital. This technological shift was most notably accompanied by a change of perception, which now associated video conferencing more closely with the domain of computing than with the telecommunication sector but equally with an extension of disciplinary perspectives. For instance, scholars who had worked on computer-mediated communication (CMC), a field that included the study of such new communicative forms as newsgroups, bulletin boards, and email, increasingly became interested in video-based forms of communication and interaction. A lot of computer science research was carried out within the Association for Computing Machinery (ACM) in research fields like computer-human interaction (HCI) and computer-supported cooperative work (CSCW) (see Furuta and Neuwirth 1994). In this line of research, computer scientists not only studied existing video conferencing solutions but aimed to overcome some of the identified deficits of "talking heads" video conferencing (Nardi et al. 1993) by experimenting with new digital interfaces (see, for instance, the contributions in Finn, Sellen, and Wilbur 1997).

Drawing on their extensive experience with digital technologies, Paul Dourish et al. (1996) argued that while conversational analysis had deepened the understanding of how technological mediation influenced conversations and interactions between individual speakers, it proved not to be very suitable for understanding (or getting into view) what people actually did within "media spaces" (see Stults 1986; Gaver 1992; Heath and Luff 1992; Bly, Harrison, and Irwin 1993)—that is, "flexible, networked, multimedia computer environments" designed to support cooperative work (Dourish et al. 1996, 33). In addition to studying face-to-face conversations in mediated environments, Dourish et al. suggested focusing on the "emerging communicative practices" (33) that coevolve over time when people and their specific work practices get in contact with networked media environments and studying these practices "in real, long-term use" (34). In other words, Dourish et al. stressed that rather than center the transfer or mediatization of "face-to-face behaviours," it seemed necessary to consider the specific circumstances, goals, and purposeful, group-centered activities that inform particular conversations and bring the people involved together in the first place (33). Rather than seeing mediated interaction as potentially inferior or less real than immediate face-to-face interaction, they insist that "the media space world *is* the real world; it is a place where real people, in real working relationships, engage in real interactions" (59)—and that taking the peculiarities of these media worlds more seriously is thus merited.

Dourish and his collaborators' insistence on considering the formation of media-specific behaviors and practices through habituation and as part of a "community of practice" (Lave and Wenger 1991) and on developing a praxeological perspective corresponds well with our own endeavor to study what we have termed *video conferencing culture* above. But it also reveals yet another gap in previous video conferencing research—namely, the fact that much work from the past five decades has focused on the here and now of video conferencing, with regard to either mechanisms of video-mediated conversations or cooperative work practices. Most studies treat users of video conferencing technologies as subjects without histories and contexts and do not consider the cultural aspects of video conferencing, such as discursive formations, sedimented practices, social norms, and political underpinnings.

Until the pandemic, and apart from a few notable exceptions (see, for instance, Otto 2013; Longhurst 2016), little work was done on cultural forms and meaning in conjunction with video telephony and video conferencing as media practices. This book represents a first step toward providing such contextualizations. To do so, zooming in on video conferencing demands us to zoom out to open the view on the background of video conferencing and its "infrastructural extensions" (Tasman 2015). Therefore, *Video Conferencing: Infrastructures, Practices, Aesthetics* asks, from a media studies perspective, what constitutes video conferencing as a media-cultural phenomenon and a constellation of particular technologies and online practices. How can we allow for the pluralization of uses and users of video conferencing due to the global Covid-19 pandemic? How can we contextualize video conferencing with respect to infrastructural conditions, use practices, and peculiar aesthetics, and what are the politics involved in video-mediated forms of remote interaction and cooperation? To put it differently, how can we take into account what lies in the background of our common experiences and everyday practice with video conferencing applications?

The Mediality of Video Conferencing

Given the rich research history outlined above, it is surprising that video conferencing is not a very well-established research object within media studies. Certainly, one reason for this is that video conferencing remained a comparatively marginal medium until the global pandemic. Our collection thus aims to fill this gap by bringing together contributions that examine the phenomenon of video conferencing from the perspective of media studies. More particularly, the volume seeks to assess video conferencing through the lens of three interrelated foci—infrastructures, practices, and aesthetics—that we take as the main aspects for delineating video conferencing's mediality. Since the 1990s, the concept of mediality has been used in media studies to stimulate the debate on the specific qualities of manifold forms, processes, conditions, and consequences of mediation. Less focused on fixed entities that then defined a *medium*, the concept of mediality aims at broader and specifically processual questions of mediation as something that is occurring, not easily grasped, and undergoing change; it is rather concerned with the processes of "becoming-media" (Vogl 2007). To that effect, it can refer to specific media forms, phenomena, and practices as well as to media in general. The "'mediality' of media," as Ulrike Bergermann puts it, "refers doubly to respective concrete individual media (formats, contents), and also to 'the media' and what they might have in common" (2016, 435). In each case, mediality is used as a concept "to call attention to what media do, to the ways in which they function as agents" (Grusin 2010, 72), so that processuality and productivity come to the fore.

The question of mediality thus responds to an inescapable conditionality: what is mediated cannot be detached from the processes of mediation, just as in speech the voice as medium always already leaves its trace (see Krämer 2015, 27–37). Against this backdrop that "there can be no neutral instance of mediation, as the medium in question will itself always shape the procedures and results of the mediation process" (Distelmeyer 2022, 51), research on mediality traces constitutive characteristics. Mediality, Sybille Krämer (2021, 88) sums up, "is to be understood as a form of generating relationality." This does not, however, determine which form of relationality is taken into consideration. Jonathan Sterne, for instance, uses "the term *mediality* (and *mediatic* in adjectival form) to evoke a quality of or pertaining to media and the complex ways in which communication technologies refer to one another in form or content" (2012, 9). But even this in other words *intermedial* understanding and focus on communication aims at a complex processuality for which different aspects and their interactions have to be taken into account—or, more concretely, "its [the medium's] articulation with particular practices, ways of doing things, institutions, and even in some cases belief systems" (10). Materiality and technologies belong to it just as much as practices and concepts as well as elementary, social, political, economic, and ecological conditions and effects.

Researching the mediality of video conferencing therefore poses a particular challenge: How can the complexity of conditions, processes, and effects be addressed and questioned? Given that the relations and interactions of human and more-than-human agencies at issue here depend on platform structures whose conditions and processes are anything but readily visible, such an undertaking poses some challenges. To examine the phenomenon of video conferencing from the perspective of media studies therefore means asking which conditions and infrastructures are at work, what kind of processes and practices appear in respect to certain technologies, and which aesthetics show up and yet allow only a part of what is effective in the process to become apparent. Therefore, we deem these questions—about infrastructures, practices, and aesthetics—essential to discussing not only the current phenomenon of video conferencing but also its history, its fundamental characteristics, and its further implications.

The concept of infrastructures specifies, especially in relation to materials and technologies, the question of conditions. Infrastructures enable and condition practices and aesthetics and are at the same time interrelated—as illustrated, for example, by the success story of Zoom in the early months of the pandemic, when increasing user numbers could be handled only by responding with infrastructural changes, as Amazon Web Services (AWS) added the performance of thousands of servers daily. Thus, especially for the phenomenon of video conferencing, the concept of infrastructure, as Lisa Parks and Nicole Starosielski have pointed out, must be understood as a dynamic category that provokes questions about "processes of distribution" and its "unique materialities" as well as "the relation between technological literacies and public involvement in infrastructure development, regulation, and use" (2015, 5). Infrastructures are not simply givens; they are in operation, are maintained and serviced, consume resources, require human and more-than-human agencies, include some and exclude others, and are usually in a state of constant flux. As Parks and Starosielski show, the interest in infrastructures challenges us "to recognize a more extensive field of actants and relations in media and communication studies" (10). In terms of video conferencing, this includes not only the software architectures of the respective services, their workers, and their servers as well as those of the third parties that provide additional support and computing power, like AWS and Oracle; it also includes the infrastructures of the internet and cloud computing—from the running protocols to the submarine cables that need to be maintained, from servers, cell towers, and air interfaces to the computers with which we ultimately carry out our video conferencing practices in those disparate spaces hidden behind the unifying term *home office*.

Although infrastructures foreground material and technological conditions, they are also closely related to practices and uses. Practices take place and become stabilized in infrastructured situations through repetition and routine. Infrastructural materialities not only enable and restrict specific forms of practice but are also transformed within and by those practices. And—as the example of video conferencing in particular shows—the infrastructures themselves also imply and rely on practices of both human (e.g., maintenance) and more-than-human (e.g., processing) agencies. The focus on practices thus emphasizes the relations users have with technologies and media. According to Nick Couldry, to ask about media practices is to ask about "what people ... are doing with media"—that is, how individual media users process and circulate meaning in everyday media practices (Couldry 2012, 6–9). Moreover, focusing on practices invites us to study "how diverse forms of *work* and *cooperation*—between different actors both human and non-human—are being constituted, stabilized, governed, and changed by and with media technologies" (Volmar 2017, 11) or, more generally, what people do with media and what media do for, to, and with people in a specific socio-historical context (see Dang-Anh et al. 2017, 7). Madeleine Akrich and Bruno Latour (1992, 259–262) have suggested the term *script* and its processual variations to describe this mutual conditionality of media and uses: technologies are conceived and implemented with an idea of specific uses and thus are designed with a script in mind. Scripts have a *prescriptive* dimension inasmuch as they delimit the potential range of actions, whereas users have to *subscribe* to these allowances and affordances or reject them—that is, develop a *de-inscriptive* stance toward designed uses (261). Moreover, scripts come along with *pre-inscriptions* (i.e., expectations on the abilities and competencies users must have to handle a specific technology) and *ascriptions* (i.e., ideas about the source of agencies, specific activity, and decision while using technologies) (261–262). Thus, the focus on different ways of dealing with scripts underlines the relationality of media and practices.

The attention to practices of different types and actors is also reflected in the gerund *video conferencing*, which gives this volume its title. The aim of this anthology is therefore not a definitional but an analytical one: it is a matter not of classifying "the video conference" as "a medium" but of exploring the mediality of that multifaceted and profoundly processual phenomenon of video conferencing. Thus, to focus on practices can mean to analyze the processes of conferencing and collaborating across spatial distances; processes of social and temporal synchronization, such as chatting and sharing screens; and practices of social and aesthetic (self-)regulation and (self-)expression, such as talking and muting. Practices of video conferencing not only have a spatial and temporal dimension but also imply forms of embodiment and ways of being located in front of a screen in domestic or professional spaces, ways of interacting with interfaces and hardware, and the diversity of embodiments supported or prevented by the respective media settings and infrastructural conditions.

Similarly, aesthetics is an important part of the relationship between media and uses. Understood as being rooted in *aisthesis*, the ancient Greek term for perception and sensation, aesthetics can even be considered as fundamental for interrelating infrastructural configurations and material conditions to the sensitivity and agency of human bodies, enabling access to media affordances and transformations of scripts. It is due to sensual perceptions that users can or cannot do something with media, whereas media also condition perceptional processes. The aesthetics of video conferencing can also be related to practices, dispositifs, and aesthetics established in older media: for example, the aesthetics of talking heads clearly connects the setting to the history and formats of television, such as talk shows and news. The experience of seeing oneself in a mirror-like way refers back to video technology, which enabled an instantaneous monitor image and a visual surveillant loop unprecedented in prior media, leading to diagnosis of the narcissistic structure of video technology (see Krauss 1976). When the video transmission is switched off, the aesthetics may resemble the telephonic setting, phenomenologically emphasizing sound. Although the term *video conferencing* emphasizes video, and thus visuality, it incorporates different media and aesthetic as well as practical regimes. Video conferencing applications combine image, speech, text, and—by way of manual interaction with interfaces—touch. The practices mentioned above (sharing screens, chatting, muting, etc.) usually rely on multiple perceptual and media modalities and registers at once. But to address the aesthetics of video conferencing implies a focus on not only appearing and making perceptible but also processes of concealment, invisibilities, and inaccessibility. This includes *anaesthetic* practices of muting and switching off the camera and the media-aesthetic conditions of the frame and off-space, as well as the basic characteristics of infrastructures and their different layers (see Schabacher 2013), which are not easily accessible for regular users.

In the attention to the interdependencies between aesthetics, practices, and infrastructures, the question of whether the visual connotations of the term *video conferencing* are misleading may even arise with regard to the mediality of video conferencing in general. The diffusion of digital (computing) devices in various forms, the proliferation of the internet, and the immensely influential sociotechnical (organizational) structure of platforms are undoubtedly among the most powerful factors that the focus on visuality cannot grasp. The video in video conferencing thus "conceals" the computer-technical as well as internet- and platform-based conditionality at play.

The Structure of the Book

The plurality of users and uses prompts us to approach the mediality of video conferencing not from the interaction between a user, a piece of software, and the resulting situation alone but by considering their extensions beyond the situation: infrastructural conditions, the embeddedness of video conferencing into the fabric of everyday practices, and how aesthetic phenomena enrich video conferencing and our understanding of it. Therefore, we address infrastructures, practices, and aesthetics as being interrelated, entangled, and even interdependent. They must be seen and discussed in relation to each other, which also has consequences for the structure of this volume. Four sections—"Teaching | Learning," "Infrastructuring | Interfacing," "Performing | Appearing," and "Working | Cooperating"—cluster texts whose perspectives on the interaction between infrastructures, practices, and aesthetics form their own focal points: perspectives on experiences and observations in the field of online education; on attention to processes of diverse levels of interfaces, ranging from user interfaces to application programming interfaces (APIs) and platform structures; on specific manifestations and strategies with which gazes are directed and corrected and artistic contexts expanded; and on different work contexts currently changing due to video conferencing and in which further historical traces of screen work can be found. The numerous cross-references throughout this volume indicate the strong interactions forged between the contributions, highlighting the mediality of video conferencing. Hence, the aim of this first anthology on the newly emerging phenomenon of video conferencing is to precisely stimulate that: debates and research that set out to explain what kind of phenomenon we are dealing with.

Teaching | Learning

It was probably in educational settings that the precarious infrastructural conditions of internet-based video conferencing and the unequal distribution of infrastructural resources in terms of, for instance, bandwidth, hardware, and domestic space became the most apparent. More than in other contexts, individual practices of "tile management"—out of personal preference or as a way to foster connection stability—caused frustration and public debate. For almost two years, the pandemic restructured our everyday lives and working routines as most parts of professional and private life moved online. Education was one of the fields massively affected by the lockdown and social distancing routines and in which a large part of the population—including teachers, children, parents, and university students, faculty and staff— had a stake. The switch to video conferencing as a technological base for instruction and learning enabled the continuation of educational work despite widespread shutdowns of schools and university campuses, although this work very often took place under less-than-ideal circumstances. The substitution of shared physical space (gathering places) for virtual "rooms" consisting of rectangles and tiles transformed the aesthetic, epistemic, and social conditions of education and stimulated the reflection on different forms of mediation. The chapters in this section shed light on the different medialities, power structures, and processes of implementation, slow habituation, and resistance involved in video conferencing.

In "A Study Abroad during Covid-19," *Kalani Michell* explores the aesthetic and epistemic affordances of the video conferencing applications Zoom and Gather.town through the prism of "virtual voyages." By taking students to Jamaica and Malaysia virtually in class, the course paralleled the experience of mediated distance, estrangement, and experimentation that accompanied the first weeks and months of online teaching in this new media environment. In her essay "Teaching into the Void: Reflections on 'Blended Learning' and Other Digital Amenities," *Donatella Della Ratta* explores the exhausting effects of video conferencing and consequential resistance techniques in order to critically engage with techno-determinist narratives and efforts of normalization and naturalization of education during the crisis. Whereas Michell and Della Ratta address the aesthetic-political dynamics of habituation and experimentation, *Andreas Weich, Irina Kaldrack*, and *Philipp Deny* focus on processes of subjectivation implied in video conferencing. Analyzing the media constellations of on-site teaching and teaching via video conferencing allows the authors to identify what kind of presence is being produced in them. It turns out that presence, despite its different configurations, serves primarily as a concept that masks practices of control used to generate specific subject positions. *Geert Lovink* also deals with specific subject effects of video conferencing and its interfaces. By addressing the experience of Zoom fatigue, he unpacks the digital restructuring of life during the pandemic, which intensifies the already existing neoliberal pressures and the mentality of "always-on," producing chronically exhausted subjects and the symptomology of media "fatigue." In the last chapter of the section, *Maha Bali* shifts the focus from fatigue and frustration to "intentionally equitable hospitality"—practices that aim to create welcoming, equitable, and pleasant experiences by starting with reflection on design choices, adaptation during facilitation, and awareness of the power within and beyond a platform.

Infrastructuring | Interfacing

The shift to the online environment also meant installing new medial infrastructures for maintaining social and professional life as well as cultivating new forms of cooperation and collaboration—a process that was accompanied by diverse challenges and disruptions. The access to resources, technologies, and know-how was unequally distributed: unstable or absent internet connections, lack of cameras or microphones, and noisy backgrounds were common experiences. Thus, a meeting usually started with phatic communication checking whether the "channel" was successfully established ("Do you hear me?" "Can you see me?" "I cannot hear you"). The infrastructures on which each of these meetings depends involve interface processes that go far beyond what becomes perceptible and available as user interfaces. Some of these even show up quite directly. For example, when a user opens BigBlueButton, a dialog box asks whether BigBlueButton is allowed to access the camera and microphone, indicating that video conferencing also relies on software-hardware interfaces that enable the video conferencing software to make use of a device's hardware. To experience, benefit from, and maybe struggle with user interfaces of, for instance, Discord, BigBlueButton, or Zoom, other forms of interfacing are always needed—like software-hardware interfaces (to allow the general purpose machine to behave as a specialized Discord or Zoom machine); software-software interfaces (to, as application programming interfaces, allow programs to interact and, as internet protocols, allow for internet traffic); or hardware-hardware interfaces (to allow the internet's submarine cables to actually make a connection). Among other things, the focus on infrastructures thus demonstrates how the interface concept, proposed by software studies in the early 2000s and continued by media studies, is fruitfully refined. Especially for the mediality of video conferencing, as the contributions in this section show, to widen the attention for interfaces toward an interface complex that encompasses different materialities and processualities, facilities and practices, resources and ideologies in equal measure becomes productive.

Jan Distelmeyer explores the traces left by the interactions between the different levels of this interface complex in the debates about and experiences of video conferencing. How these various interface processes manifest as effects, aesthetics, and practices (dealing with tiles, spaces, sounds, chats, and buttons), he argues, form new power relations and bring fundamental questions of digitality and platformization before the "user's" eyes and ears. By focusing on artistic experimentations with video conferencing interfaces, *Christian Ulrik Anderson* and *Søren Bro Pold* discuss the facial performance that becomes essential with the proliferation of video conferencing and reflect on the much larger politics of the face. They explore how the interfaces turn the face into a technical object. In his contribution, "Laws of Zoom," *Kim Albrecht* investigates the application programming interface of Zoom and the infrastructure of the hardware and software of internal and external code structures that are expressed by it. By developing an artistic method of "infrastructure imaging," he visualizes the organizing principles and structures of the Zoom API.

Performing | Appearing

This section examines how video conferencing technologies and its historical predecessors structure and produce new forms of *(in)visibilities*: of faces, bodies and private spaces, and therefore the realm of interpersonal relationships. The new visibility of the domestic, the dispositif of mirror-like self-observation, and the focus on tiled faces might be considered as three crucial aesthetic characteristics of video conferencing during the pandemic. The shifts to home office and distant learning resulted in the overlap of not only private and professional but also online and offline spaces—an aesthetics and socio-spatiality that touch upon basic questions of mediality. Being streamed in close-up, in medium shot, or without image, the tiled faces and talking heads are a basic mode of aesthetic audiovisual arrangement in video conferencing that enhances visibility and produces forms of aesthetic withdrawal. The visibility and invisibility of faces and bodies thus negotiate the social distances and proximities already implied by the entanglement of spaces. Being visible in a mirror-like structure also establishes a situation of being constantly watched by others and oneself, accompanied by modes of constant aesthetic self-evaluation and control.

The contributions in this section relate this kind of observation to the socio-aesthetic power of the gaze and its entanglement with the logics of late capitalism. The section opens with a chapter by *Laura Katharina Mücke*, which analyzes video conferencing interfaces as politicized spaces where visibility and invisibility condition social relations. By drawing on film-theoretical concepts, Mücke examines the fixed frontal view of video conferencing applications as a form of social mise-en-scène. In the next chapter, *Robert Rapoport* and *Vera Tollmann* critically interrogate the use of computer vision and generative adversarial networks for gaze correction in video conferencing applications. They argue that while gaze correction tries to collapse the difference between looking at the screen and looking into the camera, it simultaneously fails to encode and automate the social dynamics and cultural specificity of the eye contact, leading the aesthetics of optimization to prioritize user retention on platforms over the social accuracy of looking. *Martina Leeker* focuses on the use of video conferencing and virtual technology in performing arts and theater during the pandemic. By analyzing historical and contemporary artworks that engage with telepresence and the form of "distant socializing," she proposes to call this sociopolitical regime of the visible "the real virtual," thus undermining the longstanding opposition of reality and virtuality. In the final contribution to the section, *Till Baumgärtel* uses the form of the interview with the international artist Bill Bartlett, who in the late 1970s and 1980s experimented with telecommunication technologies such as satellites, slow-scan television, fax, and email in collaborative art projects. Using the technologies for making contact, collaboration, and tele-interaction, these art projects can be considered precursors of today's teleconferences and forms of communication that have since become commonplace in our use of Skype, Zoom, FaceTime, and other platforms.

Working | Cooperating

In the final section, the chapters address the usages of video conferencing in different work-related contexts. It deepens the questions and issues authors raised in the section on the educational context. The section focuses on the collaborative and cooperative dimension of practices, but it also touches upon socioeconomic consequences and sociopolitical issues of social participation and accesses enabled and prevented by video conferencing applications. Thus, the section emphasizes practices to delineate their entanglement with aesthetics, interfaces, and infrastructures. In this section, it is mainly the domestic room that constitutes the infrastructural settings of cooperative practices and conditions specific aesthetic affordances and exclusions. Access to cooperation, as several contributors show, is neither economically nor aesthetically equal.

In her chapter, *Alexandra Anikina* deals with the relation of public appearance and domestic space in video conferencing and the manifold tactics of revealing and concealing private backgrounds in mediated home office contexts. By tracing the aestheticization and professionalization of the *background*, she examines the interfaced architectures of the gaze and the economic structure of being seen, turning the act of looking and showing into a form of labor. *Winfried Gerling* traces and analyzes dispositifs of people working in front of screens as a genealogy of video conferencing in domestic environments. By revisiting historical photographs, his contribution explores how communicative relationships, working conditions, and visibilities are shaped in these screen arrangements. *Will Houstoun* and *Katharina Rein* examine the transformations that performance magic underwent by migrating to video conferencing platforms during the Covid-19 pandemic. Reflecting also on televised magic performances as precursors to those taking place online, the authors outline the mediality and the general characteristics of video conferencing. The collaborative contribution by *Tom Bieling, Beate Ochsner, Siegfried Saerberg, Robert Stock,* and *Frithjof Esch* problematizes how media participation of people with varying (dis)abilities is produced in professional settings by including the perspectives of (dis)abled people in their analysis. Writing with multiple voices, the authors discuss accessibility to video conferencing as a collaborative effort involving knowledge and access work.

Acknowledgements

We would like to thank the Collaborative Research Center "Media of Cooperation" at the University of Siegen, the Brandenburgisches Zentrum für Medienwissenschaften (ZeM), and the German Research Foundation (DFG) for supporting our research. We would like to express our gratitude to our copyeditors Jon Crylen and Sean DiLeonardi.

References

Akrich, Madeleine, and Bruno Latour. 1992. "A Summary of Convenient Vocabulary for the Semiotics of Human and Non-human Assemblies." In *Shaping Technology, Building Society: Studies in Sociotechnical Change*, edited by Wiebe E. Bijker and John Law, 259–64. Cambridge, MA: MIT Press.

Albertson, Lesley A. 1984. "Future Teleconference Meetings: Pattern and Prediction." In *The Teleconferencing Resource Book: A Guide to Applications and Planning*, edited by Lorne A. Parker and Christine H. Olgren, 394–401. Amsterdam: North Holland Elsevier Science Publishers B. V.

Argyle, Michael, Mansur Lalljee, and Mark Cook. 1968. "The Effects of Visibility on Interaction in a Dyad." *Human Relations* 21 (1): 3–17.

Armoni, Adi. 2000. *Healthcare Information Systems. Challenges of the New Millennium.* Hershey, PA: Idea Group Publishing.

Barber, P. J., and J. V. Laws. 1994. "Image Quality and Video Communication." In *Multimedia Technologies and Future Applications*, edited by Robert I. Damper, W. Hall, and J. W. Richards, 163–78. London: Pentech Press.

Bergermann, Ulrike. 2016. *Leere Fächer: Gründungsdiskurse in Kybernetik und Medienwissenschaft*. Medien'welten: Braunschweiger Schriften zur Medienkultur 25. Münster: LIT. https://doi.org/10.25969/mediarep/14841.

Birrell, J. A., and Ian Young. 1984. "Teleconferencing and Long-Term Meeting: Improving Group Decision-Making." In *The Teleconferencing Resource Book: A Guide to Applications and Planning*, edited by Lorne A. Parker and Christine H. Olgren, 278–87. Amsterdam: North Holland Elsevier Science Publishers B. V.

Blokland, Art, and Anne H. Anderson. 1998. "Effect of Low Frame-Rate Video on Intelligibility of Speech." *Speech Communication* 26 (1–2): 97–103. https://doi.org/10.1016/S0167-6393(98)00053-3

Bly, Sara A., Steve R. Harrison, and Susan Irwin. 1993. "Media Spaces: Bringing People Together in a Video, Audio, and Computing Environment." *Communications of the ACM* 36 (1): 28–46.

Cohen, Karen M. 1984. "Speaker Interaction: Video Teleconferencing Versus Face-to-Face Meetings." In *The Teleconferencing Resource Book: A Guide to Applications and Planning*, edited by Lorne A. Parker and Christine H. Olgren, 288–98. Amsterdam: North Holland Elsevier Science Publishers B. V.

Couldry, Nick. 2012. *Media, Society, World: Social Theory and Digital Media Practice.* Cambridge, MA: Polity.

Daft, Richard L., and Robert H. Lengel. 1984. "Information Richness: A New Approach to Managerial Behavior and Organization Design." In *Research in Organizational Behavior*, edited by Barry M. Staw and L. L. Cummings 6:191–233. Greenwich, CT: Jai Press.

Dang-Anh, Mark, Simone Pfeifer, Clemens Reisner, and Lisa Villioth, eds. 2017. "Medienpraktiken: Situieren, Erforschen, Reflektieren." Special Issue, *Navigationen: Zeitschrift für Medien- und Kulturwissenschaften* 17.

Dennis, Alan R., and Joseph S. Valacich. 1999. "Rethinking Media Richness: Towards a Theory of Media Synchronicity." In *Proceedings of the 32nd Hawaii International Conference on System Sciences*. (HICSS-32). Los Alamitos: IEEE Computer Society. https://doi.org/10.1109/HICSS.1999.772701.

Dietrich, Malinda. 2020. "Cultural Networks: Infrastructural Implications of AT&T's Picturephone." *Interfaces: Essays and Reviews in Computing and Culture* 1: 35–49.

Distelmeyer, Jan. 2022. *Critique of Digitality*. Wiesbaden: Palgrave Macmillan.

Dourish, Paul, Annette Adler, Victoria Bellotti, and Austin Henderson. 1996. "Your Place or Mine? Learning from Long-Term Use of Audio-Video Communication." *Computer Supported Cooperative Work (CSCW)* 5 (1): 33–62. https://doi.org/10.1007/BF00141935.

Due, Brian Lystgaard, and Christian Licoppe. 2020. "Video-Mediated Interaction (VMI): Introduction to a Special Issue on the Multimodal Accomplishment of VMI Institutional Activities." *Social Interaction. Video-Based Studies of Human Sociality* 3 (3). https://doi.org/10.7146/si.v3i3.123836.

Egido, Carmen. 1990. "Teleconferencing as a Technology to Support Cooperative Work: Its Possibilities and Limitations." *Intellectual Teamwork: Social and Technological Foundations of Cooperative Work*, 351–72. Hillsdale, NJ: L. Erlbaum Associates.

Finn, Kathleen E., Abigail J. Sellen, and Sylvia Wilbur, eds. 1997. *Video-Mediated Communication*. Computers, Cognition, and Work. Mahwah, NJ: L. Erlbaum Associates.

Furuta, Richard, and Christina Neuwirth, eds. 1994. *CSCW '94: Transcending Boundaries; Proceedings of the Conference on Computer Supported Cooperative Work, October 22–26, 1994, Chapel Hill, NC*. New York: Association for Computing Machinery.

Gaver, William W. 1992. "The Affordances of Media Spaces for Collaboration," *Proceedings of the 1992 ACM Conference on Computer-supported Cooperative Work (CSCW '92)*: 17–24. https://doi.org/10.1145/143457.371596.

Goebel, Gerhart. 1953. "Das Fernsehen in Deutschland Bis Zum Jahre 1945." *Archiv für Post- und Fernmeldewesen* 5: 259–340.

Gold, Elliot M. 1979. "Attitudes to Intercity Travel Substitution." *Telecommunications Policy* 3 (2): 88–104. https://doi.org/10.1016/0308-5961(79)90051-X.

Grusin, R. 2010. *Premediation: Affect and Mediality after 9/11*. Basingstoke, UK: Palgrave Macmillan.

Guckelberger, Olaf. 1995. *Radiologische Bildübertragung per Videokonferenz: Eine ROC-Analyse*. Berlin: Verlag für Wissenschaft und Forschung.

Heath, Christian, and Paul Luff. 1992. "Media Space and Communicative Asymmetries: Preliminary Observations of Video-Mediated Interaction." *Human–Computer Interaction* 7 (3): 315–46. https://doi.org/10.1207/s15327051hci0703_3.

Heath, Christian, and Paul Luff. 1993. "Disembodied Conduct: Interactional Asymmetries in Video-Mediated Communication."

Hiemstra, Glen. 1982. "Teleconferencing, Concern for Face, and Organizational Culture." *Annals of the International Communication Association* 6 (1): 874–904. https://doi.org/10.1080/23808985.1982.11678527.

Johansen, Robert, and Christine Bullen. 1984. "Thinking Ahead. What to Expect from Teleconferencing." *Harvard Business Review* 62 (2): 164–74.

Kawalek, Jürgen. 1997. *Unterricht am Bildschirm: der Einsatz von Videokonferenzen in EDV-Schulungen*. Vol. 570. Europäische Hochschulschriften 6. Frankfurt am Main: Peter Lang.

Kendon, Adam. 1967. "Some Functions of Gaze-Direction in Social Interaction." *Acta Psychologica* 26 (January): 22–63. https://doi.org/10.1016/0001-6918(67)90005-4.

Köhler, Stefan. 1993. "Einführung, Nutzung und Folgen von Videokonferenzen: Vergleich von 25 Unternehmen in Deutschland." Working Paper. Diskussionsbeiträge/Wissenschaftliches Institut für Kommunikationsdienste; 105.

Krämer, Sybille. 2015. *Medium, Messenger, Transmission: An Approach to Media Philosophy*. Amsterdam: Amsterdam University Press.

———. 2021. "From Dissemination to Digitality: How to Reflect on Media." *Media Theory* Special Issue: Into the Air 5 (2): 79–98.

Krauss, Rosalind. 1976. "Video: The Aesthetics of Narcissism." *October* 1: 51–64.

Kydd, Christine T. 1994. "Managerial Use of Video Conferencing." *Information & Management* 27 (6): 369–75. https://doi.org/10.1016/0378-7206(94)90017-5.

Lave, Jean, and Etienne Wenger. 1991. *Situated Learning: Legitimate Peripheral Participation*. Cambridge: Cambridge University Press.

Li, Lan, Ava Sullivan, Anwar Musah, Katerina Stavrianaki, Caroline E. Wood, Philip Baker, and Patty Kostkova. 2022. "To Zoom or Not to Zoom: A Longitudinal Study of UK Population's Activities during the COVID-19 Pandemic." *PLOS ONE* 17 (7): e0270207. https://doi.org/10.1371/journal.pone.0270207.

Lipartito, Kenneth. 2003. "Picturephone and the Information Age: The Social Meaning of Failure." *Technology and Culture* 44 (1): 50–81. https://doi.org/10.1353/tech.2003.0033.

Longhurst, Robyn. 2016. *Skype: Bodies, Screens, Space*. London: Taylor & Francis.

Mayes, Terry, and Sandra Foubister. 1996a. "Cognitive and Educational Aspects of Desktop Videoconferencing." *Interacting with Computers* 8 (2): 163–65. https://doi.org/10.1016/0953-5438(96)01024-7.

Mayes, Terry, and Sandra Foubister. 1996b. "Cognitive and Educational Aspects of Desktop Videoconferencing (Part 2)." *Interacting with Computers* 8 (3): 253–54. https://doi.org/10.1016/0953-5438(96)01024-7.

Menist, David B., and Bernard A. Wright. 1984. "Picturephone Meeting Service. The System." In *The Teleconferencing Resource Book: A Guide to Applications and Planning*, edited by Lorne A. Parker and Christine H. Olgren, 180–86. Amsterdam: North Holland Elsevier Science Publishers B. V.

Mills, Mara. 2012. "The Audiovisual Telephone: A Brief History." In *Zur ästhetischen Umsetzung von Musikvideos im Kontext von Handhelds*, edited by Henry Keazor, Hans W. Giessen, and Thorsten Wübbena, 34–47. Heidelberg: ART-Dok. https://archiv.ub.uni-heidelberg.de/artdok/2016.

Nardi, Bonnie A., Heinrich Schwarz, Allan Kuchinsky, Robert Leichner, Steve Whittaker, and Robert Sclabassi. 1993. "Turning Away from Talking Heads: The Use of Video-as-Data in Neurosurgery." In *Proceedings of the INTERACT '93 and CHI '93 Conference on Human Factors in Computing Systems*, 327–34. New York: Association for Computing Machinery. https://doi.org/10.1145/169059.169261.

New York Times. 1982. "Picturephone Service Begins." *The New York Times*, July 9, 1982, sec. Business. http://www.nytimes.com/1982/07/09/business/picturephone-service-begins.html.

Nilles, Jack M., F. Roy Carlson, and Paul Gray. 1976. *The Telecommunications-Transportation Tradeoff Options for Tomorrow*. New York: Wiley.

Noll, A. Michael. 1992. "Anatomy of a Failure: Picturephone Revisited." *Telecommunications Policy* 16 (4): 307–16. https://doi.org/10.1016/0308-5961(92)90039-R.

O'Conaill, Brid, and Steve Whittaker. 1995. *An Analysis of the Spoken Aspects of Video Mediated Communication*. Palo Alto, CA: Hewlett Packard Laboratories.

O'Conaill, Brid, Steve Whittaker, and Sylvia Wilbur. 1993. "Conversations over Video Conferences: An Evaluation of the Spoken Aspects of Video-Mediated Communication." *Human-Computer Interaction* 8 (4): 389–428. https://doi.org/10.1207/s15327051hci0804_4.

O'Malley, Claire, Steve Langton, Anne Anderson, Gwyneth Doherty-Sneddon, and Vicki Bruce. 1996. "Comparison of Face-to-Face and Video-Mediated Interaction." *Interacting with Computers* 8 (2): 177–92. https://doi.org/10.1016/0953-5438(96)01027-2.

Otto, Isabell. 2013. "Happy Birthday from Skype: Zur Darstellung von Temporalität in einer Online-Werbekampagne." *Zfm (Zeitschrift Für Medienwissenschaft)* 9 (2): 53–65.

Parks, Lisa, and Nicole Starosielski, eds. 2015. *Signal Traffic: Critical Studies of Media Infrastructures*. Urbana: University of Illinois Press.

Périn, Pascal. 1983. "Communication interactive de groupe et médiatisation." *Psychologie française* 28: 289–96.

Ruhleder, Karen, and Brigitte Jordan. 2001. "Co-constructing Non-mutual Realities: Delay-Generated Trouble in Distributed Interaction." *Computer Supported Cooperative Work (CSCW)* 10 (1): 113–38. https://doi.org/10.1023/A:1011243905593.

Rutter, D. R., G. M. Stephenson, and M. E. Dewey. 1981. "Visual Communication and the Content and Style of Conversation." *The British Journal of Social Psychology* 20 (Pt. 1): 41–52. https://doi.org/10.1111/j.2044-8309.1981.tb00472.x.

Ryan, M. D., and J. G. Craig. 1975. "Intergroup Communication: The Influence of Communications Medium and Role Induced Status Level on Mood, and Attitudes towards the Medium and Discussion." In *Meeting of the International Communications Association, Chicago, IL.*

Sacks, Harvey, Emanuel A. Schegloff, and Gail Jefferson. 1974. "A Simplest Systematics for the Organization of Turn-Taking for Conversation." *Language* 50 (4). https://doi.org/10.1353/lan.1974.0010.

Schabacher, Gabriele. 2013. "Medium Infrastruktur: Trajektorien soziotechnischer Netzwerke in der ANT." *Zeitschrift für Medien-und Kulturforschung* 4 (2): 129–48.

Schnaars, Steve, and Cliff Wymbs. 2004. "On the Persistence of Lackluster Demand—the History of the Video Telephone." *Technological Forecasting and Social Change* 71 (3): 197–216. https://doi.org/10.1016/S0040-1625(02)00410-9.

Schulte, Regine. 1993. *Substitut oder Komplement—Die Wirkungsbeziehungen zwischen der Telekommunikationstechnik Videokonferenz und dem Luftverkehrsaufkommen Deutscher Unternehmen: Mit 8 Tabellen*. Dümmlerbuch 88. Bonn: Dümmler.

Schüttpelz, Erhard. 2017. "Infrastructural Media and Public Media." *Media in Action* 1 (1): 13–61.

Schütze, Hans-Joachim. 2000. *Warum Videokonferenzen? Überlegungen und eine Experimentalstudie zur Medienvermittlung Informeller Kommunikation*. Frankfurt am Main: Lang.

Sellen, Abigail J. 1992. "Speech Patterns in Video-Mediated Conversations." In *Proceedings of the SIGCHI Conference on Human Factors in Computing Systems—CHI '92*, 49–59. Monterey, CA: ACM Press. https://doi.org/10.1145/142750.142756.

Short, John, Ederyn Williams, and Bruce Christie. 1976. *The Social Psychology of Telecommunications*. London: Wiley.

Star, Susan Leigh, and Karen Ruhleder. 1996. "Steps toward an Ecology of Infrastructure: Design and Access for Large Information Spaces." *Information Systems Research* 7 (1): 111–34. https://doi.org/10.1287/isre.7.1.111.

Sterne, Johnathan. 2012. *MP3: The Meaning of a Format*. Durham, NC: Duke University Press.

Steuer, Jonathan. 1992. "Defining Virtual Reality: Dimensions Determining Telepresence." *Journal of Communication* 42 (4): 73–93.

Storck, John, and Lee Sproull. 1995. "Through a Glass Darkly: What Do People Learn in Videoconferences?" *Human Communication Research* 22 (2): 197–219. https://doi.org/10.1111/j.1468-2958.1995.tb00366.x.

Stults, Robert. 1986. "Media Space." Xerox PARC technical report.

Tasman, David Andrew. 2015. "David Joselit | Against Representation." *DIS Magazine* (blog). 2015. http://dismagazine.com/discussion/75654/david-joselit-against-re presentation.

Unternährer, Markus. 2021. "Die Ordnung der Empfehlung." *Kölner Zeitschrift für Soziologie und Sozialpsychologie* 73: 397–423.

Vogl, Joseph. 2007. "Becoming-Media: Galileo's Telescope." *Grey Room* 29: 14–25. htt ps://doi.org/10.1162/grey.2007.1.29.14.

Volmar, Axel. 2017. "Formats as Media of Cooperation." *Media in Action. Interdisciplinary Journal on Cooperative Media* 1 (2): 9–28. https://doi.org/10.25819/ubsi/81 72.

Volmar, Axel, Charline Kindervater, Sebastian Randerath, and Aikaterini Mniestri. 2023. "Mainstreaming Zoom: Covid-19, Social Distancing, and the Rise of Video-Mediated Remote Cooperation." In *Varieties of Cooperation: Mutually Making the Conditions of Mutual Making*, edited by Clemens Eisenmann, Kathrin Englert, Cornelius Schubert, and Ehler Voss, 99–133. Media of Cooperation. Wiesbaden: Springer.

Wallbott, Harald G. 1992. "Effects of Distortion of Spatial and Temporal Resolution of Video Stimuli on Emotion Attributions." *Journal of Nonverbal Behavior* 16 (1): 5–20. https://doi.org/10.1007/BF00986876.

Wilcox, James R. 2000. *Videoconferencing & Interactive Multimedia: The Whole Picture*. New York: Telecom Books.

Williams, Ederyn. 1977. "Experimental Comparisons of Face-to-Face and Mediated Communication: A Review." *Psychological Bulletin* 84 (5): 963–76. https://doi.org/ 10.1037/0033-2909.84.5.963.

Wright, Bernard A. 1983. "The Design of Picturephone Meeting Service Conference Centers for Video Teleconferencing." *IEEE Communications Magazine* 21 (2): 30–36. https://doi.org/10.1109/MCOM.1983.1091354.

Teaching | Learning

A Study Abroad during Covid-19

Kalani Michell

In February 2021, my seminar went on a study abroad.[1] We went for free to Jamaica and Malaysia and circled back for a quick tour of our home base, Los Angeles. Our portal was a winter garden, and our pathway was lined with palms. I say my seminar "went" on a study abroad because I didn't take them to these places. What my students saw took us.

Figure 1: Abroad at home

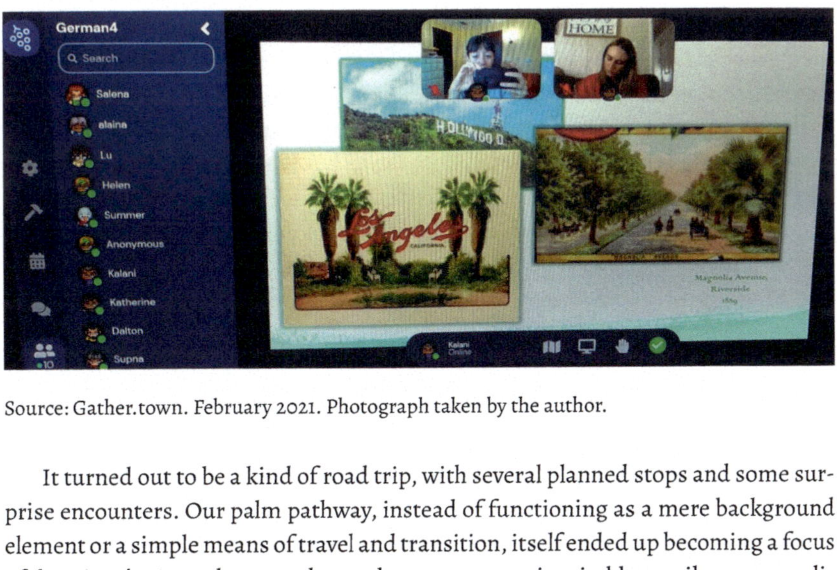

Source: Gather.town. February 2021. Photograph taken by the author.

It turned out to be a kind of road trip, with several planned stops and some surprise encounters. Our palm pathway, instead of functioning as a mere background element or a simple means of travel and transition, itself ended up becoming a focus of the trip. That's not because the road as passageway inevitably entails a more radical sense of agency or self-liberation for the driver. Roads can not only open up new possibilities but also give rise to restrictions and problems of their own (see, among

1 Many thanks to Salena Lo and Akela Morinaka for initiating this trip, as well as Donatella Della Ratta, Jan Distelmeyer, Rembert Hüser, Todd Presner, and Axel Volmar for their helpful ideas and comments.

others, Laderman 2002; Mills 2006; Archer 2016). "The psychological basis for driving seems to be deeply engrained in the pleasure of undergoing the experience of a total road situation with constraints and frustration on one hand and the flow and joy of movement on the other hand. We experience the drive as something that is not completely in our control, but rather 'draws us in' … and drives us" (Fuchs 2019, 20; see also Gadamer [1960] 1989). Instead of attempting to replicate or comprehensively recount the trip that we took in this class (and that took us), one can consider how a different means of revisiting it, through these pages, reconstructs it anew via the experience of reading. This itinerary would include routes that were planned out in advance and disorienting detours that were not: from recent changes to our home environments and overlooked histories of slide technologies to newly designed digital winter gardens with palms, hybrids, and netherworlds and opportunities to get lost—here as well as at home, in class, and on campus—along the way.[2]

Our trip took place in a seminar titled Media Environments and German Palm Tree Dreams. Environments had been on our minds that past year anyway—in terms of not only the ecological crisis on both sides of the Atlantic but also our classrooms, where we all sit around in a grid like in the opening credits of *The Brady Bunch*. Many homes still at home (fig. 1). So, we began to think about the changes taking place to our very own surroundings. Some of these emphasized the extent to which environments regulate the conditions of possibility for producing and acquiring knowledge in the first place. Institutions called the emergent shift to "remote instruction" (not "online learning") a "pivot," implying that the central pillars and values around which we work would remain the same amid this swift but temporary turn to something new.[3] Yet it soon became clear that knowledge changes when instructors "pivot" from modeling it on a whiteboard in a classroom to devising a makeshift apparatus for it out of a shower wall, magic markers, and buckets propped up on a toilet (fig. 2).

---

2 "Means of transport have the characteristic feature of integrating their user into a collective of human and non-human actants. Thus, anyone who transports or allows themselves to be transported is also inevitably transformed" (Imhof 2014, 21–22, translation KM).

3 "'Online learning' will become a politicized term that can take on any number of meanings depending on the argument someone wants to advance. … Online learning carries a stigma of being lower quality than face-to-face learning, despite research showing otherwise. These hurried moves online by so many institutions at once could seal the perception of online learning as a weak option, when in truth nobody making the transition to online teaching under these circumstances [the onset of the Covid-19 pandemic] will truly be designing to take full advantage of the affordances and possibilities of the online format" (Hodges et al. 2020).

Figure 2: Draft tripods

Source: Lorraine, Lisa. 2020. "#quarantine #teachersoftiktok #desperatetimes #biology #innovation." TikTok video, 00:09. April 7, 2020. Screenshot. https://www.tiktok.com/@lisa.tomato/video/6813185032732167430.

Other changes highlighted aspects of our physical environments we had yet to notice or fully understand. While in March 2020 we might have been searching out pro tips for controlling and curating the look of our homes during video conferencing sessions, a year later many were growing tired and skeptical of this attempt to approximate the environment of in-person learning, especially if this idea of organized space and authoritative knowledge was a fantasy to begin with. New and improved how-to videos surfaced, offering some guidance (fig. 3).

> I want my students, when they log on to my class, to feel like they're walking right into a professor's office on campus. It would be really tempting to get a

bookshelf and put it in here [this spare bedroom] and put all the books in neatly, but that wouldn't look like a professor's office at all. ... It's not just books. ... You gotta have file folders. ... It's not a book*shelf*, it's book *stack*. ... Ok, the next thing you gotta have is half a ream of printer paper. ... [A]nd you just wanna stick it there [on the bed] because sometimes you gotta use that to print quizzes right before a class. OK, so you can see how I have a whiteboard here and I've written what we're gonna do today in class? And that's great, but it'd be better if what I have written on the board looks like the ravings of a lunatic. ... We're trying to go for authenticity here. ... Listen, being in your office is not all work. ... I've got a bunch of random packets of things that imply that I've eaten many lunches in here. And we're just gonna take those and sprinkle those on the bed. ... Yeah, I think this is looking pretty good. I think this looks just like a professor's office. If a student were to log on, they'd feel like they're just right on campus, with me. (Ishak 2020)

Figure 3: Office life

Source: Ishak, Andrew. 2020. "Making Your Zoom Look More Professorial." Video, 5:46. August 13, 2020. Screenshot. https://vimeo.com/447645552.

If a real, "professorial" bookshelf is often a real mess, why try to reenact a fictional version on of it on Zoom?[4] And besides, the "asymmetry of knowledge" it once reinforced between professor and student is no longer applicable (Hüser 2020).[5]

4 Concerns about being seen in one's own messy, disorganized environment were some of the first public reactions to early forms of video conferencing, such as the *Picturephone* introduced at the 1964 World's Fair: "There were wry comments about mothers telephoning their daughters and clucking in disapproval of the disheveled state of the daughter's hair and apartment" (Fang 1997, 147). See also Katz and Crocker (2017).

5 "The discussion about whether or not a professor should have their picture taken in front of a wall of books is as old as the hills. In any case, the wall of shelves full of books no longer

Some of these changes to our home environments ultimately called for a historicization of their newness, such as the emergence of the concept of the interior in the nineteenth century and how it has long since been repurposed in various ways, or how the bed as a hybrid private/public space, media platform, and control center for networked communication was not that new.[6]

Our class discussed Zoom as a format and possible alternatives to this model of communication in times of remote learning, such as *Second Life*, Kumospace, *Minecraft*, and Gather.town, that last of which we extensively explored during one of our sessions.[7] Since presentations on Zoom usually involve a PowerPoint in "share screen" mode, I was intrigued when the first group presentation wanted to go somewhere else. A PowerPoint typically shows content and information that has already been synthesized. This was not always the fate for shared, projected presentation technologies. Joseph Licklider, for example, envisioned networked computers in specialized work settings in the 1960s as a means of modeling thinking in communal presentations rather than explaining it. Media technologies were to assist in opening up a communication process rather than projecting a finished idea or product. For Licklider,

> the computer opened up entire new possibilities: as a 'plastic or moldable medium that can be modeled [as] a dynamic medium,' it is especially suited for presenting to others how someone imagines something and, likewise in a unique way, it also allows for a common idea of something to be developed. ... [Presentations via computers in this context] were thus imagined less as a representation of static bundles of knowledge than as a flexible medium for interactive processes of cooperative model discovery. It was from this basis that

has anything to do with the asymmetry of knowledge that can be seen in earlier depictions of scholars, now that many people prefer to have their books on their USB drives. At the university, meanwhile, there is actually more of a fear of the bookshelf, and there has been for years. The university now wants to appear more relaxed to the outside world; it needs this for its Third Mission programs. A wall of books, on the other hand, has something dusty about it; no one 'lives' there" (Hüser 2020, translation KM). See also Fetters (2020a).

6 See Patton (2020, 13–23, 123–29) and Colomina (2014). "In what is probably now a conservative estimate, *The Wall Street Journal* reported in 2012 that 80 percent of young New York City professionals work regularly from bed. ... Post-industrialization collapses work back into the home and takes it further into the bedroom and into the bed itself. Phantasmagoria [of the interior] is no longer lining the room in wallpaper, fabric, images, and objects. It is now in the electronic devices. The whole universe is concentrated on a small screen with the bed floating in an infinite sea of information. To lie down is not to rest but to move. The bed is now a site of action" (Colomina 2014, 19). See also Benjamin ([1955] 1978).

7 "The ubiquity of the [Zoom] software has resulted in genericization, with many using the word 'Zoom' as a verb to replace videoconferencing, similar to 'Googling'" (Bailenson 2021).

> the problematic transition could take place, for instance, into the pedagogical context. (Pias 2020, 294, 296)[8]

Given the then-high cost of this model of thinking via a shared network of computers, and given the relative affordability of overhead transparencies, the latter led the way in pedagogical presentation practices. PowerPoint was first created as a means of printing these overhead transparencies and only later as a form of designing and projecting the presentation itself (see Pias 2020, 296–99).

Thus, although there were presentation technologies related to the history and development of PowerPoint that were originally conceived as interactive formats for modeling thinking and for work that is collaborative and creative, in the educational context of today, the default PowerPoint often displays sequentially organized material presented through bullet points, summaries, graphics, and performative oration (see Coy and Pias 2009; Robles-Anderson and Svensson 2016; Knoblauch 2008; Tufte 2003; Frommer [2010] 2012). While one person speaks, the whole group listens and epigraphically reads:

> Reading PowerPoint is wall reading, is group reading, is synchronous reading, is semipublic reading. ... In the sweaty, hormone-steeped conference room, when all eyes are on the PowerPoint presenter with his or her slides dissolving from one to the next, the emphasis is on group, consensus, team, collaboration, comprise, unity. ... The slides externalize the *truth* and allow the audience to analyze it separately, but simultaneously, from what the speaker is saying *about the same truth*. The slide is not simply an opinion, it is a written artifact on a wall owned in common by all in the room—even if, as is usually the case, the speaker wrote the words in the first place. (Gold 2002, 260)

One of the most distinctive features that we've come to learn about Zoom presentations over the past few years is how they tend to rely on prearranged interaction. With all but one face muted, the presentation doesn't even require nonverbal affirmation from the group anymore. The flow of the speaker's "verbal gloss," necessary for "illuminat[ing]" the material on the slides and for the unifying ritual of PowerPoint, is no longer interrupted by the speaker accidentally standing in front of the projected information or distracting listeners by fiddling around with the interface to move from one slide to the next (Gold 2002, 259–66). In "shared screen" mode on Zoom, the presentation interface can appear more seamless than before—as long as

8 "An illustration in one of Licklider's texts from 1968, for instance, shows a group of bridge builders (although the text itself is concerned here with tactical combat). ... One of the participants is giving a presentation on a large screen, while the others are sitting at networked computers, where they can work on things, look things up, make improvements, or also simply play around" (Pias 2020, 294).

everyone else is muted. The standard Zoom sayings about noise, disruptions, and uncertainty indicate how common such occurrences are and how accustomed to them we've become.[9] We all know that one person who always forgets to mute themselves and can imagine what would happen if this person were everywhere at once (fig. 4).

Figure 4: A friendly reminder

Source: Saturday Night Live. 2020. "Zoom Church." YouTube video, 3:08. May 9, 2020. Screenshot. https://www.youtube.com/watch?v=AYP1mXqiwqc.

> Pastor/Presenter: "The way that the Zoom machine works is that every mic is as loud as mine, so when y'all respond, I can't really hear myself preach. Amen?"
> Congregation/Audience, all with mics on: "Amen!" ...
> Pastor/Presenter: "Why are y'all still not on mute ... Stop. Answering. Me." (Saturday Night Live 2020, fig. 4)

---

9 "The titles of the sections in th[is] *not paper* are intended to remind us all of how imperfect video conferencing systems are, and of the huge limitations they impose on us" (Lindley et al. 2021). The section titles contained in this academic "not paper" include the following: "YOU'RE ON MUTE" / "PLEASE WAIT FOR THE HOST TO START THIS MEETING" / "HANG ON A SECOND ... I'LL JUST SHARE MY SCREEN" / "ARE YOU THERE? CAN YOU HEAR ME?" / "PEOPLE ARE STARTING TO ARRIVE NOW," and "THAT AWKWARD FACE PEOPLE DO AS THEY SEARCH FOR THE 'LEAVE MEETING' BUTTON" (Lindley et al. 2021). On these moments of "disturbance" within protocological networks, see Distelmeyer, in this volume, and Galloway (2004). On how such awkward "lapses" can work against teleological notions of technological progress, see Della Ratta, in this volume.

Call-and-response is not what this platform is designed to facilitate. The literal and figurative noise that interferes when Zoom faces that are meant to remain silent suddenly turn on, lighting up with a bright green outline and reconfiguring the arrangement of other faces in the room, is an indicator of the residual extralinguistic cues we usually rely on to enable the flow of verbal communication and the real incongruity that persists behind this interface function.[10] When others do get a chance to unmute and talk among themselves in Zoom, it's often when the presentation dissolves into smaller groups in sedentary breakout rooms (fig. 5). The first group presentation in our class took us out of the Zoom PowerPoint pastor-grid-slideshow that elicits predetermined outcomes, consensus, and unity and into a very different kind of grid.

Figure 5: Chatter boxes

Source: Bell, Taryn. 2021. "Do Students Hate Online Breakout Rooms?" Twitter, March 18, 2021, 1:47 a.m. https://twitter.com/tarynlbell/status/1372469644997619715.

10 See Fetters (2020b), "The Importance of Pauses in Conversation" (2017), and Bailenson (2021). The nonverbal cues received on Zoom, moreover, can easily be misinterpreted. "In a face-to-face meeting, a quick, sidelong glance where one person darts their eyes to another has a social meaning. ... In Zoom, a user might see a pattern in which on their grid it seems like one person glanced at another. However, that is not what actually happened, since people often don't have the same grids. Even if the grids were kept constant, it is far more likely the glancing person just got a calendar reminder on their screen or a chat message. Users are constantly receiving nonverbal cues that would have a specific meaning in a face-to-face context but have different meanings on Zoom. While of course people do adapt to media over time ... it is often difficult to overcome automatic reactions to nonverbal cues" (Bailenson 2021).

We met that day on Zoom, as always, having read two texts for that session: in this case, sections of Krista Thompson's *An Eye for the Tropics* and Simryn Gill and Michael Taussig's *Becoming Palm* (2017), which thematize the relationship between European colonization and the environment in Jamaica and Malaysia, respectively. Akela Morinaka and Salena Lo were presenting and let us know that they had created their own Gather.town space to explore these texts. Gather, as the platform would like to be abbreviated, is a browser-based video conferencing app, free for up to 25 people, integrated into a virtual map based on templates that you, the participant, can customize ("classroom," "rooftop party," "keynote," etc.) or freely assemble. Its 2D layout and pixilation make it feel like a retro 8-bit video game (see Metz 2020). As you move through the space via an avatar and arrows keys, you meet up with others. This is where it gets interesting. When you come within a few steps of someone, a tiny video display pops up at the top of your screen, enabling a chat between the two of you (see Jacobs and Lindley 2021). More can join if they're within reach. Or you can just pass by each other without entering a video call. You can choose with whom you want to talk, directly and outside of randomly assigned breakout rooms. You can also join a video call at designated spaces, such as a large classroom table, with anyone who happens to be seated there. This means that when this environment is employed as a classroom, for example, questions and answers no longer need to be solely directed at and channeled through the instructor as mediator as they might in a Zoom session in gallery view. Instead, this setup allows for learning opportunities to be initiated and had by multiple "players" or operators in multiple locations, even simultaneously, with or without the instructor being present (see McClure and Williams 2021, 8). This is one of the main appeals of the platform: the possibility for surprise conversations and encounters that can nevertheless remain private.[11] It's a space in which you can upload and pick up Word documents, write on a blackboard, make poster sessions, and watch films that you curate on TV sets.

I asked Akela and Salena to do what I did before I took them to Gather.town for the first time—namely, to give us the plan before we left Zoom: what our virtual map would look like; where we should meet up in the space, when, and for how long; what our activities would be, and so on. "You'll see," they said. "Just follow the palms." It wasn't long before I lost my way—and my students.

11 See Lee (2020) and Latulipe (2021). "[Gather.town] combines role-playing, game mechanics and video conferencing. It is nostalgic, contemporary and futuristic. It brings people together at a distance. It taps into our desire to share social spaces. It facilitates serendipitous encounters" (Lindley et al. 2021).

Figure 6: Environmental signage

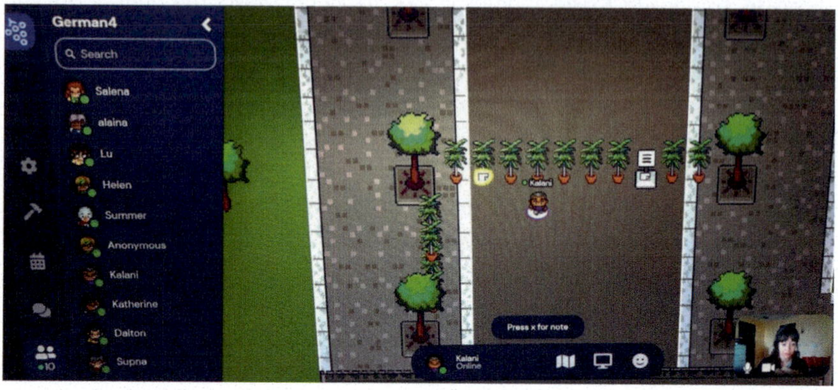

Source: Gather.town. February 2021. Photograph taken by the author.

After entering my name and turning on my mic and camera, my randomly assigned avatar transported me to a digital winter garden, where I was surrounded by a group of avatars labeled with my students' names. *Great, we all made it.* They started running in one direction, so I slowly tagged along, wanting to make sure everyone made it into the platform and knew where to go. I hit a line of palms and went straight to the building in front of me, which I figured would be our virtual classroom, filled with activities. When I entered the main hall, it was empty. *They're fast. They must be in one of the rooms. I'll catch up.* Hurriedly shifting from one arrow key to the next and holding them down in the hope that doing so would speed me up, I navigated the space room by room. Totally empty. I stepped outside and looked around. No one. Or no avatars, at least not in this part of our map. I knew my students were still in the space: I could see them all, their names and avatars, in the Gather.town interface on the left-hand side of my screen. I just didn't know *where* they were in this space. *I'll have to go back and start over. Retrace my steps.*

Returning to where I started, I was relieved to find the first visual clue I must have skipped over in a rush: a note among the palm pathway announcing with a yellow aura that it contained some information to help me navigate this new environment (fig. 6). It's not totally surprising that I missed this way-finding icon, given that "environmental signage is simultaneously there and not there—not really a 'part of' the architecture, yet indispensable to its functions. … Graphic design—signage in particular—is largely a framing activity. … Graphic design is the margins of a book, the buttons of a boom box, the friendliness of a computer interface, or the label wrapping a tin can" (Lupton and Miller 1993, 221). I walked up to the note, clicked it, and was momentarily taken out of the map and into another space (fig. 7). I read there what I had already heard Akela and Selena tell us before we entered the space:

let the palms guide you—simple, no punctuation, no bullet points underneath, an extracted line on a solid black background, my own confusion staring back at me. The moveable avatar and the world it inhabited were completely overtaken in this frame of vision: an extradiegetic space that housed a feedback loop of intrigue and contingency.

Figure 7: Palm guides

Source: Gather.town. February 2021. Photograph taken by the author.

Fortunately, I wasn't lost forever. It just took me a minute to realize that this environment required me to explore it with more curiosity than resolve. I wanted the narrative lesson of the digital environment—*Where do I need to go, and what do I need to do?*—to reveal itself without having first approached it experientially. Rather than following a singular path governed by a lock-and-key system, which requires that operators discover clues and follow prescribed steps in a certain sequence to reach a particular predetermined outcome, the gameplay built into this video conferencing app relies on principles of multiplicity, chance, and emergence.[12] The digital world was more like a sandbox game that participants are asked to wander through and,

12 "In ... emergent [video game narratives], there is no prescripted story but rather a system of existents—characters and objects—capable of various behaviors. ... When the system contains many existents capable of a variety of actions, and when these actions have side effects for other existents, the system becomes too complex to be predicable, and the stories that it produces acquire a quality of emergence" (Ryan 2016, 339–40). On the "workaround" and improvised readjustment as a response to interference in video conferencing practices from the perspective of media and disability studies, particularly in terms of how interference can be understood not only "as a problem to be solved," but as "a chance to enable situative 'crip' reorderings or productive deviations from the norm," see Bieling et al. in this volume, as well as Schabacher (2017).

if they choose, modify and cocreate using a variety of tools at their disposal. The objective was not to speed-run to the class activity but to observe ourselves and our environment while navigating, which changed how we subsequently engaged with the planned activity.[13] The long, winding path of palms functioned like the opening credits for our course, through which we can't fast-forward.[14]

I eventually found the strategically planted palms, which led me to the building in which my students, as I could tell from the presence and movements of their avatars, were already working. A note at the entrance told us to watch a series of videos that Akela and Salena had created and then enter another building to work on some Google Docs together. I held down the arrow key, walked up to the video, launched it, and saw my face show up next to my students' faces at the top; they were at the same viewing station, each with a miniature version of their avatar anchored to the bottom of their frame, reminding me who was who in the game world (fig. 8). I didn't get that I needed to mute myself upon entering the viewing station, so my sound was blasting into everyone's screens. My students quickly understood, they told me afterward. *Yeah, we just muted you*. There wasn't an assigned sequence, so each of us watched whichever video we wanted first and then moved on at our own pace. They played on loop, and we watched them as we would in a gallery: if we jumped in at the middle, we could see part of it and leave or stay until it looped back to the beginning. This meant that the filmstrip of faces with us at the viewing station was constantly in flux. While technically extracting us from the diegesis of the game world that initially guided us there as an avatar, the videos that were playing constantly referred back to it.[15] We saw each other watching, reacting, taking notes, laughing, arriving in the strip, exploring, getting bored, leaving for the next station, stretching, and taking pictures (fig. 9).

---

13 On the ways in which the hyper-performativity of the Let's Play format, in which operators simultaneously act (play, move, and make decisions) and reflect on said action, can result in epistemological contingency and productivity, see Leeker in this volume.

14 The string of palms, navigational tactics, and reading strategies at the beginning of this digital world anticipate the logic of the game world and gameplay to come, understood only later as a disordered compilation of individual aspects of this space and gamic acts. See Stanitzek (2006, 8–14; 2009, 47). On what is lost and gained in a walkthrough, see Mukherjee (2016, 64): "Often, as Garry Crawford et al. point out, 'the unavoidable consequence of playing a goal-oriented walkthrough ... is the devaluation of socially oriented play' ... and the walkthroughs that allow the player to speed through the game often make the basic narrative tools of the game, such as reading the quest descriptions, unnecessary." See also Glas et al. (2011, 149).

15 On diegetic versus nondiegetic gamic actions and instances in which these are difficult to demarcate from one another, see Galloway (2006, 5–38).

Figures 8–9: Facestrips

Source: Gather.town. February 2021. Photographs taken by the author.

The videos at these stations thematized the imaginative geography of the tropics we had read about in our texts, restaging its successive, immersive imagery and descriptive and prescriptive language. We listened to how classificatory tendencies and theories of objectivity transformed fantasies of colonial mastery into environmental realities, how they helped naturalize imperial botanical imports and imagine global origins to be organic. Jamaica, Mayalsia, Europe, Los Angeles. My mind wandered, and I thought about what it meant to be seeing this in a Gather.town configuration rather than in a PowerPoint, the objectivity of information already synthesized. I heard about the grafting and mass migrations of flora and thought about the will to order and standardization:

> Grafting is an agricultural technique that dates back to antiquity. Its major characteristic is that of intervention. Its major advantage is that it provides a short cut

> in reproduction. ... Although it [grafting] combines two pieces of plants, that is: two different bodies, it is, by contrast to hybridization, not a fusion of different genetic elements. Grafting involves "the creation of a *compound genetic system* by uniting two (or more) distinct genotypes, each of which maintains its own genetic identity throughout the life of the grafted plant." (Wirth 2014, 233–34; Mudge et al. 2009, 440)

I saw examples of how grafting and graphics create strange juxtapositions and compounds. On a recto of *Becoming Palm*, we read Taussig: "As of this writing my colleague Simryn Gill is becoming an oil palm tree along the Straits of Malacca" (2017, 37 and fig. 10). On the verso, we saw Gill-as-palm, her legs firmly planted on the ground and her hands cradling an object that refused to hold still for the camera (2017, 36 and fig. 10). As I looked at her face, replaced by lively fronds, I looked at my own and how it was framed in the video display, "all mixed up and confusing. ... now it's a mess" (Gill and Taussig 2017, 37).[16] It was stitched to a row of other images in side-by-side frames, the bodiless faces of my students in "German4," a course designator relic indicating a phase of institutional transition amid new understandings of geographical copresences and contact zones.[17] As students came and went from the viewing station, the filmstrip at the top of the screen correspondingly shuffled out one face with another, temporarily situating it next to other faces without bodies (fig. 9) or bodies replaced by the tangled limbs of a tree on one of the first British-issued Jamaican stamps (fig. 8). While we learned about and observed the historical transformation of the environmental objects onscreen, we did so while simultaneously observing ourselves and our transformations as pandemic video conferencing participants. I thought about supposedly unadulterated, stable states in pre-this moments and post-that eras and other markers of time that might try to neatly organize its complexity. I listened to how this image on the stamp stirred up debates about whether the environment seemed too Welsh and too universal and how to isolate and typify the landscape. Thought about grafting as an instrument of writing, as a description of the process of writing itself.[18] As a remediation that rejects the

16 "[A] basic characteristic of frame analysis [is] that 'discussions about frame inevitably lead to questions concerning the status of the discussion itself, because here terms applying to what is analyzed ought to apply to the analysis also'" (Stanitzek 2005, 34–35; Goffman 1974, 11).

17 See Abraham (2021): "The new department [of European Languages and Transcultural Studies at UCLA] brings together the existing departments of Germanic languages, French and Francophone studies, Italian and Scandinavian. ... The term 'transcultural' emphasizes shared European roots and an expanded focus on the perspectives of filmmakers, writers and theorists from Africa, Asia, the Caribbean, Central and South America, and elsewhere."

18 "That is how the thing is written. To write means to graft. ... The graft is not something that happens to the properness of the thing. There is no more any thing than there is any original text. Hence, all those textual samples provided by *Numbers* do not, as you might have been tempted to believe, serve as 'quotations,' 'collages,' or even 'illustrations.' They are not being

belief that content can be presented in new formats and packaging, in a new light, without being transformed and transforming, since

> each grafted text continues to radiate back toward the site of its removal, transforming that, too, as it affects the new territory. Each is defined (thought) by the operation and is at the same time defin*ing* (think*ing*). … Inserted into several spots, modified each time by its exportation, the scion eventually comes to be grafted onto itself. The tree is ultimately rootless. (Derrida [1972] 1981, 355–56)

Figure 10: Grafting

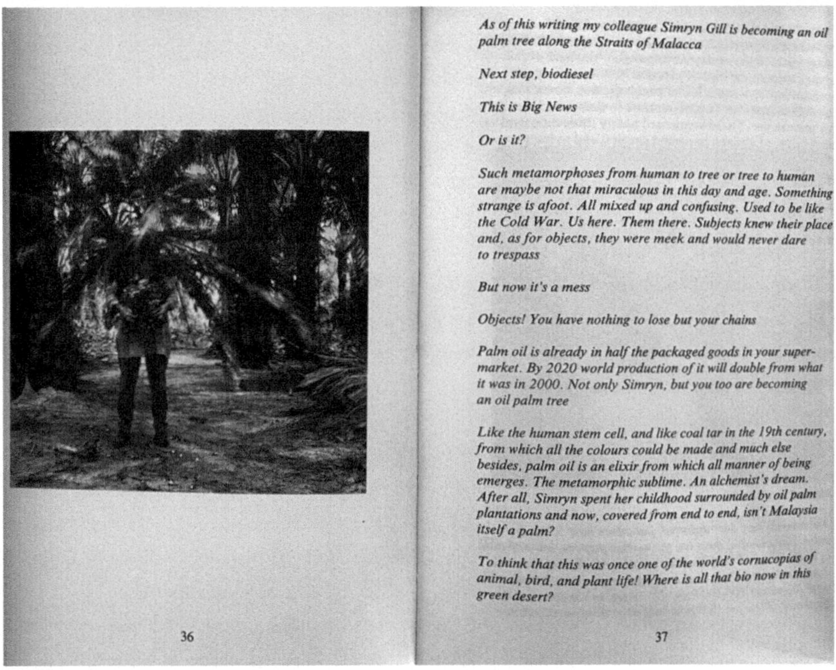

As of this writing my colleague Simryn Gill is becoming an oil palm tree along the Straits of Malacca

Next step, biodiesel

This is Big News

Or is it?

Such metamorphoses from human to tree or tree to human are maybe not that miraculous in this day and age. Something strange is afoot. All mixed up and confusing. Used to be like the Cold War. Us here. Them there. Subjects knew their place and, as for objects, they were meek and would never dare to trespass

But now it's a mess

Objects! You have nothing to lose but your chains

Palm oil is already in half the packaged goods in your supermarket. By 2020 world production of it will double from what it was in 2000. Not only Simryn, but you too are becoming an oil palm tree

Like the human stem cell, and like coal tar in the 19th century, from which all the colours could be made and much else besides, palm oil is an elixir from which all manner of being emerges. The metamorphic sublime. An alchemist's dream. After all, Simryn spent her childhood surrounded by oil palm plantations and now, covered from end to end, isn't Malaysia itself a palm?

To think that this was once one of the world's cornucopias of animal, bird, and plant life! Where is all that bio now in this green desert?

36

37

Source: Simryn Gill and Michael Taussig. *Becoming Palm* (2017), 36–37. Image: Simryn Gill, from *Vegetation* (2016).

The pathway of palms had taught us that objects could be actionable. This was a key argument in the texts we had read beforehand. The line between the palm as a background, scenic, and/or picturesque representational image and the palm as a living, botanical, exotic, and/or cultivated object had long been blurred:

---

applied upon the surface or in the interstices of a text that would already exist without them" (Derrida [1972] 1981, 355; see also Sollers 1968).

> Photographs presented in the [colonial] lectures, books, and postcards formed a visual grammar that over time, through successive reproduction and repetition, defined what was characteristic or representative of the island. ... In the case of Jamaica at the turn of the twentieth century, although imagers of the New Jamaica concerned themselves with crafting a modern Jamaica, they did so through a purposeful use of images from the past; namely, eighteenth-century plantation paintings and nineteenth-century naturalist representations. (Thompson 2006, 34)

The palms I first confronted in this new Gather.town environment were, in fact, scenery (they, as objects, were not actionable). But embracing the possibility that they might offer themselves up for acts beyond that of looking and recognizing that I am the active agent within this process is the only way to find the actionable object within them: the note. So, while the palm as scenery is not in and of itself actionable, approaching the palm as pure scenery will prevent one from progressing. Palms were already singled out in the texts we read as objects that were historically *made to be scenery* rather than objects that were essentially and only that, so it makes sense that, if one wants to think about how to put into practice the lessons from the texts, one must first think about how to position and reconfigure these objects anew.[19] The challenge was to not only show the history of palm tree dreams, of palm fantasies and sites of projection, but also to appropriate and put into a new practice the speculation that these dreams once embodied, while considering the different instruments at my disposal to do so and their visual politics of pedagogy, the epistemic values inscribed into these new foreign environments of remote learning.

It was an experience of estrangement that felt familiar—a dizzying groundlessness that takes over in a completely new territory, almost like living abroad for the first time. Since study abroad programs were put on hold during the Covid pandemic, there's been talk about to how to make them more accessible (see Durden 2020). Mobility as a precondition for in real-life study abroad has meant that only "2 percent of all undergraduates and 16 percent of those who earn a bachelor's degree" in the U.S. take part, leading some institutions to experiment with virtual exchange as a way to address this equity barrier (Fischer 2021). But it's not really the *what* of study abroad that is key, the Google Earth VR technology that shows students the iconic monuments and museums they know from postcards (see Redden 2020). It's also the *how* of study abroad—how experiences of estrangement in a radically different environment can make students question what they thought they already knew. "'What matters isn't place but what happens in that place'" (Larry A. Braskamp

19 On "the conversion or re-purpose ... of things ('Zweckentfremdung der Dinge') contrary to their original design intention," see Bieling et al. in this volume as well as Schüttpelz (2006) and Brandes and Erlhoff (2006).

in Fischer 2015). It's normal for a temporal lag to accompany some of the most vivid and meaningful experiences from studying abroad. Something in this new environment felt disorienting and uncomfortable in the moment, and after returning home, students kept reflecting on it, mulling it over, learning from it, and incorporating it into their everyday perspectives and worldviews (see Fischer 2013). "The most common type of memories [from study abroad] were ones that caused anxiety related to confronting difficult, stressful situations of being in a different culture. Alumni reported these types of experiences were still significant in providing meaning to their lives, even decades later. Can virtual education abroad reproduce these long-lasting effects?" (Whalen 2020). Or, for those of us (still) teaching remotely: Can the strange video conferencing landscapes and landmines we have found ourselves navigating make us more flexible when it comes to finding ways of fostering these long-lasting effects in our own classroom cultures, particularly given that these environments were usually as new for instructors as they were for students?

David Davies writes about this in his account of how he accidentally buried his virtual anthropology class. In their first *Minecraft* session in January 2021, Davies noted that "students gave examples of how they *couldn't make sense* of what was going on. They had problems with learning the controls. They couldn't orient themselves. ... They didn't know *how to comport themselves* in the new context" (2021). It gets interesting when Davies's next experiment with the platform fails, and the instructor is then no longer the tour guide. A key portal in their virtual classroom is obstructed, leaving the class, including the instructor, dispersed, trapped, and confused in the dangerous underworld dimension of the game, the Nether.

> If I, in any way, considered the Minecraft experiment comparable to study abroad, I was failing on our first day and I was getting panicked. I'm used to having control over class time and the general order of things in the classroom. Yet, at that moment, I had no control. My class was spread across two dimensions in a virtual world during a pandemic. And, we still had to get to the day's assigned text! What about course content?! (Davies 2021)

What about course content? We're used to thinking of media technology in the classroom as a delivery method for course content. When it doesn't function in this way, when students disobey the media devices policy in the syllabus by filling their laptops and smartphones with content unrelated to the course, we think of it as an adversarial screen. The second screen debate brings many ideological beliefs about the relationship between pedagogical practices and media technology to the surface, from "highbrow" versus "lowbrow" content and "active" versus "passive" engagement in "public" versus "private" spheres, to information "directly" related to course content versus information that "distracts." Already in 1971, when the Open University in the UK first experimented with broadcasting course content to its students via radio and television, lowbrow associations with these medial forms presented a problem as well as a potential for knowledge disseminated and authorized by a university.

> If universities and museums have traditionally been given the job of establishing and maintaining what Raymond Williams called "the selective tradition" of so-called high culture ... then television became one of the most powerful technologies for questioning the unassailable consecration of the masterpiece and the genius. A television-based university was therefore caught in an ambivalent position between forces that sought to maintain a hierarchy of cultural production and those for whom culture was everywhere and ordinary. (Highmore 2018, 182; Williams 1977, 115)[20]

Conventional approaches to "lowbrow" medial forms that outright disregard or diminish their epistemic values and potential for knowledge production and dissemination at the university level are often informed by hierarchical beliefs about learning modalities, such as "the prospect of the mass of passive listeners who were gullible, suggestible, malleable, impressionable, compliant. ... assum[ing] that the default position of the listener was one of passivity out of which they had to be jolted" (Lacey 2013, 114). In the formative era of broadcasting, "distracted listening" was something that men, based on the gendered correlation of the private sphere with feminine domesticity, had to try to actively resist.[21] The history of the concept of distraction itself across various media reveals similar fluctuations about "good" versus "bad" forms of distraction according to class-based and gendered ideologies.[22]

---

20 "The OU [Open University] was born from an insistence that technology mattered, that technology shaped our perception of the world. As an institution, it shared [John] Berger's sense of seriousness as well as his desire to question the assumptions that presumed the natural superiority of one mode of cultural production over another" (Highmore 2018, 183; Berger 1972). This bias against broadcast media perpetuated despite the many instances at the OU when, for example, "a teaching exercise [in television] turned into research" (Benton 2018, 95). "You might think that, between writing texts and doing television, radio, or radiovision programmes, the media would be more lightweight, but in fact, in my experience, it was always the other way around. A lot of research that I later developed and that others who were involved in the course [A305: History of Architecture and Design 1890–1939] later developed, came out of television and radio, because every programme was primary research" (Benton 2018, 96). See also Moreno (2020).

21 "The assumption here [even by those who were championing radiogenic artforms, such as Rudolf Arnheim] that the listener is male is not just a product of the linguistic bias of the age. It also has to do with the pervasive but unspoken alignment of distracted listening with the domestic and therefore feminine sphere. The male listener has to *struggle* against the feminine surroundings to achieve mastery over the acoustic environment" (Lacey 2013, 129).

22 "The actual hype of a deep-attention reading is, seen from a media archaeological perspective, not simply nostalgic. It forgets its 'dark side,' as it was seen in the civil cultures of the 18th

While, before Covid, we could still try to fool ourselves into believing that our policies on (against) media devices in the classroom worked for us and were necessary on our syllabi to prevent temptations, distraction, and mischief, we now have to return to and complicate the original *Minecraft* question: What about course content when the second screen is not only inevitable but also, as in the case of video conferencing, an essential part of the means of delivery and interaction?[23] How can our mandatory, synchronized foray into virtual environments that experiment with the second screen present an opportunity for us to better understand, experience firsthand, and test out the complex relationships between technology and pedagogy in these historical media developments and practices, such as the history of remote university education via television and radio and the gendered lineage of "active" viewing and "distracted" listening? Such experiments would first have to acknowledge that the second screen we pay so much attention to in times of remote instruction was always already more than a means of delivery, and this goes for instructors and students alike. After all, for whom is the experiment in the experimental humanities intended?[24] And how might we use these new video conferencing platforms to set up experimental classroom systems that are deliberately not clearly defined from the outset, so we can generate new questions altogether (Rheinberger [2001] 2006, 33)? If research begins with the choice of a system, it might be worth considering how to make getting lost an essential part of the process of returning to campus too.

---

and 19th century, when bored middle-class women were accused of being addicted to reading novels and were condemned for escaping into exciting dream worlds. Deep concentration was regarded as dangerous then, because it leads to absentmindedness and even mental confusion, making individuals unusable, particularly for a capitalist economy. ... I was surprised to read in *Dialectics of Enlightenment* that, according to [Theodor W.] Adorno and [Max] Horkheimer, a total excess of distraction comes close to art in its extremity. ... In this passage of their book, Adorno and Horkheimer are saying ... that an accumulation and intensification of distraction is able to fulfill the task of negation that was originally dedicated to art, because it alters the state of the subject in the world completely. With this thought in mind it would be really funny and, at the end much less elitist, to speculate on what Adorno would say about the Internet" (Löffler 2013, 552, 554–55). See Löffler (2014, 315–16) and Horkheimer and Adorno ([1947] 1996, 143–50).

23 See Kanth (2020).

24 "Most of us [in the humanities] still tend to be nonexperimentalists. We stick with what already exists, seeing our objects of study as finished products, faits accomplis ... works of literature created three hundred years, thirty years, or three years before we turn our attention to them. Completed before our arrival and summoned now only to be observed and critiqued, these antecedent objects stand at an input-discouraging distance. ... We don't dream of collaborating with these texts, nor do we design experiments to test their behavior under altered circumstances" (Dimock 2017, 243).

References

Abraham, Melissa. 2021. "UCLA establishes Department of European Languages and Transcultural Studies." *UCLA Newsroom*, February 8, 2021. https://newsroom.ucla.edu/releases/new-department-european-languages-transcultural-studies.

Archer, Neil. 2016. *The Road Movie: In Search of Meaning*. New York: Wallflower Press.

Bailenson, Jeremy N. 2021. "Nonverbal Overload: A Theoretical Argument for the Causes of Zoom Fatigue." *Technology, Mind, and Behavior* 2 (1). https://doi.org/10.1037/tmb0000030.

Bell, Taryn. 2021. "Do Students Hate Online Breakout Rooms?" Twitter, March 18, 2021, 1:47 a.m. https://twitter.com/tarynlbell/status/1372469644997619715.

Benjamin, Walter. (1955) 1978. "Louis-Philippe, or the Interior." In *Reflections: Essays, Aphorisms, Autobiographical Writings*, edited by Peter Demetz, translated by Edmund Jephcott, 154–156. New York: Harcourt Brace Jovanovich.

Benton, Tim. 2018. "A305 Course Team Chair." Interview with Joaquim Moreno on July 20, 2017. In *The University Is Now on Air: Broadcasting Modern Architecture*, edited by Joaquim Moreno, 90–97. Montréal: Canadian Centre for Architecture and Jap Sam.

Berger, John. 1972. *Ways of Seeing*. Television series. London: BBC

Brandes, Uta, and Michael Erlhoff. 2006. *Non Intentional Design*. Cologne: Daab.

Colomina, Beatriz. 2014. "The Century of the Bed." In *The Century of the Bed*, edited by ARGE curated by_vienna, 18–23. Vienna: Verlag für moderne Kunst.

Coy, Wolfgang, and Claus Pias, eds. 2009. *PowerPoint. Macht und Einfluss eines Präsentationsprogrammes*. Frankfurt am Main: Fischer Taschenbuch Verlag.

Davies, David. 2021. "So, I Buried My Anthropology Class Alive…" *Museumfatigue.org*, January 30, 2021. https://museumfatigue.org/2021/01/30/so-i-buried-my-anthropology-class-alive.

Derrida, Jacques. (1972) 1981. *Dissemination*, translated by Barbara Johnson. Chicago: University of Chicago Press.

Dimock, Wai Chee. 2017. "Experimental Humanities." *PMLA* 132 (2): 241–249.

Durden, William G. 2020. "An Expanded Notion of Study Abroad." *Inside Higher Ed*, July 28, 2020. https://www.insidehighered.com/views/2020/07/29/people-who-practice-international-traditions-and-cultures-within-united-states-must.

Fang, Irving. 1997. *A History of Mass Communication: Six Information Revolutions*. Boston: Focal Press.

Fetters, Ashley. 2020a. "Washington's Secret to the Perfect Zoom Bookshelf? Buy It Wholesale." *Politico*, December 26, 2020. https://www.politico.com/news/magazine/2020/12/26/books-by-the-foot-washington-dc-covid-books-440347.

Fetters, Ashley. 2020b. "We Need to Stop Trying to Replicate the Life We Had." *The Atlantic.com*, April 10, 2020. https://www.theatlantic.com/family/archive/2020/04/why-your-zoom-happy-hour-unsatisfying/609823.

Fischer, Karin. 2013. "Your Brain on Study Abroad: The Experience Changes Lives, and Neurons, a Scholar Says." *The Chronicle of Higher Education*, May 31, 2013. https://www.chronicle.com/article/your-brain-on-study-abroad-the-experience-changes-lives-and-neurons-a-scholar-says.

Fischer, Karin. 2015. "Why a Global Education Doesn't Have to Mean Going Abroad." *The Chronicle of Higher Education*, August 12, 2015. https://www.chronicle.com/article/why-a-global-education-doesnt-have-to-mean-going-abroad.

Fischer, Karin. 2021. "The Year Without Study Abroad." *The Chronicle of Higher Education*, April 1, 2021. https://www.chronicle.com/article/the-year-without-study-abroad.

Frommer, Franck. (2010) 2012. *How PowerPoint Makes You Stupid: The Faulty Causality, Sloppy Logic, Decontextualized Data, and Seductive Showmanship That Have Taken Over Our Thinking*, translated by George Holoch. New York: New Press.

Fuchs, Mathias. 2019. *Phantasmal Spaces: Archetypical Venues in Computer Games*. New York: Bloomsbury.

Gadamer, Hans-Georg. (1960) 1989. *Truth and Method*. Translated by Joel Weinsheimer and Donald G. Marshall. New York: Crossroad.

Galloway, Alexander R. 2004. *Protocol: How Control Exists After Decentralization*. Cambridge, MA: MIT Press.

Galloway, Alexander R. 2006. *Gaming: Essays on Algorithmic Culture*. Minneapolis: University of Minneapolis Press.

Gill, Simryn. 2016. *Vegetation*. Photograph. Silver gelatin print.

Gill, Simryn, and Michael Taussig. 2017. *Becoming Palm*. Berlin: Sternberg Press.

Glas, René, Kristine Jørgensen, Torill Mortensen, and Luca Rossi. 2011. "Framing the Game: Four Game-Related Approaches to Goffman's Frames." In *Online Gaming in Context: The Social and Cultural Significance of Online Games*, edited by Garry Crawford, Victoria K. Gosling, and Ben Light, 141–58. London and New York: Routledge.

Goffman, Erving. 1974. *Frame Analysis: An Essay on the Organization of Experience*. Cambridge, MA: Harvard University Press.

Gold, Rich. 2002. "Reading PowerPoint." In *Working with Words and Images: New Steps in an Old Dance*, edited by Nancy Allen, 256–270. Westport, CT: Ablex.

Highmore, Ben. 2018. "Televarsity: At Home with The Open University." In *The University Is Now on Air: Broadcasting Modern Architecture*, edited by Joaquim Moreno, 177–184. Montréal: Canadian Centre for Architecture and Jap Sam.

Hodges, Charles, Stephanie Moore, Barbara Lockee, Torrey Trust, and Aaron Bond. 2020. "The Difference Between Emergency Remote Teaching and Online Learning." *EDUCAUSE Review*, March 27, 2020. https://er.educause.edu/articles/2020/3/the-difference-between-emergency-remote-teaching-and-online-learning.

Horkheimer, Max, and Theodor W. Adorno. (1947) 1996. *Dialektik der Aufklärung. Philosophische Fragmente*. Frankfurt am Main: Fischer.

Hüser, Rembert. 2020. "Der Professor als Zeremonienmeister." *FAZ*, April 25, 2020. https://www.faz.net/aktuell/karriere-hochschule/erste-woche-online-uni-der-professor-als-zeremonienmeister-16740870.html.

Imhof, Maria. 2014. "Automobil." In *Wörterbuch kinematografischer Objekte*, edited by Marius Böttcher, Dennis Göttel, Friederike Horstmann, Jan Philip Müller, Volker Pantenburg, Linda Waack, and Regina Wuzella, 21–23. Berlin: August.

Ishak, Andrew. 2020. "Making Your Zoom Look More Professorial." Vimeo video, 5:46. August 13, 2020. https://vimeo.com/447645552.

Jacobs, Naomi, and Joseph Lindley. 2021. "Room for Improvement in the Video Conferencing 'Space.'" Paper presented at *AoIR 2021: The 22nd Annual Conference of the Association of Internet Researchers*, September 15, 2021. https://doi.org/10.5210/spir.v2021i0.12188.

Kanth, Tanmay. 2020. "when the professor who banned phones and laptops in class now has to learn how to teach online." Facebook, March 11, 2020. https://www.facebook.com/groups/underfunded/permalink/3522278767787552.

Katz, James E., and Elizabeth Thomas Crocker. 2017. "Visual Interpersonal Communication in Daily Life: Skype as a Precursor of Perpetual Visual Contact." In *Between the Public and Private in Mobile Communication*, edited by Ana Serrano Tellería, 60–76. Abingdon: Routledge.

Knoblauch, Hubert. 2008. "The Performance of Knowledge: Pointing and Knowledge in Powerpoint Presentations." *Cultural Sociology* 2 (1): 75–97.

Lacey, Kate. 2013. *Listening Publics: The Politics and Experience of Listening in the Media Age*. Cambridge, UK: Polity.

Laderman, David. 2002. *Driving Visions: Exploring the Road Movie*. Austin: University of Texas Press.

Latulipe, Celine. 2021. "A CS1 Team-Based Learning Space in Gather.Town." *SIGCSE '21: Proceedings of the 52nd ACM Technical Symposium on Computer Science Education*, March 5. Page number 1245. https://doi.org/10.1145/3408877.3439587.

Lee, Isaac. 2020. "Penn Students Use Digital Platform Gather to Imitate In-Person Office Hours." *The Daily Pennsylvanian*, October 22, 2020. https://www.thedp.com/article/gather-town-penn-cis-virtual-office-hours.

Lindley, Joseph, David Philip Green, Miriam Sturdee, and Hayley Alter. 2021. "This Is Not a Paper: Applying a Design Research Lens to Video Conferencing, Publication Formats, Eggs ... and Other Things." *Extended Abstracts of the 2021 CHI Conference on Human Factors in Computing Systems (CHI EA '21)*. Association for Computing Machinery, New York, NY, USA, Article 16, 1–6. https://doi.org/10.1145/3411763.3450372.

Lorraine, Lisa. 2020. "#quarantine #teachersoftiktok #desperatetimes #biology #innovation." TikTok video, 00:09. April 7, 2020. https://www.tiktok.com/@lisa.tomato/video/6813185032732167430.

Löffler, Petra. 2013. "The Aesthetics of Dispersed Attention: An Interview with German Media Theorist Petra Löffler." Interview by Geert Lovink. *NECSUS: European Journal of Media Studies* 2 (2): 545–55.

Löffler, Petra. 2014. *Verteilte Aufmerksamkeit: Eine Mediengeschichte der Zerstreuung*. Zürich: Diaphanes.

Lupton, Ellen, and J. Abbott Miller. 1993. "Critical Wayfinding." In *The Edge of the Millennium: An International Critique of Architecture, Urban Planning, Product and Communication Design*, edited by Susan Yelavich, 220–232. New York: Cooper-Hewitt, National Museum of Design, Smithsonian Institution, and Whitney Library of Design.

McClure, Colin D., and Paul N. Williams. 2021."Gather.town: An Opportunity for Self-Paced Learning in a Synchronous, Distance-Learning Environment." *Compass: Journal of Learning and Teaching* 14 (2): 1–19.

Metz, Rachel. 2020. "Forget Zoom: You Can Do Meetings on a Website That Looks Like a Retro Video Game." *CNN Business*, October 6, 2020. https://www.cnn.com/2020/10/06/tech/gather-zoom-video-game-chat/index.html.

Mills, Katie. 2006. *The Road Story and the Rebel: Moving Through Film, Fiction, and Television*. Carbondale, IL: Southern Illinois University Press.

Moreno, Joaquim. 2020. "Being Open." *e-flux Architecture*, May 2020. https://www.e-flux.com/architecture/education/332216/being-open.

Mudge, Ken, Jules Janick, Steven Scofield, and Eliezer E. Goldschmidt. 2009. "A History of Grafting." *Horticultural Reviews* 35: 437–493.

Mukherjee, Souvik. 2016. "Videogames as 'Minor Literature': Reading Videogame Stories through Paratexts." *GRAMMA: Journal of Theory and Criticism* 23: 60–75.

Patton, Elizabeth. 2020. *Easy Living: The Rise of the Home Office*. New Brunswick, NJ: Rutgers University Press.

Pias, Claus. 2020. "Overhead Projector." Translated by Valentine A. Pakis. In *The Oxford Handbook of Media, Technology, and Organization Studies*, edited by Timon Beyes, Robin Holt, and Claus Pias, 286–300. Oxford: Oxford University Press.

Redden, Elizabeth. 2020. "Study Abroad Faces a New Reality." *Inside Higher Ed*, April 27, 2020. https://www.insidehighered.com/news/2020/04/28/study-abroad-field-faces-altered-reality-and-unknowns-about-when-regular-programs.

Rheinberger, Hans-Jörg. (2001) 2006. *Experimentalsysteme und epistemische Dinge: Eine Geschichte der Proteinsynthese im Reagenzglas*. Frankfurt am Main: Suhrkamp.

Robles-Anderson, Erica, and Patrik Svensson. 2016. "'One Damn Slide After Another': PowerPoint at Every Occasion for Speech." *Computational Culture* 5, January 15. http://computationalculture.net/one-damn-slide-after-another-powerpoint-at-every-occasion-for-speech.

Ryan, Marie-Laure. 2016. "Narrative." In *Debugging Game History: A Critical Lexicon*, edited by Henry Lowood and Raiford Guins, 335–41. Cambridge, MA: MIT Press.

Saturday Night Live. 2020. "Zoom Church." YouTube video, 3:08. May 9, 2020. https://www.youtube.com/watch?v=AYP1mXqiwqc.

Schabacher, Gabriele. 2017. "Im Zwischenraum der Lösungen. Reparaturarbeit und Workarounds." *ilinx—Berliner Beiträge zur Kulturwissenschaft* 4: XIII–XXVIII.

Schüttpelz, Erhard. 2006. "Die medienanthropologische Kehre der Kulturtechniken." *Archiv für Mediengeschichte* 6: 87–110.

Sollers, Philippe. 1968. *Nombres*. Paris : Éditions du Seuil.

Stanitzek, Georg. 2005. "Texts and Paratexts in Media." Translated by Ellen Klein. *Critical Inquiry* 32 (1): 27–42.

Stanitzek, Georg. 2006. "Vorspann (titles/credits, générique)." In *Das Buch zum Vorspann: "The Title is a Shot,"* edited by Alexander Böhnke, Rembert Hüser, and Georg Stanitzek, 8–20. Berlin: Vorwerk 8.

Stanitzek, Georg. 2009. "Reading the Title Sequence (Vorspann, Générique)." *Cinema Journal* 48 (4): 44–58.

"The Importance of Pauses in Conversation." 2017. *The Economist*, December 14, 2017. https://www.economist.com/books-and-arts/2017/12/14/the-importance-of-pauses-in-conversation.

Thompson, Krista A. 2006. *An Eye for the Tropics: Tourism, Photography, and Framing the Caribbean Picturesque*. Durham, NC, and London: Duke University Press.

Tufte, Edward R. 2003. *The Cognitive Style of PowerPoint*. Cheshire: Graphics Press.

Whalen, Brian. 2020. "Education Abroad in a Post-COVID-19 World." *Inside Higher Ed*, April 13, 2020. https://www.insidehighered.com/views/2020/04/14/how-covid-19-will-change-education-abroad-american-students-opinion.

Williams, Raymond. 1977. *Marxism and Literature*. Oxford: Oxford University Press.

Wirth, Uwe. 2014. "Between Hybrid and Graft." In *From Literature to Cultural Literacy*, edited by Naomi Segal and Daniela Koleva, 232–249. New York: Palgrave Macmillan.

Teaching Into the Void

Reflections on "Blended" Learning and Other Digital Amenities[1]

Donatella Della Ratta

> "Tried to save myself but my self keeps slipping away
> Tried to save myself but my self keeps slipping away
> Tried to save myself but my self keeps slipping away
> Tried to save myself but my self keeps slipping away
>
> Talking to myself all the way to the station
> Pictures in my head of the final destination"
> —Nine Inch Nails, "Into the Void"

> "Think of those humans who are attractive for the primary reason of how the presentation of their body is impenetrable or brooding or fierce
> or impassive with brooding or fierceness.
> This category of desire is simple, slightly mechanistic:
> to penetrate the brooding, fierce, impenetrable presentation."
> —Ann Boyer

Preface

This essay combines the experiences of teachers and students[2] in the turn to "blended learning" during the Covid-19 pandemic, to build prefatory theories of new, emergent phenomena—from the exhaustion produced by constant "self-gazing" and its role in "zoom fatigue," to the anxieties and oppressions of the dominant interfaces (The Tyranny of the Rectangle) and their

---

1 "Teaching into the Void: Reflections on 'Blended' Learning and Other Digital Amenities" was originally published as part of the Institute of Network Cultures' Longforms on January 6, 2021, and can be found online here: https://networkcultures.org/longform/2021/01/06/teaching-into-the-void. The postscript was added in April 2022, and it appears for the first time in this anthology.

2 All interviews were conducted with students and teachers of John Cabot University in Rome unless otherwise noted. Those who wished to be interviewed anonymously remain unnamed.

consequential resistance techniques (Camera On, Or Camera Off: That is the Question). Through anecdotes and personal accounts, we will enter a founding critique and examination that aims to walk these interactions and isolations with both compassion and criticality, toward a counter-politics reliant on exposure, as well as poesis. A counter-politics that finds and forms itself in the aural rather than the visual (Intimacy Out of the Void), one that is most present (and most potent) in the "awkward moments" of lags, lapses, glitches, bandwidth failures, and frozen frames (In Praise of Awkwardness).
The emphasis lies throughout on withstanding naturalization and remaining in defiance of the "normalization" efforts and narratives that pertain to all of the above. There is an urgency here.

This is a direct account, examination, and rebuttle of what it means to be a teacher and a student during the techno-determinist moment of "the Corona Crisis"—a crises of education, embodiment, and alienation, in more ways than one.

"Shall I look over *here*? Over *here*, or over *here*?" Igor "Yes Men" Vamos asked his clueless and appalled students in a hilarious parody[3] of our Zoom/Teams/Meet daily tragedy. Vamos's skit has become a sad, accurate reproduction of the everyday experience of both students and teachers, now forced to study under digitalized, remote conditions resultant of the Covid-19 pandemic. This crisis of Corona has produced innumerable crises of and for education. These crises span the troubles of sociality and substitutive interfaces, to the ails of the isolated individual bodies and minds—troubles which arise simply from being a human subjected to these new environments in their various forms of virtuality and isolation.

To begin with the semantics, there is an entry point provided by the introduction of what has come to be known as "blended learning." As a teacher, before being obligated to move to this environment (aka "hybrid") where we would teach both online and on site—which we can see was and is an attempt to achieve ubiquity and maintain the myth of multitasking as noble, wise, and possible—those words, "blended learning," sounded sweet and promising.

Digital capitalism has a talent for name-selection and (re)branding flawed substance with bombastic labels. For me, "blended learning" initially evoked images of British chaps drinking impossible amounts of scotch in crowded pubs, a motif that my domesticated post-pandemic imagination no longer even dares to resurface without triggering fears of the (self)surveillance and (self)sanctions that the virus has made us all so familiar with since. Instead, what my Proustian *petite madeleine* now brings forth, is an aseptic nightmare of wires—a rather coherent Covid-free sanitized environment, liberated from people drinking, sweating, and shouting aloud together.

---

3 https://www.youtube.com/watch?v=JcHOKNy8Omo.

"Blended learning" is not a whiskey brand. As pompously stated[4] by one of the thousands of learning platforms that have mushroomed out of the shadows of Covid-19, it is a "formal education program where students learn at least through some online delivery of content and instructions with some elements of students control over time, path, place and pace; and at least in part at a supervised brick and mortar location way from home."

This is what we're all supposed to do after measures such as social distancing, forced lockdown and curfews, have been implemented. The question is, within this, what are we really "in control" of? What is it that makes it a "personalized learning experience" and "unique to your students," as the platform promotes and attempts to sell? How can we use words such as "individualization," "personalization," and "differentiation"—"tailored to meet each individual student's needs"—to describe an experience that looks increasingly similar, as Sofie Smeets[5] writes, to a *séance*?

"Can you hear me?"
"Are you there?"
"If you can hear me, can you give me a sign?"

Most of the platforms we use on a daily basis for work and leisure have names that suggest collectivity and togetherness. They hint toward shared spaces that do not exist, and the hinting begins from the very moment we call these spaces into being, from the moment we engage in the performance of their sociality. We say, "let's meet on Zoom" or "See you on Skype," as if we are meeting on stools at a bar, on benches in public squares, or on the dance floor awaiting a concert. When in reality, we grab a drink and sip it in front of the gallery view. We watch our friends resemble the *Brady Bunch* opening credits, looking like a virtual stamp collection in their tidy, repetitive, rectangular spaces. Throughout this charade, the oddest element is that we spend the social hour not so much looking at the gathering, rather we sit there instinctively watching, monitoring ourselves.

No matter whether one is casually drinking with friends, holding an online birthday celebration, showing up to a work meeting, or teaching a class—the self-reflecting gaze is always on, ever-present whenever the camera is on, too. It is a Narcissus-like situation, except instead of loving yourself to death, you develop a constant sense of inadequacy and anxiety: Do I look professional enough? Do I look good enough? Do I look fresh enough? *Fix your hair, apply a "resting smile," adjust the camera angle to look less puffy, less tired. Just. Look. Happier.*

---

4 https://www.christenseninstitute.org/blended-learning.

5 https://www.uu.nl/en/news/are-you-there-can-you-hear-me-the-impact-of-covid-19-on-higher-education.

Tending to the Face: From Surgery to Ring Lights, We're All Influencers Now

The cosmetic surgery market is booming in the time of the pandemic. In the Italian newspaper *Repubblica Federica*, a "life coach" from Bergamo confessed that once having to switch to "smart" (online) working because of the lockdown, she began noticing flaws on the skin of her face, which she had not perceived to be there before. Valeriano Vinci, a surgeon from Milan, says "(the spike in cosmetic surgery) is a bit like what happened with selfies, now it's because of video conference calls that we are obliged to gaze at our own image on the screen."[6]

> "How frequently do you look in the mirror? Does your face please you?
> Are you disgusted to detect familial features? Do you worship or hate your ancestors?
> Do you consider your image erotic? Do you pretend that you are a star's child?
> If you squint, does your reflection become abstract? Is abstraction a transcendental escape from identity or a psychotic spasm of depersonalization?"
> —*Wayne Koestenbaum*

If surgery seems a little extreme, why not try a ring light? The ring light is a face-flattering "doughnut of light." A miracle lighting device that "cleans" your face of all shadows by delivering brightness in a circle rather than from a single point. The ring light was a nerdy thing before March 2020. Prior to the mass-conversion to time spent in front of webcams, ring lights were part of the tool belt of Instagram influencers, YouTubers, TikTokers, and new/next-generation digital "creators" (the masters of online, facial performances). Pre-pandemic, ring lights were mostly known and used by those who were livestreaming from inside their home aquariums of kitchens, bathrooms, or bedrooms, implementing their illuminating doughnuts to impress followers, gain more subscribers, and increase their likes and shares. An army of YouTubers, Twitch streamers, OnlyFans creators, Chaturbate cam models are now sharing their hack for eternal beauty with office employees, businessmen, fitness instructors, and us—teachers.

"Americans had spent the past decade mastering the momentary muscle movements of a good selfie, but starring in a high-quality live video in front of co-workers or romantic prospects for hours at a time is a different beast entirely," writes Amanda Mull in *The Atlantic*.[7] "People had no idea how to contend with broadcasting their own face—weird shadows, awkward backdrops, and under-the-chin shots from low-slung laptops abounded." How to make the flattening gallery view look

---

6 https://www.ilprimatonazionale.it/cronaca/chirurgia-estetica-lockdown-177223/.

7 https://www.theatlantic.com/technology/archive/2020/11/ring-lights-for-all/617143/.

alive again, as if people were moving in an actual place, crowded and swarmed with life? How to render the boring *aesthetics of the rectangular* into an object glowing out from the uniformity and numbness of the on-life?

Suddenly, everyone was looking for a tip to just look better, less boring, and more professional. The ring light is the revenge of Gen Z against Boomers and white-collar workers. "Those kids" who spent hours dancing in front of their laptops, inventing silly choreographies, live commenting as somebody else played a video game, and making DIY tutorials for pretty much anything and everything, are now teaching the global working class a lesson. In the time of a pandemic we are all, whether we like it or not, living the backend, glamourless life of influencers. We all work for ourselves and with ourselves, from inside of our bedrooms and kitchens. Thus, the ring light has become a must-have investment, no matter your profession. It is the most wanted gadget of the work-from-home mandate.

Unlike an ergonomic chair, a specialist microphone, the latest HD webcam, noise-canceling headphones, or an oversized office printer that we had never considered to own before, the ring light is frivolous. It is a fierce reminiscence of the "good life," disco culture, glamour parties, and crowded photo shoots, all bundled into one. The ring light does not originate from boring corporate settings, rather from the realms of make-up, TikTok dances, and sex cams. It glows in the dark, illuminates your visage, frees you from the visual constraints of the desk and the desktop, and most importantly, makes your skin look younger and positively radiant. Move over Bill Gates, the essential office products are now brought to you by James Charles and Charli D'Amelio. The ring light is the ultimate gift for a locked-down and curfewed Christmas. It gives all of the glamor, excitement, and frivolity one surely deserves during these dark times.

> Something I would have never thought that I could have done before is dancing in front of thousands of people.
> —*Charli D'Amelio*

The ring is said to give one "manga eyes." These are the peculiar, enlarged style of eyes drawn for characters in Japanese comics. In front of a ring light, eyes become huge and perfectly round with tiny pupils and no iris. Manga eyes convey "a cute, delighted look" and symbolize "extreme excitement," Wikipedia says.[8] Perfect.

> How interesting ... what could it mean for history, that a face is wrong for itself in a time in which all is also so wrong. The animals sit forlorn or ride subways into city centers. The water has become poison. The old behave like the young, and the young are too worried to move. Pilotless weapons have the names of birds,

8 https://en.wikipedia.org/wiki/Manga_iconography.

> so why shouldn't faces, also, lead away from the facts? To the lovers of the contradictions, these faces are a perfect account of our time: the poetry of the wrong.
> —*Anne Boyer*

With this facial upgrading-capability, and when compared to the surveillance-looking, CCTV cameras that have been installed in many educational settings, the ring light looks more like an illuminated path toward liberation. Unlike the old-school, top-down Foucaldian *dispositif*, which makes the student/teacher/subject feel constantly observed and under the control of a panopticon-esque institution (continuously reminding those present that they are part of a surveillance project and performing their own role in it), the ring light juxtaposes lightness with freedom of movement. You are your own *metteur-en-scène*. The camera is in front of you, it glows in the dark and makes you glow return. A hovering portal that sits afore like the entry into Wonderland. There are many roles available: Humpty Dumpty, Alice, the Mad Hatter ... or the Cheshire Cat.

Here, There: Any-Space-Whatever

In his essay from 1995, "The Exhausted," Gilles Deleuze mentions the Cheshire Cat in a description of what he calls "Langue III"; a stage of language that comes after that of names and voices, of rationality and memory, of objects and representation. It's the language of an image that is rendered into a *process* rather than into *content*, divorcing it from the thing represented, disengaging and transforming it into "a possible event that doesn't even have to realize itself in the body of an object any longer." Like the Cheshire Cat's disembodied eyes and smile in Lewis Carroll, images appear and disappear in and out of thin air. They dissolve into space, *a* space, that Deleuze writes can be "any-space-whatever." Here we are, from this *any-space-whatever*, all glowing, appearing, disappearing, and reappearing, in the shape of floating doughnut ring lights, coming to you live straight out of Wonderland.

The CCTV camera spies on us. It suffocates. The selfie stick triggers us too hard on what we've lost—the travels, the freedom of movements, all that hanging out. But the ring light stares at us *with warmth*. It glows before us and invites us to enter the wonder of its circle. The ring light is, in fact, our Wonderland. We are the Cheshire Cat, disembodied, disconnected from our own flesh, from the others, and from any possible representation. With big eyes, clear skin, and appeasing grin, we are still on the search for all (and any) possible connections, new algorithmic combinations to be done and undone, anything that can be liked and shared, streamed, or streaked. In this never-ending set of variables and possibilities that can be tried and untried, "all order of preference and all organization of goals, all signification" is lost.

As Deleuze said in the title: "Exhausted." Radiantly puzzled by where the tiredness is coming from.

It is here that the semantics and tactics start to bleed, blend, and *become* the exhausting performance. This is the real welcome into the world of blended learning—where the same greeting is given into this hall of mirrors as it is to the mise-en-scène where "collectivity" is in nonstop (re)enactment. Students and teachers alike are busy gazing at themselves, rather than at the "Teams" in front of them. During a video lesson, we are not immune to the constant drive to look at one's self. To the minute camera adjustments in search of the better angle. Worried about the personal image far more than the content you are about to deliver. In the physical classroom we might have a live performance crisis, but in the hybrid environment surely the experience of crisis is one of *the gaze*—not of the Other, but of the self. "Shall I look over *here*? Or over *here*? If you guys all look at your computer, all of you, including those who are in the physical classroom, then I can look at everyone on the screen. I think this is the way to do that ... How do they do it in other blended classes?" Like Igor Vamos in his tragicomic video, all of us have uttered these words—usually in a panic. This is the crisis of the gaze: Where do I look? *Here*, *here*, or *there*?

Where is it *here*, and where is it *there*? Does *here* suggest the physical presence, and *there* hint at the remoteness held by boundaryless and immaterial cyberspace? A student majoring in international affairs voices her frustration toward this imaginary line that divides the two spaces, which (unwillingly but irremediably) becomes a way to assess the "quality" of students.

> There was a general assumption by some professors that the students who were online were not paying attention, just because we were online. Things like "Oh, they are probably away having lunch right now" or, "I wonder if I have anyone's attention from the remote students . . ." were said several times during my classes. Professors need to understand that being online is not like being in a classroom and answering a question in the physical classroom is much easier than doing so remotely—at least it is for me. If we do not answer, it does not mean that we are doing something else. Maybe we just simply do not feel comfortable speaking in that moment.

Meanwhile, on the other side of the screen, teachers are required to behave like jugglers—which is not the easiest thing to do whilst also experiencing performance anxiety. Antonio Lopez, a teacher of communication and media studies, explains:

> When the majority of students were physically present, it seemed to be smoother because most of my attention could be in the classroom and I could incorporate the online students easier. The dynamic reversed itself once the majority of students were online and only a few were in the physical classroom. At that point it just felt like a farce.

Behind the curtain, in the pantomime
Hold the line
Does anybody want to take it anymore?
—*Queen, "The Show Must Go On"*

The Tyranny of the Rectangle

With professors and students both lost, the interfaces we use simply do not help us overcome the geographical confusion to find our bearings. The gallery view reduces the environment to a sequence of geometrical forms and quantified tiny spaces, putting each student into a little cage, forcing the entire education experience into the aesthetics of the rectangle. This is what a "team" looks like in the age of the pandemic: a grid of isolated box shapes confined to unbreakable bounds. You cannot cross the borders of your rectangle, just as you cannot go out of your apartment. Spaces are limited and delimited, even in the limitless domain of the digital. "It's very difficult to create a space of connection, a space of togetherness within this aesthetics of the rectangular," a communications senior student expresses.

The rectangular is what drives attention toward yourself rather than diverting it to others. There is no real *Meet* happening in *Teams*. In the solitudinal aesthetics of the rectangular, the only thing to *Zoom* in on, is yourself. Going beyond yourself means engaging with floating icons that most likely only carry initials (camera off), as if they were patients under medical observation whose identities need to be kept anonymous.

The once innocent, banal rectangular has turned into a powerful self-containment device and peer-surveillance tool. "When online, there's no one to hide behind when all the squares are lined up next to each other, and I'm much more aware of how clean my apartment is in the picture behind me," says Madison, a grad student in art history, "I'm guilty of spending a lot of time looking at my own square when the camera is on. I'm constantly checking to make sure I'm still in the frame and that I'm not making any faces/reactions that I wouldn't make in class." "It is weird," adds Briana, a student in communications, "[...] when physically in class I am completely able to focus my attention on both watching and hearing; online, I was not able to do these two things at once. I was distracted by looking at other people's faces, or my own."

A truth is that in physical presence we are never confronted with our own gaze, neither as teachers nor students. We do not get to see the resting grimace on our face, or our eyes blinking as we explain a concept, just as students do not have the chance to check whether they have a clueless stare or a glimmer of curiosity toward what is being said. Blended learning has put us in the awkward situation where we have to not only sit in constant confrontation with our face, but find the best way to check (and adjust) our expressions to make them convey a certain deliberate emotion, from interest and curiosity, to visible passion and participation.

> The impassive face has its rival: the face that can never hold still. The face is kinetic, elastic ... the curse of digital photographers and bio-informationists who must try to fix, in data, what is in its very form unfixable.
> This face provides an onrush of information which comes so quickly it almost evades processing: this face is prolific, a human comedy of feeling.
> —*Anne Boyer*

Moving from the hypertrophic self-reflecting gaze of the "selfie," which was widely interpreted as a mere sign of the empty aesthetics of pure narcissism, video-calling platforms have added an unprecedented (self)disciplinary dimension. The environment constantly reminds you of your presence and of the Others." Ironically, it is precisely when acquiring a disembodied status that we are reminded of our flesh and bones, skin and muscles—by this enduring and endless mirror gaze.

"If the camera is on, I often look at myself, making sure I am 'presentable' (something I would not care about during a live in-person lecture, as I would not be able to see myself)," tells Giulia Villanucci, who is enrolled in a master's program in the United Kingdom. For Briana, the sense is more a matter of "being watched": "Although I knew it was not the case, it made me stressed." Matjia, a communications senior who has been remote from Serbia for the entire semester, echoes the same concerns, adding that "it seems as if all eyes are on you constantly and you're being surveilled. This, of course, is both good and bad at the same time."

> YouTube videos and practice have taught me all I know
> —*James Charles*

"I enter light and it's from the gaze that ... I am photo-graphed." The camera appears to have finally acquired the magical power of "looking at you." In the anecdote of the sardine can that Lacan recalls in *Seminar XI*, a young Lacan is out at sea on a boat with a group of fishermen when one of them, who goes by the name of Little John, points to something floating in the water: "You see that can? Do you see it?" Little John says to young Lacan, who becomes anxious in a state of discomfort, "Well, it doesn't see you!" Sorry, Little John, but in fact, it does see you. It does see us, now the object gazes back. The mere fact of the built-in webcam reminds us that we are constantly gazed upon, being watched, and under surveillance. "In the depths of my eye the picture is painted, but the subject is not in the picture ... if I am anything in the picture, it is always in the form of the screen ... the stain, the spot," later wrote older Lacan.

Camera On, or Camera Off: That Is the Question

If hypervisibility implies hyper (self and peer) surveillance, then how do you call presence into being in the absence of bodies, in the lack of a physical shared space and its light, smell, and shape? "Camera on" has come to mean "presence" in the blended learning environment. "Why don't you guys switch your cameras on, at least at the beginning of class, so I can see you, you can see each other" is the ubiquitous plea of teachers. A begging and desperate attempt from their side to not teach into the void in preference for staring at other people's faces alongside their own—even if the other faces are reluctant and disinterested. By now this request is frequently met with "Sorry professor, I have a bad connection," "Unfortunately my camera does not work" or "I am connecting from my mobile, I don't have enough bandwidth to turn my camera on atm," to which the teacher is obliged to answer: "Sure, no problem."

Lately, there are increasing suggestions to enforce the use of cameras as a proof of attendance, in the hope that it will encourage students' participation. "Sleeping with camera on: classic! I much prefer it to the mute rectangle," a professor in public speaking tells me, who is strongly in favor of having all cameras on during class time. Giulia Martinez-Brenner, who studies in the Netherlands, writes:

> I cringe when the teacher asks the class *those questions*. The ones that aren't even proper questions, just ones to check and see if everyone is there and following. I stare at the sea of initials and think about my own, GM, that are currently hiding our pitiful conditions. I'm slumped over my chair, obviously still in my pajamas, resembling some sort of slimy invertebrate more than anything else. I have Instagram open in one hand, a coffee in the other, and I can't help but give a small smirk. *I can get away with this?* Without doubt it's a childish satisfaction. I try to convince myself it's freedom ... but is it? The initials stay silent. The professor asks again, and my anonymity hits me with a surge of guilt.

Carlos, a student in international business, is firmly convinced that "cameras should always be on. It is a sign of disrespect to the teachers and fellow students if the camera is off while they are speaking or interacting and I believe 'awkward' is not enough of a word to express how it feels." Briana also empathizes with the professors, noticing that "it was a bit sad when they would gaze at our initials because no one had their camera on, since they probably felt they were talking to nobody." And yet she adds: "This, though, was controversial for me because, while I did want to keep the professor's company, I knew I was going to be less engaged if I turned my camera on."

The shadow of anonymity covers up something other (or more) than lack of interest. If one is embarrassed by their physical presence, they might very well opt for not turning the camera on. A teacher in my communication department tells me that he was enforcing the use of the camera. He eventually gave up on this practice, realizing—only after a long email exchange—that one of his students who had never shown up on camera had a problem with their body image, feeling overweight and hated their reflection so much that they did not even allow mirrors inside their apartment.

Besides the reflection of the self, many students, as well as teachers, do not have the luxury of living in beautiful, spacious apartments. There is a huge variety of housing situations—from living in a foster home to having only a bed in a shared space with many others or even being a grown-up still living at home with Mom and Dad—and this is where "camera off" hints at embarrassment rather than laziness or lack of commitment. "Camera on" has turned into a luxury option that underprivileged people, migrants, proletarians, or anyone experiencing a physical or mental disorder (including stress and anxiety), simply cannot afford. Rather than helping class participation or facilitating the learning experience, it has become an exclusive, divisive, and ultimately oppressive, tool.

Before being tool to surveillance and discipline, the camera is a technology of the self. The camera carries the quasi-magical power of individuation, the quality of putting the self into being. "My camera is off. I started turning it off because I hated catching glimpses of my face in the corner, and I hated even more the thought that other people were seeing the same thing," Giulia Martinez-Brenner writes. If we opt toward merely sanctioning the option of camera-off, without understanding why some prefer it in the first place, we risk losing track of this powerful aspect of the construction of the self.

From a student's perspective, the "always on" camera might generate the endless anxiety of having to restlessly perform the self and put it into being in front of others. This anxiety is, however, not only a "student thing." During this busy semester of webinars all over the place, I heard more than once the solemn declaration being spoken by organizers: "It's nice for our speakers if they can see/hear other people in the panel, it's not just our students who are shy." "It would be nice if folks can turn their cameras on or unmute themselves, so panelists can feel the energy of the crowd... can feel that they are, in fact, speaking to an actual crowd." And so on.

Theaters and cinemas may be shut down, but never has there been so much space for performance of the self. In endless *Zoom-Teams-Meets* we are confronted on a daily basis with the absence not only of the body, but also of faces and voices, performing for the self rather than for others, continuing to maintain the veneer that things are "kind of" normal and business goes on "as usual"—that all of this has simply moved to a different space, and only temporarily (or so, we kid ourselves). We take to this performance both for ourselves, and on behalf of a system. For an infrastructure of life that we desperately want to keep intact while we are well aware of its collapse. Here it lies, what we already know: our daily performance is nothing but a farce.

Intimacy out of the Dark

And yet, sometimes, in these disembodied communications launched into the void, something "intimate" might come to emerge from the darkness. A communications senior reports on a teacher who plays soft music in the background while talking over the slides and sharing the screen: "This voice over with the light music in the background creates a certain intimacy, a shared space, a space for connection." Another teacher uses the same word, *intimacy*, to describe the space created by the faceless sounds he encounters daily in his blended teaching. He is also convinced that sound might be a more powerful dimension—and far more suitable to our internet bandwidth—than video sharing. It appears that the sound dimension can be a liberation from the self-centered (and hyper performative) gaze, providing a way out of the aesthetics of the rectangular.

I cannot help but be reminded of "songs playing from another room," a YouTube microgenre initially made up of pop songs from the 80s and 90s that have been remixed so as to sound as if they were, in fact, being played from another room. This genre now includes hundreds of thousands of remixes from contemporary artists, ranging from Tame Impala to Lana del Rey, and beyond. Jack Wilson, an expert in YouTube videos featuring songs playing from another room," writes that "while these videos are vectors for a number of affects—nostalgia, recalling memories (real and imagined), mourning—and while these affects are diverse, the point from which they emerge is the same: a synthetic sense of something occurring that we are not a part of."[9]

This "something" we are not part of (or no longer part of, at least for the time being) is the overcrowded room. The one too hot in summer and too stuffy still in winter. The one with too much noise coming from the street below. This is the same one with the blackboard that's always scribbled upon from previous classes, the one with a desktop computer that's barely working. The disembodied sound dimension, and its potential for a freeing-up from the anxiety of the hyperconscious self-gaze, might help create an opening for spaces of intimacy and in defiance of connectivity—with its lo-fi, low-tech, glitchy sound, indifferent to the glossiness and emptiness of the high-res camera. Connectivity does not always equal connection—and vice versa.

Teacher, Call Center Operator, Food Courier: One and the Same?

A colleague tells me that he thinks about teachers as similar to call center operators. Weird hybrid bodies, somewhere in between the exploited workers of the con-

9 https://networkcultures.org/wp-content/uploads/2020/06/VV_ReaderIII.pdf.

temporary service industry and the vintage romantic image of switchboard operators from the 1950s. On the one hand, they are helping people connect and get in touch; on the other hand, they are playing into the myth of efficiency and multitasking—headphones on, responding to queries no matter the time, place, or condition, they're coming from. We are service operators in isolated cells. No doubt feeling alienated ourselves, how is the operator-come-teacher able to create empathy with others while experiencing the same lonely seclusion?

Teaching "on demand" alludes to educators as not-so-different to food couriers. Customers(/students) open their apps at any time, selecting their goods, ordering and paying, to a 24/7 temporality. Then, someone will deliver it wherever it's needed, day or night. Eventually, a consumer will fill their bellies, which in turn should make the courier happy, as they have fulfilled their "mission." In March 2020, during the most severe phase of the Italian lockdown, one of the few working categories allowed to move freely around the city were the heroic bike riders, the food couriers. Just like them, we teachers fight the starvation of souls and provide emergency care, delivering "food for thought," one could say.

Are we stuck into this suffocating binary, somewhere between being quick-response call center operators and valorous food couriers? Are we entrapped in the multitasking, efficiency, on-demand game? Are we condemned to be switchers and bikers without having our own switchboard, our own bike?

Many among students and teachers think that asynchronous is a better solution for the challenges of our times. Record a class whenever and wherever you want, upload it, leave the students consume it at whatever time suits them, which eventually ends up lowering the participation rate even further. Briana speculates that "the fact that classes were recorded was perhaps an additional reason to why people did not ask questions or comment on the material. They were uncomfortable knowing that their comments could be heard over and over or could stay forever on the chat."

In Praise of Awkwardness

"Can you hear me?"
"Are you there?"
"If you can hear me, can you give me a sign?"

Recalling the séance situation of the online classroom: "A séance is an attempt to communicate with spirits."[10] We sit in the blended session attempting to call forth the presence of the Others, simultaneously trying to make our own presence present, while remaining in the absence of both. *Séance* is a freakily fitting word for

---

10 https://en.wikipedia.org/wiki/Séance.

the palaver, coming from the French word for "session" and the Old French *seoir*, "to sit."

These séance-calls of endless fine-tuning procedures take forever. This is to the utmost annoyance of everyone involved, both teachers and students. "Blended learning" mostly sounds like "Can you guys hear me?... You need to unmute yourself. Can you see my PowerPoint?" "This is how any and every student presentation started this semester," a senior student in art history tells me. Antonio Lopez also shares this frustration:

> It's very difficult to maintain a natural flow of weaving discussion and lecture while at the same time having to manage all the technology. Under normal circumstances, I just launch Youtube videos from the class Moodle page and we watch and discuss them together in class. But I could not stream the videos to online students, so they had to watch them on their own using the Moodle links. It was a hope and a prayer that when I showed the video in class, the students at home were also watching it on their own and that our timing synched up.

Madison also reflects on the amount of time spent on making the tech work.

> On days that were truly "hybrid" (in that some of us were in class and others at home), we almost always had to start class late because of technical issues. From the perspective of a student in the classroom, it felt like the distanced students didn't have to engage as much. From a distanced-student perspective, it felt like it was harder to gauge the classroom (e.g. I couldn't tell if someone in the class was trying to speak since the camera faces the board), so I didn't feel as comfortable participating as I normally would.

The discrepancy between the promise of seamless real-time delivery, efficiency, and speed that comes embedded with assumptions of the digital, and the much more messed up and messy reality made of glitches, interrupted sound bites, frozen frames, and abrupt disconnections, creates what anthropologist Rebecca Stein calls *lapse*: "An analytics of lapse writes against theories of technological progressivism and dreams of techno-modernity by spotlighting sites and examples where technology fails, somehow, to deliver on its promise." While the concept originated in the context of the Israeli-Palestinian conflict, it gives insight for reflecting upon what is happening in our classrooms, kitchens, living rooms, and lives.

Lapse is the hiatus, the gap, the abyss that separates the bright, utopian dream of techno-solutionism and the cruel reality of techno-dysfunction. The lapse is that awkward moment when the teacher has prepared the perfect PowerPoint, adjusted the camera and the light to catch the best angle of the face (wrinkles and frowns carefully expunged from the frame), found a brand-new background that makes one look like they're speaking from the outer space (deleting the reality of the tiny, chaotic, sad studio)—and then, all at once, everything goes awry. The bandwidth doesn't fulfill its expected presence and performance. We all know this moment: Suddenly you're unable to display the PowerPoint and your face simultaneously (*It worked earlier today, I swear!*) and have to make a real-time decision amid the flurry of embarrassment.

And so, you give up on everything. From the special background to that perfect camera angle. One ends up doing a solo performance to the screen; forsaking your face for the bright PowerPoint, now simply speaking into the void.

Regardless of the amateur antics that arise, Madison writes: "I think that overall I prefer in person learning, but the 'hybrid-learning' experience made the best out of the situation at hand." I, too, feel that no one would disagree. We were forced into this blended, hybrid (call it what you will) environment because of an emergency. And surely, we are lucky to have what we have—lapses, glitches, frozen frames, floating icons of initials, un-called for cats wandering across screens, and all. These flaws and irritations are better for making do with than nothing.

Recently, a video work[11] I made with one of my former students was part of an (online) exhibition called "*no-longer-being-able-to-be-able*."[12] Hang Li, the brilliant young curator, took inspiration from Byung-Chul Han's *The Burnout Society*. Hang Li writes that the exhibition

> questions the contemporary life immanent in excessive positivity and information. In such a society, feelings of exhaustion, anxiety, and disorientation, are growing within obese bodies that are indoctrinated with neoliberal fantasies. The pandemic, which could be a pulse of the myth of endless positivity, expansion, competition, and exploitation, has seemingly turned out to be revealing and reinforcing the "normal" way of being (in both the art field and societies-at-large). In response to the neoliberal norm of being-able-to-be-able, *no-longer-being-able-to-be-able* attempts to unpack the culture of excessive positivity, ceaseless expansion, and overaccumulation. It also attempts to discuss the potential ways of expression and resistance amid the overloaded, oversaturated, and overdrawn existence.[13]

During the first (Zoom) meeting to get to know the other artists involved and exchange views, Hang Li timidly apologized: "I wanted to play the sound of the video I am showing, but I am not being able to, because my partner is working in the same room." Suddenly, the meeting turned into lots of people apologizing as they wanted to switch the camera on but were "not being able to" as they didn't have enough bandwidth or had other people in the room that were also trying to get stuff done.

---

11 https://vimeo.com/437631123.

12 https://www.skelf.org.uk/S_Q/Hang/Hang_Section2%20(hang).html.

13 https://hang-li.net.

This "not being able to" was precisely the point. We could have declared that we were doing our best to cope with the situation and produce art, learn, and teach. That was not always possible though, as expectations only sometimes met wishes and desires. We collectively found that "blended learning" and other bombastic names, myths of efficiency and hyper-connectedness, hypertrophic self-oriented gazes and high-res surveillance cameras, were the rubbles of the neoliberal moment lying behind us. It was within this that we (silently, and together) came to decide that it was just fine being, and feeling, awkward.

Many students told me that emojis sent in chats during online sessions might not make the teaching and learning process more efficient, but they do make it more fun and less tense. Antonio Lopez described how "for the last day of class one student used a background image of a nuclear explosion, which made the whole class laugh." With GIFs, hearts, smileys, and even raised hands—all of which are useless from a teacher's perspective as you are not able to monitor the chat while you're trying to juggle between camera, PowerPoint, connection problems, etc.—elements can be utilized that are "merely" something fun and that help release the tension of overly intimidating and frozen atmospheres. In the absence of a body, an emoji can help do the job. When internet connection fades in and out, a simple round of chuckles might make up for the fiber's failures.

Acknowledging the awkwardness of the situation, whether it being face-to-face, hybrid, or entirely online, can be of great solace. The aesthetics of contemporary online platforms tend to optimize sound and video in order to render a seamless experience of presence. However, human awkwardness irremediably makes its way back in the video call, within all the glitches and frozen frames, within interrupted sounds and unasked-for echoes. Geraldine Snell, an artist who is part of the *no-longer-being-able-to-be-able* exhibition, says "We should let go of the shame ... We shouldn't be ashamed of our real-life background, of the noises at our place, of the people we live with." Recounting a Zoom experience she had with some BFA students, Anna Frijstein, another artist from the group, emphasizes how they had explored collective care through vulnerability and shame. "It was 'awkwardness' that was the overarching feeling for both the participating students, and definitely for myself as the performing teacher. Feeling vulnerable, exposed, and awkward because of the virtual 'voidness.'"

If any lesson should be learned from the past "blended" year, it would be to welcome—and celebrate—awkwardness. As Franco Berardi writes in *Futurability: The Age of Impotence and the Horizon of Possibility*, competition and its predatory rules have resulted in reducing our "ability to feel compassion and to act empathically. The hyper-stimulated body is simultaneously alone and hyper-connected, subject to psychological and emotional exhaustion." We are inculcated with the myth of performativity and efficiency, when instead we get loneliness and depression.

Rather than perpetuating the farce of performing the self, we could get acquainted with the awkwardness inherent in this very farce. Instead of ignoring that "awkward moment," why not celebrate it? "That awkward moment when someone farts in the car …," "That awkward moment when you have already said 'what?' three times and still have no idea what the person said, so you just agree," "That awkward moment after you've already taken a bite and someone starts praying for the meal," "That awkward moment when you wave to someone, but they don't wave back …" The internet is full of awkwardness, of which it seems only meme culture is in constant celebration of.

Poet Wayne Koestenbaum writes in his book *Humiliation*: "Performers spawn performers, an intergenerational saga of distress. Liza (in the eyes of a shame-hungry public) is humiliated by inability to reach her mother's pinnacle, or by inability to reach her own former pinnacle. Past triumphs rise up to humiliate the present self." *Humiliation* is a tribute to that "abasement of pride"—as Wikipedia defines the feeling—"which creates mortification or leads to a state of being humbled or reduced to lowliness or submission. It can be brought about through intimidation, physical or mental mistreatment, or trickery, or by embarrassment if a person is revealed to have committed a socially unacceptable act."[14]

Right there, on the avenue of embarrassment, is where humiliation meets awkwardness. It thrives upon the clumsiness of your well-rehearsed professional performance that stumbles on impact with bad bandwidth, slow connection, frozen frames, a kid screaming in the background, the family dog's impromptu barking, and the cat purring right in front of the screen—not to mention the merciless close-up of a pimple growing bigger and bigger from the stress and anxiety of it all. Why don't we take the shame element out of the picture (so to speak), and embrace the awkwardness as a vaccine against hyper-performativity and self-burnout? To paraphrase Koestenbaum: Imagine a society where awkwardness is essential, "as a rite of passage, as a passport to decency and civilization, as a necessary shedding of hubris." Why not make a space for potential flourishing out from our own self-acknowledged impotence and collective discomfort? How about we just make the best, make maximum awkwardness, make shameless *fun*, of this one massive "awkward moment" that we're living in?

Acknowledgements

I want to thank Student Government and all the students and professors at John Cabot University in Rome who have given me insights and feedback on their "blended learning" experience.

---

14 https://en.wikipedia.org/wiki/Humiliation.

Thank you also to Giulia Villanucci, Giulia Martinez-Brenner, Hang Li, Anna Frijstein, Geraldine Snell, Lyric Dela Cruz, Jess Henderson, and Geert Lovink.

Postscript

I reread this text in spring 2022, a little more than a year since the original essay was drafted. "Teaching into the Void" broke out from an urgency at the end of the sinister year 2020, my own urgency to make sense of what had happened to us human beings and to the ways in which we used to communicate with each other. At the end of fall 2020, the first term during which we were forced into a new environment merging the online and on-site, I put together my own auto-ethnographic notes with some students' and colleagues' reflections on how the experience of teaching and learning at a distance had looked like. Despite the screen fatigue and the approaching end of the term with the nightmare of final exams, many of them answered enthusiastically to my questions in a sort of cathartic release, as to let the accumulated tension go. Then I sewed together these accounts (including my own) of fear, anxiety, and uneasiness about our hyper-mediated life in the pandemic with a personal journey into today's digital culture, finding resonances between the promised wonderland of influencers' and content creators' glowing ring lights and Deleuze's nonstop performance of images resulting in exhaustion, between Charli D'Amelio's playing and replaying her TikTok self and the Cheshire Cat's grin without a face, and so on.

A year later, looking at this fear and loathing of the online self, lost in the hall of mirrors of the video conferencing self-view, not much has changed. Well, now you can hide yourself and blur your background. You no longer have to gaze at your own grimace; you can focus on the grimaces of others. No longer must you be in a position for others to guess how big your bedroom is, how many times you change the sheets, or how many books lie on your shelves, maybe left unread in the dust.[15]

At the same time when they become "choices" in the select menu, though, hiding and disappearing are turned into platform affordances. They are changed into tech-mediated performances. You need to tap the "blur" option if you want to set your surroundings in the dark. You need to select the "hide self-view" button if you want to disappear completely from other people's gazes and, mostly, from your own. As Jan Distelmeyer points out in this volume, the connections and encounters put into being by today's platforms might be described as "programmatic," both in the sense of the fundamental computing condition of "programmability" and in terms of con-

15 Here I am thinking of the "credibility bookshelf" evoked by Anikina's text. See Anikina, this volume.

figuring a new mode of the social that determines what is possible and how it is with regard to human relationships at large. Finally, everything becomes "Zoomable."[16]

Whether we use them for learning, working, or socializing, today's platforms are not so much, as Andersen and Pold emphasize in their chapter, about constructing a bunch of narcissist selves gazing at their own images or just lurking, as exercising power and control "by way of subjectivation."[17] They become tools for the administration of the everydayness, forms of "corporeal management"[18] regulating the visibility and visuality of our faces and bodies as well as the aspiration to become invisible. There's no hiding, no disappearing from tech. The techno-determinist moment of the early corona days has taken over, becoming pervasive. We seem to need technology more than ever, even to perform basic functions that were once the domain of the non-technological. In the end, escaping tech does need some tech these days.

In the space that has become the "new normal," whether called hybrid or blended or whatever is apt to suggest a "merger" (a word so dear to neoliberal capital) between places, sociality might have been reduced to an interface, a platform affordance, or an option in the select menu. With more and more isolated individuals and cities, is there any sociality left that is not tech-driven, that stays purely bodily? And yet, does it even make sense to think about the body as separate from technology? Isn't the body, as Marcel Mauss wrote back in 1934, the first and most natural technical instrument?[19] That is even more so now, in the moment when all things human are being (re)thought algorithmically and becoming digitally integrated, a Cronenbergian undistinguished blend of machine and mind, body and wires.

It is in this new form of sociality that we have to redefine our thinking, teaching, and learning experiences. This should not be thought in terms of binaries—bodies versus machines, nature versus technology—yet articulated around an idea of technology that does not stand as certainty and perfection against the fallacy of the body. We need to think about technology as something allowing glitches, lapses, obsolescence, and decay—that opens up to and celebrates awkwardness.

---

16 See Mücke, this volume.

17 See Andersen and Pold, this volume.

18 See Andersen and Pold, this volume.

19 See Andersen and Pold, this volume.

References

Berardi, Franco. 2017. *Futurability: The Age of Impotence and the Horizon of Possibility*. Brooklyn: Verso Books.

Boyer, Anne. 2018. "Erotology III: Categories of Desire for Faces." *A Handbook of Disappointed Fate*. Berkeley, CA: Small Press Distribution.

Deleuze, Gilles. 1995. "The Exhausted," transl. by Anthony Uhlmann, *SubStance* 24 (3): 3–28.

Han, Byung-Chul. 2015. *The Burnout Society*. Redwood City, CA: Stanford University Press.

Koestenbaum, Wayne. 2011. *Humiliation*. London: Picador.

Lacan, Jacques. 1998. *The Seminar Book XI: The Four Fundamental Concepts of Psychoanalysis*. Edited by Jacques-Alain Miller. Translated by Alan Sheridan. New York: W. W. Norton & Company.

Mull, Amanda. 2020. "Americans Got Tired of Looking Bad on Zoom." *The Atlantic*, November 19, 2020. https://www.theatlantic.com/technology/archive/2020/11/ring-lights-for-all/617143.

Paoletti, Ilaria. 2020. "Chirurgia Estetica, Boom di Ritocchini. Tutto 'Merito' del Lockdown." *Il Primato Nazionale*, December 15, 2020. https://www.ilprimatonazionale.it/cronaca/chirurgia-estetica-lockdown-177223.

Smeets, Sofie. 2020. "Are You There? Can You Hear Me? The Impact of Covid-19 on Higher Education." *Utrecht University*, November 20, 2020. https://www.uu.nl/en/news/are-you-there-can-you-hear-me-the-impact-of-covid-19-on-higher-education.

Stein, Rebecca L. 2017. "Go Pro Occupation: Networked Cameras, Israeli Military Rule, and the Digital Promise." *Current Anthropology* 58 (S15): 56–64.

Wilson, Jack. 2020. "(playing from another room)." *Video Vortex Reader III: Inside the YouTube Decade*. Edited by Geert Lovink and Andreas Treske. Amsterdam: Institute of Network Cultures.

Presence in Video Conferencing in Teaching Contexts as a Means for Positioning Subjects

Andreas Weich, Irina Kaldrack, and Philipp Deny

Before the Covid-19 pandemic, video conferencing was rarely used in school or university teaching settings. Lessons generally took place "in person." As a result of the restrictions on social contact and the closure of schools and universities that have been repeatedly implemented, to varying degrees and with differing scope, since spring 2020, video conferences have become a cornerstone of "Emergency Remote Teaching" (Hodges et al. 2020) and consequently part of the "new normal" in school and university teaching, at least for a certain period of time. One supposition of this paper is that video conferences were initially employed as a media format to reproduce the "presence" of in-classroom teaching in the form of distant teaching.

Notions of copresence can already be found within the concepts that we use to talk about video conferencing: we "attend a meeting" in a "meeting room" or "join a conference." We experience a kind of simultaneity—at least, when the bandwidth is sufficient and the connection is stable. Educational discourses conceptualized (and in part still do) video conferencing as an attempt to "translate" (Macgilchrist 2020) on-site teaching scenarios and to constitute some kind of "remote presence" as a translation of "physical presence" as this was seen as a cornerstone of successful teaching and learning.

But research on video conferencing and hybrid settings in teaching contexts shows that on-site "physical presence" and the "remote presence" of video conferencing settings do not differ significantly in terms of learning outcomes (Raes 2022). So fears of a lack of cognitive comprehension and missing cognitive learning objectives without being physically copresent seem to be unfounded. However, on-site and video conferencing settings do differ in terms of engagement, with the latter suffering significant drop off (ibid.). Teachers also struggle to "activate" their students (Malewski, Engelmann, and Peppel 2021). On the one hand, this shows that "engagement" and "activation" are relevant categories and objectives in educational research and practice, and that problems in this sense can arise, especially when switching to video conferencing in teaching contexts. The learning subject is supposed to be "engaged" and "activated" and is addressed as such. On the other hand, it shows that when we talk about presence we are predominantly referring to a specific subject position known and expected from on-site teaching that is difficult to "translate" into video conferencing in teaching contexts.

In this paper, however, we do not aim to further perpetuate this kind of research but rather to take it as another point of departure when deconstructing the notion of physical presence as a precondition for successful learning. We will focus more on media-analytical questions such as: which type of presence is generated by which kind of media constellation, and which kind of subject positions evolve from this process. We will first briefly introduce the concept of media constellations as a heuristic concept enabling us to analyze the elements and relations that constitute those aspects of presence and the expected/desired result, as well as divergent subject positions in in-classroom and video conferencing situations in school and university teaching. We will then outline a number of semantic layers included in the term presence, its connections to media history, and how presence can be linked to the heuristics of media constellations. In the following analysis, it will become clear that physical copresence is not a means in itself, but forms the basis of instructional practices and modes of subjectivation that affect demeanor, focus, and conversational behavior. The "remote presence" in video conferences is not only characterized by material differences, that is to say spatial and technical arrangements, but also by different perceptible content, practices, knowledge, and subject positionings. The media constellation that evolves from this interplay does not so much reproduce the physical copresence as such but establishes a shared perceptual space, with mutual representation and availability, that is supposed to address the students in their subject position known from in-classroom teaching. So the question is, what exactly changes in terms of presence and subjectivation when comparing the media constellations of in-classroom teaching and teaching in video conferences; what is, or is attempted to be, kept stable in favor of a certain "normality" (Boys 2022)?[1]

Media Constellation as an Analytical Heuristic

In order to analyze the constitution of the different aspects of presence as well as of subject positions, we will apply the media constellation model (Weich 2020a and 2023, Weich et al. 2023) as an analytical heuristic. This model conceptualizes media not as objects with a certain mediality but sees mediality as a product of the meaningful (in terms of constituting meaning) interplay between materialities (e.g., hardware, spaces/architecture, bodies of people), knowledge, practices (cultural and

1 The research that this paper is based on was partly conducted within the Leibniz-Science-Campus—Postdigital Participation—Braunschweig that was funded by the Leibniz-Association and the Ministry of Science and Culture of the German State of Lower Saxony.

discursive elements), content (the perceptible elements that signify the constituted meaning), and subject positions (requirements of human actors, interpellations).[2] The media constellation approach indicates that the notion of "translating" presence from in-classroom teaching to teaching in video conferences is kind of misleading, as changing even one element automatically reconfigures other elements and relations and subsequently the entire media constellation.[3] It allows us to map out how video conferences in teaching contexts constitute subject positions by examining the interplay between the different elements and how this is connected to presence. The focus on subject positions means that we will reconstruct how the teaching and, first and foremost, the learning individuals are situated and "placed" by the materiality of their private spaces, the hardware, infrastructures, and their own physical bodies; how they are positioned by them as well as addressed by the visual and auditory contents of the media constellation as certain subjects; what practices are enabled by this constellation in terms of agency; and which practices are expected from the learning individuals due to a shared knowledge of teaching scenarios and mutual interpellations. In comparing in-classroom teaching to teaching in video conferencing, we will see how the elements and the interplay between them changes and what role different concepts of presence play within this interplay.

---

2 In contrast with other media studies approaches such as the *dispositif* (in terms of an apparatus), this approach avoids the assumption that *the* video conferencing *dispositif* exists in the same way as there is *the* cinema *dispositif* as a medium, in favor of a more differentiated perspective and the opportunity to conceptualize complex and variable interconnections. In contrast to a broader and more Foucauldian understanding of a *dispositif*, it also avoids the suggestion that there is a general applicability or a strategy and an urgency as Foucault did for sexuality, for example (although the question of urgency seems promising in this case as well). An Actor-Network-Theory approach would not allow subject positioning to be taken into account due to a different underlying ontology. Framing video conferences as situations, on the other hand, would not address the question of media or mediality as such. At the same time, the term mediality itself remains abstract as it only addresses the distinctions and distinctiveness of certain media without addressing what things to look at when analyzing it. The media constellation approach offers groups of elements and relations that can be used for a heuristic analysis of this or their specific mediality, which can be understood as the specific constellation of materialities, knowledge/practices, content and subject positions that provide the conditions for the production of meaning.

3 It is, of course, necessary to drastically simplify, as there is no one type of in-classroom teaching. Rather there are many forms, which we will not go into here in favor of a (stereo)typical conception based on our own experiences and observations in schools. Two prototypical scenarios could be a discussion-based form of teaching, as is perhaps customary in media studies tutorials or in German or politics lessons in schools, and teaching forms focused more in imparting information, such as lectures or school math lessons. There are of course differences between media constellations and forms of presence at school and at university, but these are outweighed by their similarities, which is why we will address them together and only allude to differences when it appears necessary.

Presence (*Präsenz*)

To be present and to maintain a presence, (of a person) being present and to present (an object): the semantics of "presence" in German[4] range from existing, being there, physical attendance, and present tense, as well as show and give, introduce, depict or illustrate, and realize. The German dictionary offers two meanings for "*Präsenz*" (presence): "attendance, [conscious and deliberate] actuality or reality" as well as "physical bearing."[5] People "have presence" when they make a (largely positive) impression on others and draw their attention. The sense of a temporal and spatial "here and now" and the perception of the same have been understood under the term "*Präsenz*" since the word was adopted in German as a loan word from the French in the seventeenth century (Pfeifer 1989, 1312/1313). As an adjective, "*präsent*" (present) means "to be in a particular place, existing or occurring now, currently available." This last definition is connected to the meaning of the Latin participial adjective "*praesens*," which is also used in the sense of "momentary, immediate, urgent, effective" (ibid.). These aspects of "being to hand," in the sense of available and useable, and of efficacy linger in the use of "demonstrating presence." This term is frequently used in instances of state authority: police or the military mark their presence and their ability to deploy or to act. The expression "show presence" tends to resonate, in this sense, with the ability and preparedness to immediately intervene anywhere (in an entire city or district, etc.).

This first semantic examination of what presence can be, therefore includes aspects of (a) current existence or availability, which is (b) perceived by someone. The term also implies (c) certain access to available resources and the opportunity to act. In the sense of introduction and depiction the lexical field also includes (d) representation. These aspects can be found, to differing extents, in the conceptual history of the word and its root. They also repeatedly emerge in the discourses surrounding media, where their meaning sometimes shifts over time.

From a media studies perspective, it is evident that different aspects of the concept of presence outlined here, have regularly accompanied the introduction and implementation of "new" media throughout history. Accordingly, discourses of presence in media studies repeatedly gain certain attention and readings.

Thus, early media of image and writing have been studied in terms of their relationship to presence. Writing, according to Assmann (2006) for example, functions as both memory and utterance and is thus especially indebted to memory and voice.

---

4 As we live and work in Germany and the observed video conferences take place in German-speaking countries, we have focused on the etymology and semantics of concepts of presence in the German language. We assume that the German semantics are also of interest to international readers.

5 https://www.duden.de/rechtschreibung/Praesenz, last accessed 6.12.21.

Already in ancient Egypt, writing is thematized in relation to the voice of individuals. Funerary inscriptions are formulated as a "call to the living," thus pointing to the idea of being present in a medium and, in a sense, communicating (Assmann 2006, 186).

This amalgamation of existence, representation, and perception is in turn fundamental for scripture-based (monotheistic) religions, insofar as God reveals himself and the truth through the holy scripture (Nordhofen 2009). Both the transcendence and presence of God merge in scripture, which consequently becomes an infinitely referential structure, both carrying and suspending transcendence and presence in equal measure (Derrida 1976). Scripture's differential structure as a referential system demands an absent origin that only shows itself in scripture, thus establishing the absent presence as transcendence.

Image media, for their part, are committed to (tele)presence in two ways. Grau (2001) underlines two salient ideas about images: First, there is a long (ritual and religious) tradition of cult images that function, in a sense, as a medium of transmission. Images of saints and/or icons have a certain effect, and such images can in turn be acted upon, which also has an effect. On the other hand, since the "invention of central perspective" (Schmeiser 2002), images have been negotiated as perceptual impressions. While in the first tradition the represented enter the here-and-now, in the second tradition the recipient enters the space of the represented.

Other media such as photography and early cinematography connect to these discourses of presence and update the relationship between existence, perception, and representation. The telephone has only successively been understood as a medium that enables a common communication space and personal proximity. It was initially viewed as a variation of the telegraph: used more in a formal context for the transmission of information and instructions. In the private sphere, the telephone was initially used by the upper classes to communicate with servants within the house (König 2004). The telegraph and telephone made it possible to act at a distance, in the form of orders and instructions, and thus connect with the concept of presence we outlined above.

Media and presence intersect anew at the emergence and discursivation of media-based public spheres. According to early mass psychology, media enable dispersed masses to perceive the same thing, thereby facilitating transmissions between individuals that bring those individuals together into something collective. Furthermore, the distribution and circulation of content through mass media creates communication between those individuals that can result in a shared opinion. The television discourses of the 1950s and 1960s follow on from these discourses and result in television becoming a medium of perception as well as a simulator of liveness and presence.[6]

In the 1990s, with the rise of the Internet and the phenomenon and concept of telepresence and virtual reality (or immersion in virtual realities), media studies discourses on presence and immersion gained relevance and were prominently discussed by Roy Ascott, for example. Since then, the question of presence and agency has been a recurring theme in game studies. Discourses on presence currently play a significant role in terms of the distribution and acceptance of each new media type.

For our purposes it is important to establish which specific forms of presence are applied to and created through the context of video conferences in teaching situations. In order to structure our observations of the relevant aspects, we apply the concept of media constellations as outlined above. Presence comes into play, here, on different levels:

i) as knowledge within the media constellations and the discourses on presence in in-classroom teaching and teaching in video conferences which are inscribed in the communication in and about them and include the expected ways of a) being existent and available, b) perceivable, c) able to act, and d) represented as well as a product of practices that take place in relation to those expectations.

ii) as a material factum of a) being existent and available, b) perceivable, c) able to act, and d) represented in certain ways.

ii) as a product of perceivable content that signifies a) being existent and available, b) perceivable, c) able to act, and d) represented.

iv) as subject positioning in terms of a set of expectations toward the individuals to be a) existent and available, b) perceivable, c) able to act, and d) represented in certain ways.

Presence in Media Constellations Employed for Teaching

We will begin by briefly outlining the media constellation typical for traditional in-classroom teaching, and the accompanying forms of presence, since initially the switch to using video conferencing for teaching attempted to "translate" existing teaching concepts (see above).

6 Kaldrack and Röhle (2016) traced these genealogical lines with regard to Facebook. In particular, they argued that Facebook Open Graph reorders the constellation of transmission, perception, and communication into a technical infrastructure that is staged as a simulator of presence. In the process, presence and participation are configured medially, even if they remain unoccupied in terms of content.

A Typical Media Constellation for In-classroom Teaching

The most obvious characteristic of the media constellation of in-classroom teaching is the materiality of a shared physical space, in which the teacher and students are simultaneously physically present. In terms of the aspects of presence outlined above, typical (even stereotypical) in-classroom teaching—regardless of institutional location and individual implementation—can be said to follow this very general pattern: (a) a group of copresent individuals in (b) a situation of mutual perceptibility, where (c) mutual influence enables joint reference to (d) representations of teaching content.

In institutional settings the materiality of the room—its architecture and furnishings—are already geared toward optimizing mutual perception, through a certain seating arrangement, for example. Such conditions can, to a certain extent, be viewed as infrastructural requirements that enable or hinder different kinds of mutual availability and influence as well as enabling certain content to be presented. The various arrangements can be assigned visibility regimes that stand in vague relation to institutional framings. The spatial arrangement and distribution of those present in the classroom is designed to prioritize the teacher's view of the students, while lecture theatres in universities focus on centrally positioned content and seminar rooms are designed to dispense with hierarchies in terms of lines of sight (see Pongratz 1990). The hierarchies and power relations that are more or less implicit here are thus repeatedly (re)produced performatively within the space and the specific practices and rituals (Wulf et al. 2007, Wulf et al. 2011). This can be exemplified by a ritual, most likely well known by many, that signals the start of the lesson. As soon as the teacher enters the room students immediately move to their usual seats, which results in them generally all facing the same direction. Conversations are abruptly ended and other activities also generally cease. The teacher then greets the class by saying "good morning," for example. The students generally answer in unison (sometimes over-emphasizing every syllable) by saying "Good morn-ing Miss/Mr X-Y." In terms of the media constellation model, we see here the alignment of the heterogeneous elements of the in-classroom constellation.[7]

Teachers in their subject position are commonly required to structure lessons and thus also media practices in certain ways, that is to say, to structure conversations regarding the content of the media constellation in the form of presentations or discussions. They also organize the perceptions of the participants according to conventional practices within, and in addition to, the existing power structures. The key mechanism is to attempt to prevent interruptions in concentration.

---

7 While this precise scenario does not usually occur in a university context, the underlying function of the ritual nevertheless persists, albeit more informally.

Teachers demonstrate their presence by their material position in the room and invoke the subject position of the learners as students through targeted looks and gestures and by coordinating interaction and communication and linking them together. These coordination practices also refer to the different technical media employed, and their content, such as when the board or textbook, spoken content and non-verbal gestures, relate to one another. In other words, teachers aim to establish or maintain the presence of their students through these practices as well as to coordinate the relationship between specific content represented within the constellation. Established courses of action are available: one of which is to confront the students with questions and observe their reactions (in terms of body language, gestures, facial expressions, signals, whispering with their neighbor, etc.). The students are thus called upon in their subject position and can now react (which becomes one aspect of the content of the outlined media constellation) visibly and, if necessary, audibly: reactions may range from disinterest to the urge to communicate. The teacher's task is generally to keep communication going. S/he can look specifically at one particular student and check their reaction, or address a student individually and prompt a specific reaction. The teacher can also emphasize this by moving around the room, toward a particular person, for example. Control and disciplinary practices may also be connected with such actions, aimed at regulating unwanted behavior.

Bureaucratic practices that check or document attendance are also relevant in this regard to attest the existence of students within the given situation. In schools, attendance is generally checked by the teachers, who see the person in the room and enter a name in a list (a type of material record as content of the media constellation). In universities lists are sometimes passed around in which students are expected to register their own attendance. While attendance in school is the legally stipulated norm, compulsory attendance for university courses must be specifically justified. Institutional practices, and the knowledge on which they are based, therefore differ greatly in this respect.

So, the subject position of the student within this media constellation can be characterized by being present in the sense of being materially/physically there, seen, heard, smelt in terms of "perceptible content," discursively represented as content in lists, addressed by initiation-practices and ongoing discursive and spatial practices as attendant, available, and able to act within the communicative practices initiated by the teacher.

A Typical Media Constellation of Video Conferences in Teaching Contexts

Presence, as mentioned above, emerged as a problematic issue in discourses on video conferencing in teaching constellations when the pandemic started. So, what are the elements, relations, and related aspects of presence in video conferencing in teaching contexts? We will again refer to (stereo)typical characteristics of types of teaching and learning, taken from our own experiences and project work in schools. And clearly video conferences are not purely a technical "tool" to be "used," rather it is a case of producing media constellations related to video conferencing systems within teaching situations (Weich 2020b; see also Dang-Anh et al. 2017).

Decentralized Physical Existence

One aspect of in-classroom teaching that is continued in video conferences is synchronicity. In order for the video conference to function, all participants must enter the media constellation at the same time and put themselves materially and physically in a specific situation. They are physically present but individually and in dispersed locations. The organization of one's physical being does not involve physically moving to or within the seminar room, classroom, or lecture theatre, but remaining in a (generally) private room. The sense of "here and now" in the shared physical space is no longer possible and becomes instead a sense of "there and now" to the extent that the subjects are in decentralized locations and variously situated spaces and do not have to focus on the space in which they are currently residing, but rather on the interface with its presentation and interactive possibilities, which determines the mutual availability. Ideally, the room is warm and light enough, and sufficiently self-contained to allow the occupant to be able to focus on the content of the video conference. Rooms in educational establishments are designed to fulfil such requirements, but the functional reality during video conferences frequently diverges from the ideal. Many find themselves in material surroundings that undermine their focus and concentration as elements of presence. The spatial situation in private surroundings makes it much more likely that practices and content of the lesson have to compete with the pupils' leisure practices and content as well as their leisure subject positions and activities (see below). In short: the physical aspects of presence as well as the subject positions in video conferencing differ massively from those of in-classroom teaching due to the different media constellation.

Mutual Perception Via the Interface

Although the practices described above, related to movement in the room, no longer apply, parts of the audio-visual perceptibility of the bodies involved are "translated." A process that requires complex interplay between different materialities: in addition to cameras, monitors, and the end devices to which they are attached, there are also LAN cables, wireless routers, broadband and fiber optic modems, the cables to the infrastructure providers and between the infrastructure elements themselves, including mobile network masts if using mobile devices, servers, etc. Teachers and students require sufficient bandwidth in their private space and a suitable device. The participants can still meaningfully perceive one another but in contrast to in-classroom teaching, only as the audiovisual content of a media constellation and as technically disseminated presentations. Copresence in this sense becomes a collection of icons, names and/or video images, shared speech, and other interactions, which address the individuals behind the accounts and verify the simultaneity of their being logged in, as well as their focus on the learning content. The possibilities for mutual availability and influence also shift. Central to this are directed interactions via a complex interface (Distelmeyer 2021, 53–97), that have similar goals to those in in-classroom teaching, but function differently. Depending on whether the cameras are switched on or off, the media constellation and therefore also the aspects of presence and subject positioning differ fundamentally.

With Cameras Off

In many of the video conferences we observed, students frequently turned their cameras off, thus removing many possibilities and processes of perceptibility. We observed many reasons for students not to use their cameras and were told others in conversations and interviews—material issues such as not having a (working) camera, not having sufficient broadband speed, not wanting to feel under surveillance, wanting to protect one's private space or to continue with competing activities in that private space, which is possible, largely unnoticed, without a camera image. Activities such as cooking, cleaning, hanging out, playing games, sleeping, or playing with pets or children, were just some of those mentioned by students. The subject positions created by the greeting ritual in in-classroom teaching, outlined above, cease to exist and are now in competition with domestic subject positions and practices arising from the private environment and the established routines within it. The new media constellation enables students to shape parts of their subject positioning compared to the media constellation of in-classroom teaching, in which they have no control over their visibility. The materiality allows for practices to "eliminate" some of their representation as content of the media constellation. At the same time, certain aspects of copresence are lost.

When cameras are not switched on, students are frequently only identifiable by names or icons on the user interface provided by the video conferencing software, such as symbols indicating whether the microphone is on or off, or the symbol to raise your hand. In this case, perceptibility is established by visual content on screen without bodily representation. The teacher therefore lacks the optical confirmation that the students are correctly positioned within the room and have assumed an attentive posture or attitude (physical and cognitive), appropriate to their subject position, as this is not represented as content. This also means, however, that the associated positioning of the subjects as students or pupils is initially omitted and that the obligation to call up the associated practices and bodies of knowledge is made more difficult. Against this backdrop, substitute practices of control emerge. During our observations we noticed what can be interpreted as attempts to actively interpellate the students, in order to refer them to their subject position within the lesson.

Some teachers, for example, called the names of each participant at the beginning of the lesson thus employing the bureaucratic form of the attendance list known from in-classroom teaching in order to check the presence of the subjects in terms of their availability and their ability to (re-)act. A further set of practices involved continually checking participation, encouraging everyone to write comments in the chat, or to take part in questionnaires.

The chat function, available in most video conferencing platforms, is a central feature for establishing perceptibility and the ability to (inter-)act. Notification signs and practices become established in video conference chats, such as posting ! (information), ? (question), h (raised hand), m (message), as well as written questions, which are either answered verbally via audio or directly in the chat. This creates a written log of the meeting's progression. Practices from other media-cultural spheres become integrated in the lesson through the chats, such as emojis, "written oral language," and the use of capital letters or repeating letters. In our observations we frequently encountered abbreviations being used such as "afk" (away from keyboard) or "brb" (be right back), which introduces, on one hand, abbreviations established in other spheres as content and, on the other hand, the associated practices from these areas: whilst briefly being absent from an online game or chat is not an issue, this is—at least in the context of schools—not generally allowed in a teaching situation and requires an explanation as it conflicts with the expected subject position. Presence and absence are being negotiated under the premise of the new media constellation and not that known from in-classroom teaching. In some cases, the domestic practices mentioned above play an important role (the doorbell rang), and can justify some absences. Yet another layer of communication is constituted. In most "private chats" a bilateral copresence and a kind of "private" subject position occurs at the written level, which can be linked with reactions visible from the camera image for all others to see. This is to some extent similar to whispering with the person you are sat next to, or to passing around notes in the classroom, however, a rather peculiar kind of copresence emerges in the case of video conferences.

Visibility by Camera Images

In many cases, if not done voluntarily, the students are asked or forced to turn their cameras on (at least at the beginning and end of the meeting), in order to be seen and to greet each other. Then, as they are addressed as visible subjects and their visual representation is constituted as the content of the media constellation, they become present in terms of their mutual visual perceptibility. The image produced by the camera generally depicts the head and shoulders, and sometimes the hands as core visual content of the media constellation. This changes the familiar perception of a person's entire physical appearance in in-classroom teaching in favor of a limited audio-visual representation (touch and smell are also lost). The resolution and the usual size of the image box on the screen make it difficult to identify facial expressions and gestures as content of the media constellation that signifies presence in terms of perceptibility, availability, and symbolic agency. Particularly pronounced or over-exaggerated movements have become established in practice, such as exaggerated nodding or shaking of the head, or wide-open eyes to compensate that fact. So, the subjects stage themselves as visual content in order to show that they are "present" in this respect and that they fit in the subject position desired by the teacher. At the same time, nobody can see who is looking at whom, so no one can be sure whether the teacher or other learners are checking one's actual subject status. Subjectivation functions here more similarly to a panopticon (where the deviants can never be sure if they are being watched or not) than to a direct interpellation by the teacher's gaze, for example. A particular feature of the technical infrastructure is that if one wishes to give the impression that one is looking at a particular person, or all the others in the meeting, one can look directly into the camera. The teacher's gaze that looks at all students at once becomes the visual content that would never be possible in on-site teaching. But this also makes it impossible to observe the other participants at the same time, or to perceive how other participants are looking at your image. This rules out parts of the practices of conversation management, control, and discipline described above. In addition, it is much harder to recognize and interpret the reactions of others. Our observations show how the functions of such glances and looks are being newly configured through the use of emoticons/icons and particularly through written comments in chat functions, which can become a central "log" of reactions as a new form of content.

One of the specific visual effects of video conferences is that participants are not only able to see the camera images of others, but that they are also confronted with their own image as content of the media constellation. On this level, there is the potential for continuous self-monitoring, which acts as a cue for each person to fit into their assigned subject and can lead to practices of self-discipline. At the same time, there is the danger that one's own presence (image) as content, and the sense of one's awareness of and influence upon it, distracts from the content on which one should be focusing. One student described this as "irritating, distracting and takes some getting used to," as effective demands on one's own outward appearance (hair, clothes, posture) in a public space are subject to permanent self-monitoring. The knowledge of one's own visibility can, however, also be used proactively. It makes it possible to control the presentation of oneself as described above through gestures, clothes, accessories (such as gaming headsets), and also through conscious presentation of the space in which one finds oneself (from being in front of bookshelves to being in a whirlpool) as the content of the camera image. Digitally inserted backgrounds create yet another content category, which can be highlighted using practices and technology, particularly known from films, TV series, or video productions.

Audio

Although we refer to *video* conferencing, audio plays a crucial role, too. The material requirements for audio as content of the media constellation—besides the infrastructures already mentioned—are a microphone and speakers or headphones, as well as the physical ability to speak and hear. Mutual speaking and hearing are core aspects of presence in in-classroom teaching. In video conferences teachers and students are primarily talking in their respective location to themselves. Speaking and hearing one's voice in the room reaffirms the sense of self-presence, and discerning reactions from the other participants proves one's ability to act. Knowing that one's own voice is a shared part of the content of the media constellation and hence heard by the others, as well as hearing the others speak, establishes a level of mutual perception. A central challenge of video conferences in this context consists of establishing content and practices at the visual level which can be coordinated with the audio channel and can in turn integrate or contextualize the visual and written contributions in speech. The technical and material conditions do not allow simultaneous speech which sometimes fulfils a coordinating function in in-classroom situations through the changeover from one speaker to another. Visual markers that in in-classroom situations would signal that one would like to speak, such as body language or facial expressions, are generally not adequately visible. The technical and material functions of the "noise gate" discount as potential content signals that are too quiet, meaning that sounds that indicate one would like to speak (clearing one's throat) are also filtered out. The slight delay in transmission of sound and pictures also presents a challenge in terms of temporal coordination. Another factor is that the practices of addressing and prompting one another through the use of looks, pointing, gestures, and position within the room cannot be implemented in this media constellation. Alternative strategies are calling people by name, the "keep quiet until someone sacrifices themselves" approach familiar from in-classroom teaching, which usually does not work well in video conferencing, especially when the cameras are off.

Shared Perceptions

The sense of presence in video conferences remains linked to the aim, that those logged in will tend to perceive the same things, or that the interface will present them with the same content. There are, however, a series of factors to consider that differ according to the individual situations of those "gathered," which can lead to differences in perceptible content. In addition to different material conditions (space, end devices, infrastructure) which can affect sound and picture quality, for example, there are also generally differences in presentation: different versions of the application (in the browser or app) may organize content differently, different settings in the interfaces (tile view or gallery, making certain fields visible or hiding them) allow participants to have different views, and individual participants can also switch between other windows or applications that they have open at the same time. The arrangement of one's own display and showing or concealing one's own space (use of background pictures, positioning oneself in front of a bookcase, etc.) are elements of individual practices and knowledge bases, which in turn can affect the established aspects of presence in the teaching setting in diverse ways. When using Zoom, participants in a video conference can use the speaker view which facilitates their focus on the person speaking by hiding or minimizing the representations of other participants. In contrast, it can be useful for the speaker to select the gallery view, in order to be able to see the greatest number of participants, perceive their postural and facial reactions, and adjust their speech accordingly. In comparison with traditional in-classroom teaching, much more complex arrangements and equipment are involved in adopting a position of paying attention and making it visible, such as adequate lighting, selecting a suitable picture detail, and the position of the end device to obtain a decent camera angle. The material arrangement means that aspects of mutual perception and the awareness of what others can perceive, as well as the associated possibilities and practices of specifically reacting to and influencing one another, no longer function through the act of seeing and being seen since they are no longer part of the media constellation.

Invisible Presence

Finally, it is important to note two levels on which participants or subject positions are involved/present but not usually perceived or even not perceivable in the media constellation. The first is that in contrast to in-classroom teaching, video conferences always raise the question of who owns which parts of the infrastructures used and how we become present for whom—in the form of our data. In in-classroom teaching, data are only recorded in a rudimentary fashion, in the form of attendance lists or class register entries for example, as long as no digital devices are used, and it is generally clear who has access to that information. In video conferences a wide range of data is accrued as "hidden content," and it is difficult for participants to know what happens with that information. This scenario involves new subject positions with economic interests behind the media constellations used for teaching, which are fulfilled for example by companies such as Zoom or Webex that put the learning and teaching subjects in the position of a data resource.

On the second level there are occasionally people present in video conferences who are neither audible nor visible: teachers in schools have for example reported that parents of students have listened and observed, without the teacher's knowledge, and have later spoken to them about their teaching style. This behavior oversteps boundaries that exist in in-classroom teaching and opens up new visibility arrangements and control practices. In other scenarios it is possible for siblings, house or flat-mates, or partners (of students or teachers) to be present. This results in the new subject position of the undetected observer or listener.

In reality, the levels and constellations analyzed here, and the associated spaces in which one is present, overlap in highly complex ways. In some cases, the overlaps are functionally related to one another in terms of the teaching objective, and in others they are the expression of the fact that media constellations are also "contested," insofar as individuals appropriate the intended subject positions and attempt to quietly subvert them, or attempt to reshape the entire constellation.

Conclusion and Desiderata

Stemming from the observation that the use of video conferences as a substitute for in-classroom teaching in the first lockdown was a much discussed issue and closely connected to the concept of presence, we have used the heuristic analysis of media constellations to examine the materialities, knowledge and practices, content, and subject positions that emerge from in-classroom teaching and video-based synchronous teaching. The idea of presence with its dimensions of (a) current existence or availability, (b) being perceived by someone, (c) certain access to available resources and the opportunity to act, as well as (d) representation has then been connected to the media constellation analysis as different aspects of presence have been identified as part of the knowledge, product of practices, material factum, and content of the media constellation and especially related to the positioning of subjects. This made it possible to determine more clearly what the function of the much-vaunted presence actually is in in-classroom teaching and video conferencing.

Following this rather simplified characterization leads us, first, to conclude that copresence is the basis of a specific set of control practices in in-classroom teaching that lead students to take a certain subject position. This, in turn, ensures the "correct" body language and conversational posture, which mark receptivity as well as a willingness to talk about the (learning) content and allow the students to exercise practices specific to subject positions and learning goals, e.g., information intake, knowledge acquisition, presentation, collaboration, discussion, or argumentation. In the media constellation of video conferencing, we see a different set of materialities, knowledge, practices, and content that try to "translate" certain aspects of presence known from in-classroom teaching to the video conference. But new elements and interrelations also arise that employ different aspects of presence in order to (re-)establish a similar subject position to that of in-classroom teaching. Video conferences configure presence as the simultaneous gathering of representations of addressable but physically decentralized participants, who should perceive the teaching person, each other, and the respective information presented as well as having an exchange about this information. However, the traditional practices of having influence upon one another and offering mutual availability only function to a limited degree in video conferences. Instead of using position within the room, looking, pointing, and sanctioning undesirable behavior as disciplinary and motivational practices, visualizations, audio, and datafied control practices come to the forefront. This also brings into view the fact that presence in teaching particularly serves the performative embodiment of attentiveness and concentration, as well as the practice of appropriate communication practices and modes of subjectivation.

If one follows the description and analysis of in-classroom teaching and video conferencing in teaching contexts presented here, then the key aim behind the longing for presence is not the (re)establishment of presence in the sense of "here and now" (a), but rather the creation of certain communication and control practices and subject positions, which represent the preliminary basis of institutional hierarchies and didactic objectives. Further levels of presence, such as shared perception (b), mutual influence (c), and the presentation of content (d), are not aims in themselves, but rather a means to an end. The analysis of the creation of presence in video conferences in teaching contexts allows us to draw conclusions about the function of presence in traditional in-classroom teaching insofar as the differences, but especially the similarities, between the two media constellations reveal that teaching is primarily concerned with subject positioning and engaging in practices that correspond with these positions.[8] The analysis of this specific kind of presence, and the media constellation through which it is generated, provides a basis for reflection on the design of teaching with and in the context of video conferences. It is possible in this way to surpass the limitations of the basic "translation logic." It is important to reflect upon the function that presence fulfils and which combination of materiality, knowledge and practices, content and subject position could or does fulfill the same role.

A range of further research questions could emanate from this brief analysis and subsequent speculation. On one hand it would be appropriate to systematically analyze works from pedagogy and educational sciences, and (subject-specific) didactic works on presence, visibility, addressing, and subjectivation in terms of their relationship with the media studies analyses developed here. In light of current developments, it also seems promising to continue this historical line of questioning by considering presence and other relatively new media constellations, such as those associated with virtual reality (VR) environments, which have (again) attained a prominent role in the discourse, due for example to the recent metaverse imaginings of

8 Building upon our analysis and rationale, we can conclude that physical copresence in in-classroom teaching is not a means in itself, but is rather a foundation for (habitualized) practices related to demeanor, focus, and conversational behavior. Copresence appears to be the basis for control and disciplinary techniques directed at correct concentration and conversational behavior. Interface-based copresence in the media constellation of video-supported synchronous teaching is not able to fulfil this function in the same way and so there is a partial shift toward other disciplinary and control practices.

Mark Zuckerberg. The respective underlying media constellations in teaching contexts could be analyzed, and the production of the different kinds of presence could subsequently be related to those constellations. It would also be interesting to examine how the relationship between presence and mediality is renegotiated within the historical timeline.

References

Assmann, Jan. 2006. "Evolution durch Schrift." In: *Evolution und Menschwerdung* (Nova acta Leopoldina NF 93, Nr. 345) edited by Harald zur Hausen, 181–193. Halle: Leopoldina.

Bolter, Jay David, Richard Grusin. 1999. *Remediation. Understanding New Media.* Cambridge. Massachussets: The MIT Press.

Boys, Jos. 2022. "Exploring Inequalities in the Social, Spatial and Material Practices of Teaching and Learning in Pandemic Times." *Postdigit Science and Education* 4: 13–32. https://doi.org/10.1007/s42438-021-00267-z.

Dang-Anh, Mark, Simone Pfeifer, Clemens Reisner, and Lisa Villioth. 2017. "Medienpraktiken: Situieren, erforschen, reflektieren. Eine Einleitung." In: *Medienpraktiken: Situieren, erforschen, reflektieren. Navigationen—Zeitschrift für Medien- und Kulturwissenschaften*, Vol 1/2017: 7–36.

Derrida, Jacques. 1976. *Of Grammatology*. Baltimore & London: Johns Hopkins University Press (trans. Gayatri Chakravorty Spivak).

Distelmeyer, Jan. 2021. *Kritik der Digitalität.* Wiesbaden: Springer VS.

Grau, Oliver. 2001. "Telepräsenz: Zu Genealogie und Epistemologie von Interaktion und Simulation." In: UNSPECIFIED Frankfurt am Main: Suhrkamp. 39–63.

Hodges, Charles, Stephanie Moore, Barbara Lockee, Torrey Trust, and Aaron Bond. 2020. "The Difference Between Emergency Remote Teaching and Online Learning." *EDUCAUSE Review*, March 27, 2020. https://er.educause.edu/articles/2020/3/the-difference-between-emergency-remote-teaching-and-online-learning.

Kaldrack, Irina and Theo Röhle. 2016. "Creating Subsets of the Masses. Taxonomies, Orders, and Crowds in Facebook's Open Graph Protocol." In *Social Media—New Masses,* edited by Inge Baxmann, Timon Beyes, Claus Pias,75–100. Zurich Berlin: diaphanes.

König, Wolfgang. 2004. "Telegraphie, Telefonie, Funk. Kommunikation und Technik im 19. Jahrhundert." In *Die Medien und ihre Technik. Theorien, Modelle, Geschichte,* edited by Harro Segeberg, 56–70. Marburg: Schüren (Schriftenreihe der Gesellschaft für Medienwissenschaft (GfM) 11). https://doi.org/10.25969/mediarep/14231.

Macgilchrist, Felicitas. 2020. "Digitale Bildungsmedien: Diskurse und deren Wirkkraft in der Corona-Pandemie." *Lernende Schule* 91: 14–19.

Malewski, Sarah, Sabrina Engelmann, and Lars Peppel. 2021. "Erleben, Herausforderungen und zukünftige Lehrszenarien in der Online-Lehre. Eine Mixed-Method-Studie zum Covid-19 Sommersemester 2020 aus Sicht von Lehrenden." *MedienPädagogik* 40 (CoViD-19): 97–117. htt-ps://doi.org/10.21240/mpaed/40/2021.11.12.X.

Nordhofen, Eckhard. 2009. Kultbild – Schrift – Körper – Kunst. Zu einer elementaren Medientheorie des Monotheismus. *TEXT UND ZEIT – Magazin für Literatur und Gesellschaft*, http://www.text-und-zeit.de/rel/christ007.html.

Pfeifer, Wolfgang. 2019. *Etymologisches Wörterbuch des Deutschen*. H-P. Berlin: Akademie-Verlag.

Pongratz, Lugwig A. 1990. "Schule als Dispositiv der Macht – pädagogische Reflexionen im Anschluß an Michel Foucault" *Vierteljahrsschrift für wissenschaftliche Pädagogik* 66/3: 289–308.

Raes, A. 2022. "Exploring Student and Teacher Experiences in Hybrid Learning Environments: Does Presence Matter?" *Postdigital Science and Education* 4: 138–159. https://doi.org/10.1007/s42438-021-00274-0.

Schmeiser, Leonard. 2002. *Die Erfindung der Zentralperspektive und die Entstehung der neuzeitlichen Wissenschaft* München: Wilhelm Fink.

Weich, Andreas. 2020a. "Digitale Medien und Methoden: Andreas Weich über die Medienkonstellationsanalyse." *Open-Media-Studies-Blog*, 16.06.2020. https://zfmedienwissenschaft.de/online/open-media-studies-blog/digitale-medien-und-methoden-weich.

Weich, Andreas. 2020b. "Hervorbringung von Medienkonstellationen statt Nutzung didaktischer Werkzeuge. Versuch einer medienkulturwissenschaftlichen Didaktik der Bildungsmedien am Beispiel von Videokonferenzen als Unterrichtsform." *Medienimpulse* 2/20, Nähe(n) und Distanz(en) in Zeiten der COVID-19-Krise: 32p. https://doi.org/10.21243/mi-02-20-20.

Weich, Andreas. 2023 (online) / 2024 (print). "Medienkonstellationsanalyse." In *Handbuch Digitale Medien und Methoden*, edited by Sven Stollfuß, Laura Niebling, and Felix Raczkowski, Wiesbaden: SpringerVS.

Weich, Andreas, Philipp Deny, Michael Friedewald, Marvin Priedigkeit, and Ina Schiering. 2023. "Let's figure it out. Participatory means for reflecting educational media in a postdigital world." In *Postdigital Participation in Education. How Contemporary Media Constellations Shape Participation*, edited by Andreas Weich and Felicitas Macgilchrist. London: Palgrave Macmillan.

Wulf, Christoph, Birgit Althans, Gerald Blaschke, Nino Ferrin, Michael Göhlich, Benjamin Jörissen, Ruprecht Mattig, Iris Nentwig-Gesemann, Sebastian Schinkel Anja Tervooren, Monika Wagner-Willi, and Jörg Zirfas. 2007. *Lernkulturen im Umbruch. Rituelle Praktiken in Schule, Medien, Familie und Jugend*. Wiesbaden: SpringerVS.

Wulf, Christoph, Birgit Althans, Kathrin Audehm, Gerald Blaschke, Nino Ferrin, Ingrid Kellermann, Ruprecht Mattig, and Sebastian Schinkel. 2011. *Die Geste in Erziehung, Bildung und Sozialisation. Ethnographische Feldstudien.* Wiesbaden: SpringerVS.

The Anatomy of Zoom Fatigue

Geert Lovink

> Humankind is so resilient. For example, I have acclimated to Microsoft Teams.
> —*Ian Bogost*

> Word of the day is "clinomania": the excessive desire to stay in bed.
> —*Susie Dent*

> Poverty of hermeneutics today: Post-it Notes, Miro, tag clouds, a search bar, infinite scrolling recommendations.
> —*Geert Lovink*

This is it. During the COVID-19 pandemic, the internet came into its own. For the first time ever, it experienced a sense of completion.[1] Glitches were common. Video calls lagged then froze. Laptops or routers had to be restarted. In those bright early days of the first lockdown (March–April 2020), few dared to complain. Almost overnight we saw a mass migration to Zoom. And oh, what freedom! To paraphrase Marx and Engels, it was now possible to teach class in the morning, attend a conference in the afternoon, and socialize after dinner—while never leaving the fucking screen. We hadn't yet arrived at the feeling of being trapped in a virtual prison. In fact, as we tweaked and improved our online personas, in-person meetings began to feel strange or secretive. Somehow, we became trapped in a *Videodrome* future, a scenario that suggested some very dark outcomes.

From mid-2020 onward, I began collecting evidence on the trending topic of "Zoom fatigue." Needless to say, experiences of this kind are not just limited to Zoom but extend to Microsoft Teams, Skype, Google Classrooms, GoToMeeting, Slack, and BlueJeans—to name but a few of the major players. In our pandemic era, cloud-

1 German: *Vollendung*—completion, perfection.

based video meetings became the dominant work/life environment in not only education, finance, and health care but also the cultural and public sector. Every stratum of management withdrew into new enclosures of power. The same environment was adopted by high-flying business consultants and precarious freelancers. While their lives were very different, they had one thing in common: they worked very long hours.

Zoom has multiplied work, expanded participation, and systemically devoured any time we might have once had for writing, thinking, leisure, and relations with family and friends. Excessive screen time takes a toll. Body mass index levels have increased. Affective states and mental health have taken a hammering. Spatio-motor coordination has suffered. Video vertigo is a peculiar condition that also prompts more widespread forms of disorientation. Minka Stoyanova teaches computer programming and spends 20 hours per week on Zoom: "My ability for non-work-related social-distancing encounters has gone down greatly," she confessed. While some people "schedule Zoom cocktail parties and birthday meet-ups, I dread having to log back into the interface."[2]

It is a question of strategy. Should we resist this new normal and go on strike? Should we refuse to deliver online classes, hold management meetings, or offer virtual medical consultations? This is easier said than done. Paychecks are at stake. At first, staying at home felt like a privilege. We even felt a little guilty when others had to venture out into the pathogenic world. Now, many fear that video calls are here to stay. "Companies big and small, all over the world, are transforming themselves into a business that is more digital, more remote, and more nimble,"[3] observes *Fast Company*. Expensive real estate can be sold off, expenses dramatically reduced, and discontented staff neatly isolated, preventing any communal organization.

The video dilemma is intensely personal. "If work exhausts my video call time, I intuitively cut informal video calling with allies, friends, possible collaborators," designer Silvio Lorusso observes. "This makes me sad and makes me appear rude. It's a self-preserving attitude that leads to isolation." The debate should not be about hanging out on FaceTime or Discord with friends for a game night, doing karaoke, holding a book club, or watching Netflix together. Video time is part of the advanced post-Fordist labor regime, performed by self-motivated subjects who are supposed to be doing their jobs. But then you drift off while pretending not to. Your eyes hurt, your concentration span diminishes, multitasking is a constant temptation, and that physically, psychically uncomfortable feeling hums in the back of your head ... You've heard it all before.

---

2 Private email exchange with Minka Stoyanova after a public call on the nettime mailing list, July 3, 2020.

3 https://www.fastcompany.com/90558734/this-one-concept-will-transform-the-future-of-work-post-covid.

In 2014, Rawiya Kameir defined internet fatigue as the state that follows internet addiction: "You scroll, you refresh, you read timelines compulsively and then you get really, really exhausted by it. It is an anxiety that comes along with feeling trapped in a whirlwind of other people's thoughts."[4] Philosopher Nigel Warburton echoed this fatigue with his Twitter post that asked, "Does anyone have a plausible theory about why Zoom, Skype, and Google Hangout meetings are so draining?"[5] He received 63 retweets, 383 likes, and a few replies. The responses closely mirrored popular diagnoses and advice now offered across the web. So what were the main drivers of this exhaustion after a Zoom meeting, this post-screen slump? Responses included the brain's attempt to compensate for the lack of full-body, nonverbal communication cues; a sense of constant self-consciousness; engagement in multiple activities with no real focus; and a consistent tugging temptation to multitask. Suggested remedies are predictable: take breaks, don't sit for too long, roll your shoulders, work your abs, hydrate regularly, and integrate plenty of "screen-free time" into the day.

Living in Video Space

Isabel Löfgren lives in Stockholm, but Zoom has become her official place of residence. Her office is now located in that sleek black rectangle in her pocket, her mobile device. "Our living rooms have become classrooms," she states. "Does it matter what is on display behind you? What does it say about you? If you have a bookshelf in the background, or your unfolded laundry in a pile on the chair behind you, it's on display and up for scrutiny. What is personal has become public." Zoom sets up shop in the private space of the home, becoming another room in the house. Which theorists or philosophers predicted this strange scenario? Gaston Bachelard certainly didn't in *The Poetics of Space*. Neither did Georges Perec in *Life: A User's Manual*, as he failed to include a screen in his fictitious apartment block.

Actually, Czech philosopher Vilém Flusser anticipated this state of affairs, predicting "the technical image as phenomenology." Technical implies something state-of-the-art, a technology that is both smooth and sophisticated. Yet as Löfgren notes, Zoom's functionality is surprisingly simplistic or even crude: "You can raise your hand and clap like a preschooler, chat like a teenager, and look at yourself in your own little square as if peering at a mirror."[6] In fact, in many cases, Zoom fails to work altogether. Lorusso chronicles a long litany of dysfunctions in his first days of use.

4 https://www.complex.com/pop-culture/2014/03/is-internet-fatigue-ruining-your-life.
5 https://twitter.com/philosophybites/status/1252148409672380424.
6 Private email exchange with Isabel Löfgren, June 26, 2020.

> I couldn't install Microsoft Teams, my camera wouldn't activate, and, worst of all, the internet connection had hiccups. The connection was neither up nor down; every other attempt it just became super slow. Let me help you imagine my video calls: all would be smooth for the first five minutes and then decay took over—frozen faces, fractured voices, reboots and refreshes, impatience and discouragement. A short sentence would take minutes to manifest. It was like being thrown back to the times of dial-up connection, but within today's means of online communication.[7]

Zoom was broken, but we used it anyway. All-too-quickly, it became the new normal. Video calling moved from a global experiment to a foregone conclusion. We adjusted to a new interpassive mode. That was it. Completion achieved.

"I am utterly zoomed out and exhausted," Henry Warwick writes from Toronto. "Between watching the nation of my birth (the United States) commit a long slow political suicide and having friends die of COVID and working like a dog while on what is de-facto nine months of bio-house arrest, I'm not in a great mood." Henry's summer was spent making video bits and preparing for the delivery of asynchronous class material, which he describes as

> not really a university education—it is a step above a YouTube playlist. Sitting in front of a Zoom window makes it difficult to forge those friendships and networks, and it's certainly a buzzkill for adventure. In addition, there is the issue of Internet Time as I have students all over the world. It's hard for them to attend a two-hour lecture when it's 2.00 a.m. where they are. It's utter madness. Making these videos was a serious time drain. I refuse to give Adobe my money, and Apple screwed Final Cut Pro so badly that I am editing my videos in DaVinci Resolve, which has the benefit of being free-ish. I have never used Resolve, so the learning curve was not insignificant.[8]

Long before the recent pandemic, philosopher Byung-Chul Han was already observing in *The Burnout Society* that we lived no longer in a disciplinary society but in one defined by performance. This performance is not spectacular or intense but a kind of mundane repetition. Spending hours in virtual conferences doesn't feel like being in a paranoid panopticon—but neither is it a celebration of the self. We are not being punished—but we also aren't feeling productive. We aren't subjected—but we can't say we're activated either. Instead, we are hovering, waiting, pretending to watch, trying to stay focused, wondering when we might squeeze in a lunch break or recharge with a caffeine hit. Much like the seemingly endless pandemic, we are being asked to endure never-ending sessions on Zoom. The Outlook Calendar is the

---

7 https://www.platformbk.nl/en/remote-work-demand-dail-up/.

8 Private email exchange with Henry Warwick, October 1, 2020.

new jail warden. This is not a brief sprint, where we emerge sweaty and uplifted, but a marathon that leaves us drained and depleted. What's wearing us out is the *longue durée*.

Tired subjects perform badly. Screen-time apps and MyAnalytics summaries now tell us precisely how many minutes of our lives are being wasted as we calibrate our productivity and efficiency to collaborate with colleagues. It's hard not to wonder whether the IT sector isn't about to get into bed with big pharmaceutical companies. The society of synthetic performance enhancement is now prime for a dramatic expansion. There is no hope that this simulacrum of life can ever protect us from accelerating economic and social collapse. Despite the guilt trips, we are allowed to admit that we're not achieving much.

In response, the system has turned emphatic and switched to worry mode about our mental state. Soon after the introduction of lockdown, with quarantine in place, the authorities set about investigating whether their pitiable subjects were still coping. With society on hold, it is the waiting that tires us out. A few years ago, David Wojnarowicz tracked in *Close to the Knives, A Memoir of Disintegration* how another disease took its toll on the body, noting the disintegration that resulted from his encounter with the "fatality, incurability and randomness of AIDS ... so powerful and feared."[9] Now we witness our own version of disintegration, watching as our lives fray at the seams. Trapped in the waiting room, we are being asked—very kindly—to stay in survival mode, to press on despite our burnout.

Pressing on means mastering anger and numbing intense emotion. What we experienced during the shock of lockdown was *aesthetic flattening*: a highly reductionist substitute for human interaction, as Cade Diehm and Jaz Hee-jeong Choi described our online social life in that period: "a core source of angst in reflections on the hypermodern vulgarity inherent in the same software being used for everything from professional meetings to remote birthdays to funerals, or the absurdity of rushed, voyeuristic on-screen pedagogical endeavors, marred by limited support and buffer for failures."[10] According to Diehm and Choi, video calling is an unsatisfactory low-resolution audiovisual interaction. This is paired with the reduction of body and identity from three dimensions to two, facilitated by universalist design thinking for smooth-surfaced dumb terminals that cannot accommodate performative or deeply immersive interactivity and self-expression.

Zoom has become the universal client, the software suite for everything. It seems to tick off a giant list of use cases: social media, work, entertainment, food orders, gaming, watching Netflix, seeing how family and friends are doing, and livestreams to observe those in hospital. In the context of a global lockdown, it offers some level

---

9 See also https://blogs.ethz.ch/making-difference/2019/05/09/introduction-posthuman-bodies-judith-halberstam-and-ira-livingston/.

10 https://newdesigncongress.org/en/pub/aesthetic-flattening.

of telepresence, allowing us to steer clear of buses, trains, and airplanes. But what a sad form of teleportation. What happened to the future? We need to go back to early science fiction novels, to revisit those far-fetched dreams. Utopia and dystopia seemed to merge in 2020. All we want is to re(dis)cover the body. We demand instant vaccines. We want less tech. We long to go off-line, to travel. We want to leave the damned cage behind.

Zoom Doom

Some weeks into lockdown, the question arose as to why video conferencing was so exhausting. Zoom fatigue is "taxing the brain,"[11] people complained. Why are classes and meetings on Skype, Teams, and Google Hangout so draining? This was expressed not as some sort of interface critique but as an actual existential outcry. Popular articles on Medium name it as such. Common titles include variations of "Do You Have Zoom Fatigue or Is It Existentially Crushing to Pretend Life Is Normal as the World Burns?" and "The Problem Isn't Zoom Fatigue—It's Mourning Life as We Knew It."

It took just days for the Zoom fatigue trope to establish itself, a sure sign that internet discourse is no longer controlled by the "organized optimism" of the marketing lobby. Managerial positivism has been replaced by the arrival of instant doom. According to Google Trends, the term made the rounds back in September 2019 and reached its peak in late April 2020, when the BBC reported on it.[12] "Video chats mean we need to work harder to process non-verbal cues like facial expressions, the tone and pitch of the voice, and body language; paying more attention to these consumes a lot of energy," stated one expert. "Our minds are together when our bodies feel we're not. That dissonance, which causes people to have conflicting feelings, is exhausting. You cannot relax into the conversation naturally." Another interviewee describes how on Zoom "everybody's looking at you; you are on stage, so there comes the social pressure and feeling like you need to perform. Being performative is nerve-wracking and more stressful."[13] Maybe Han's performance prediction was correct.

"I usually stand and move around when lecturing, sometimes making large gestures," states Michael Goldhaber, "just sitting at a desk or wherever is sure to be fatiguing. Doing this in a non-fatiguing way will require fundamentally re-thinking

11 https://www.nationalgeographic.com/science/2020/04/coronavirus-zoom-fatigue-is-taxing-the-brain-here-is-why-that-happens/.

12 https://trends.google.com/trends/explore?q=zoom%20fatigue.

13 https://www.bbc.com/worklife/article/20200421-why-zoom-video-chats-are-so-exhausting.

the system of camera, mic and screen with respect to participants."[14] The sad and exhausting aspect of video conferencing can be attributed to the "in-between" status of laptops and desktop screens. They are neither mobile and intimate, like the smartphone and FaceTime, nor immersive, like Oculus Rift–type virtual reality systems. Zoom fatigue arises because it is so directly related to the "bullshit job" reality of our office existences. What is supposed to be personal turns out to be social. What is supposed to be social turns out to be formal, boring, and (most likely) unnecessary. This is only felt on those rare occasions when we experience flashes of exceptional intellectual insight and when existential vitality bursts through established technological boundaries.

As programming teacher Stoyanova noted, the ability to see oneself—even if hidden in the moment—creates a tiring reflective effect, the sensation of being surrounded by a hall of mirrors. Educators feel that they are constantly monitoring their own demeanor, while simultaneously trying to reach through the interface to their students. In a blog post, L. M. Sacasas describes the effect of paying so much attention to oneself:

> We are always to some degree internally conscious of ourselves, of course, but this is the usual "I" in the "I-Thou" relation. Here we are talking about something like an "I-Me-Thou" relation. It would be akin to having a mirror of ourselves that only we could see present whenever we talked with others in person. This, too, amounts to a persistent expenditure of social and cognitive labor as I inadvertently mind my image as well as the images of the other participants.[15]

It is like practicing a speech in front of a mirror. When speaking to yourself, you experience a persistent cognitive dissonance. In addition, there is the lack of eye contact—even if students have activated their video—which also makes live lectures more difficult to conduct. "Without the non-verbal feedback and eye-contact one is used to, these conversations feel disjointed."[16] Curiously enough, speaking into the void nevertheless kickstarts the adrenaline glands, which certainly isn't the case when rehearsing in front of a mirror. We have entered a strange mode of performance that aligns with predictive analytics and preemption. Even though the audience might just as well not be there, performing on Zoom still activates biochemical responses in the body.

And yet if Zoom is a mirror, it is a delayed and distorted one. Online video artists Annie Abrahams and Daniel Pinheiro point to the rarely discussed effects of delay.

---

14 Michael Goldhaber, nettime mailing list, July 7, 2020.

15 https://theconvivialsociety.substack.com/p/a-theory-of-zoom-fatigue.

16 Private email exchange with Minka Stoyanova, July 3, 2020.

> We are never exactly in the same time-space. The space is awkward because we are confronted with faces in close up for long time spans. We first see a face framed like when we were a baby in a cradle as our parents looked down upon us. Later it became the frame of interactions with our lovers in bed. This makes it so that while video-conferencing, we are always connected to something very intimate, even in professional situations.

Abrahams and Pinheiro also observe that it is impossible to detect much detail in the image we see.

> Video conferencing is psychologically demanding because our brains need to process a self as body and as image. We lack the subtle bodily clues for the content of what someone tells. Our imagination fills the gaps and makes it necessary to process, to select what to ignore. In the meantime, we are continuously scanning the screen (there is no overview and no periphery). We are never sure we are "there," that the connection still exists, and so we check our own image all the time. We hear a compressed mono sound, all individual sounds are mixed into one soundscape.[17]

The result of all this compression and distortion is an impoverished interface, a crude simulacrum of social interaction. Isabel Löfgren responds that we should think of Zoom as a "cold medium," one that demands more participation from the audience, according to Marshall McLuhan's concept of cold and hot media: "The brain needs to fill in the gaps of perception, which makes our brains (and our computers) go on overdrive." In terms of camera angles, Löfgren adds that we are constantly looking at a badly framed medium shot of other bodies: "We have no sense of proportion in relation to other bodies ... emotional closeness to the subject on the other side of the camera is eliminated with the lack of eye contact," leaving us with no "pheromonal connection." In this sense, "Zoom terminology is correct," she notes; "our experiences of others occur in 'gallery mode.'"[18]

Caught in the Grid

The Zoom regime keeps the subject locked in, on task and on track. *Keep your eyes on the camera*, our digital alter ego whispers through our earphones. According to Sacasas, video conference calls are a "physically, cognitively, and emotionally taxing experience as our minds undertake the work of making sense of things under

17 https://doi.org/10.16995/jer.67.

18 Private email exchange, October 3, 2020.

such circumstances. We might think of it as a case of ordinarily unconscious processes operating at max capacity to help us make sense of what we're experiencing."[19] We are forced to be more attentive; we cannot merely drift off. Multitasking may be tempting, but it is also very obvious. The social (and sometimes even machinic) surveillance culture takes its toll. Are we being watched? Our response requires a new and sophisticated form of invisible daydreaming, absence in a situation of permanent visual presence—impossible for students, who are not afforded their grades unless the camera stays on.

Video conferencing software keeps us at bay. Having fired up the app and inserted name, meeting number, and session password, we see ourselves appear as part of a portrait gallery of disappointing personas that constitutes the Team. Within seconds you are encapsulated by the performative self that is you. Am I moving my head, adjusting myself to a more favorable position? Does this angle flatter me? Do I look as though I'm paying attention? And this professional image is often disrupted by the distractions of "real life"—partners who walk into the room, a passing pet, needy kids, and the inevitable courier ringing the doorbell. "Thanks to my image on the screen, I'm conscious of myself not only from within but also from without," notes Sacasas. He describes the experience as a double event that the human mind experiences as if it were real.

Why do I have to be included on the screen? Don't I have the right to be invisible? I want to switch off the camera and become a ghostly half-presence. I want to be a voyeur, not an actor. I long to be frozen like an ancient marble bust, neatly standing in a row with other illustrious figures, brought to life with a click like the figures in *Night at the Museum*. But no, it's too late, I've already joined the call and appeared on stage. The software lords have decided otherwise and gifted the world with the virtue of visible participation. They demand total contribution. The set is designed to ensure that we stay focused all the time, making the fullest possible contribution, expending maximum mental energy. You hate dressing up for that video call (but you do it anyway). Bored and tired of the emotional labor, you change your background to a tropical beach, a paper-thin paradise to inject some cheer into the situation.

Writing for *Artforum*, Paula Burleigh observes that "the most pervasive of COVID imagery has little to do with the actual disease: it is the digital grid of people congregating virtually on Zoom for 'quarantini' happy hours, work meetings, and classroom instruction."[20] The grid Burleigh describes as a hallmark of minimalist design and modernist art, "conjures associations with order, functionality, and work, its structure echoed on graph paper and in office cubicles." In his two-part "History of the Design Grid,"[21] Alex Bigman describes how the system of intersecting verti-

---

19 https://theconvivialsociety.substack.com/p/a-theory-of-zoom-fatigue.

20 https://www.artforum.com/slant/paula-burleigh-on-the-zoom-grid-83272.

21 https://en.99designs.de/blog/tips/history-of-the-grid-part-1/.

cal and horizontal lines was invented in Renaissance painting and page layout. This led to the development of graphic design. The assumption that images are more dynamic and engaging when the focus is somewhat off-center is something video conferencing designers have yet to take on board.

The grid cuts through any rational divisions between boxed-in subjects. Individuals are unable to spill over into the space of others except when they gossip on a back channel. Let's commemorate the guilty pleasures of the Zoom bombers who, early on in lockdown, carried out swarm raids on open sessions they found on websites and social media.[22] For some, spraying management meetings with graffiti and workshops with porn was seen as annoying puerile male behavior. Others got the joke, understanding how this anarchist gesture disrupted the platform's regime of squared tiles and perfect order. As Burleigh concludes, "the grid is rife with contradictions between what it promises and what it delivers."[23] Individualized squares are the postindustrial equivalent of a Le Corbusier housing nightmare: we are sentenced to live in our very own utopian prison cells. Here one finds tragic normalcy, punctuated at moments by deep despair.

We're alive but are being slowly caught in the grid, trapped inside existential reality. Its insistence on 24/7 mindfulness can only lead to a regressive revolt, an urge to take revenge. How can we blow up the social portrait gallery, with its dreadful rectangular cutouts? Jailed inside the video grid, you zone out, drifting away from the management meeting and entering a virtual version of Velázquez's *Las Meninas* (1656). You move on to the next room, the Kazimir Malevich 1915 Suprematist exhibition. You snap back to attention only to realize the depressing reality: you're back inside your own sad version of *The Brady Bunch*'s opening credits. You're on Zoom, not roaming inside some artwork.

The body gets depleted, bored, and distracted and ultimately collapses. No more signals! Please provide less, turn the camera off. The number-one piece of popular advice on combating Zoom fatigue is simply "do it less"—as though that's even an option. There is an imperative here, and it's about productivity and efficiency, not software—in this sense, the "you don't hate" aphorism can also apply here: "you

22 https://en.wikipedia.org/wiki/Zoombombing. In a private email from September 18, 2021, Donatella Della Ratta points at the pleasure of disrupting the official reality of videocasting out of the private sphere, for instance in a 2017 BBC broadcast two small children walk in (https://www.youtube.com/watch?v=Mh4f9AYRCZY). "How far we are from that cheerful atmosphere, the first time we have an incursion of 'real life' into screen life, kids playing and screaming in the background, the guy trying to be serious and professional, the interviewer laughing, and millions of people cheering while watching this. How far that cheerful atmosphere is from the incursion of real life we have to experience on a daily basis when we work, teach, do business meetings in constant fear that the cat will jump on the keyboard or someone will ring our door bell or our little kid will start screaming from the other room."

23 https://www.artforum.com/slant/paula-burleigh-on-the-zoom-grid-83272.

don't hate Zoom, you hate capitalism."[24] Should we be designing indicators of group sentiment? In what way can we fast-forward real-time team meetings? More back channels, perhaps, and less ongoing visual presence. But wait, isn't there already enough multitasking happening? If anything, we long for intense and short virtual exchanges followed by substantial off-line periods.

The Zoomopticon

Zoom watches you. The video filter that adds a mask, a funny hat, a beard, or a lip color demonstrates that Zoom is watching you through face-tracking technologies. Søren Pold, Danish interface design researcher, observes that Zoom provides only "a slight overview and control of the sound you're receiving and transmitting." This Zoomopticon, as Pold calls it,

> is the condition in which you cannot see if somebody or something is watching you, but it might be the case that you're being watched by both people and corporate software. Zoomopticon has taken over our meetings, teaching and institutions with a surveillance capitalistic business model without users being able to define precisely how this is being done.[25]

How can we respond to this surveillance regime and its pressure to be professional? In her *Anti-Video-Chat-Manifesto*, digital art curator Michelle Kasprzak echoes the understanding of Zoom as a surveillance tool. She calls out this eavesdropping, identifying other individuals and agencies on the call: "Hello NSA, hello Five Eyes, hello China, hello hacker who lives downstairs, hello University IT Department, hello random person joining the call." In response to this regime, Kasprzak calls on us to turn off our video cameras:

> DOWN with the tyranny of the lipstick and hairbrush ever beside the computer, to adjust your looks to fit expectations of looking "professional." DOWN with the adjustment of lighting, tweaking of backgrounds, and endless futzing to look professional, normal, composed, and in a serene environment. DOWN with not knowing where to put your eyes and then recalling you need to gaze at the camera, the dead eye in your laptop lid.[26]

---

24 See, for instance, this 2021 study that compares to have cameras on and off during virtual meetings: "Fatigue affects same-day and next-day meeting performance" (https://doi.apa.org/fulltext/2021-77825-003.html).

25 Private email exchange with Søren Pold, October 6, 2020.

26 https://michelle.kasprzak.ca/blog/writing-lecturing/anti-video-chat-manifesto.

She calls upon us to "refuse fake living in an IKEA showroom with recently-coiffed hair, refuse to download cutesy backgrounds which take up all our CPU, and refuse to fake human presence."

Social Media as Medicine?

Zoom takes its toll on our physical and mental well-being. London-based cultural anthropologist and research consultant Iveta Hajdakova writes,

> Last week I had three nightmares, all related to remote work. In one, I was fired because of something I said when I thought I was offline. In the second, my colleagues and I were trying to get into an office through a tiny well. We were hanging on ropes and one of them became paralyzed, which I think was a dream version of a Zoom freeze. The third nightmare was about me losing track of my tasks. I woke up in panic, convinced I had forgotten to send an important email.

In the early days of lockdown, she struggled with headaches and migraines. Luckily, she writes, these have gone,

> perhaps due to a combination of factors, having a desk and a more ergonomic setup, being able to get out of the flat, limiting non-essential screen and headphone time, and adopting lots of small changes to my routine. The head and the ears are feeling much better now, but something isn't quite right, as the nightmares signal. I've started feeling disconnected and I think this is not merely a result of social isolation but of a more profound sense of disorientation.

As a result, Hajdakova is noticing a growing sense of confusion and uncertainty: "I feel like I am losing the ability to anchor our interactions in embodied human beings and shared physical environments."

Zoom is on its way to becoming a social environment, a strange remediation of office life gone by. "In the beginning, recreating the office experience over video calls worked because all of us still had the shared reference point," Hajdakova continues. "But the more we're removed from the office in space and time, the more I'm forgetting what it is that we're imitating. We're creating something new, a simulacrum of the office." And yet this simulacrum is a pale imitation, reducing a worker and her rich personality to a collection of chat handles and cutesy icons. "I don't want to be just a face and voice on Zoom calls, an icon on Google docs, a few written sentences, I want to be a person ... Social media helps so I've been posting on social media a lot." Friedrich Nietzsche once noted, "When we are tired, we are attacked by ideas we conquered long ago." When Facebook is experienced like a panacea, we know something must be deeply wrong. But why is this feeling of discontent so hard to pin down? The inert state is essentially regressive.

Proving our own existence is like running on a hamster wheel. "The more I try to be a real person, the more I'm getting trapped in the simulation of myself," Hajdakova says. "I'm communicating and sharing just to remind people I exist. No, it is to remind myself that I exist ... Like McLuhan's gadget lover, like Narcissus, staring at his own image." We are not only losing a sense of reality, memory, and confidence, Iveta argues,

> but also losing a sense of understanding for other people. Just knowing that they feel X or Y but having no way of connecting with them through some kind of mutual understanding. In general, Zoom is traumatizing for me because of the way my mind works—I need physical things, shared environments etc., otherwise, I lose not only confidence but also memory and motivation.[27]

No Diagnosis, No Cure

After surviving the COVID-19 siege, we've earned the right to wear the T-shirt: "I survived Zoom." Is a different kind of Zoom possible? We have found the experience we've undergone draining, yet coming together should empower us. What's wrong with these smooth, high-resolution user interfaces accompanied by low-resolution faces due to shaky connections? It's been a delusional dream televising events and social interactions, including our private lives. Is the "live" aspect important to us, or should we rather return to pre-produced, watch-them-whenever videos? In education, this is not a marginal issue. There is a real, time-honored tension between the exciting "liveness" of streaming and the detached, flat coolness of being "online."[28] How can we possibly reverse the Zoom turn?

Already we've seen some formulaic solutions being offered. In 2021, Stanford researchers published four causes of Zoom fatigue and proposed, in tired Silicon Valley fashion, "four simple fixes."[29] The obligations of students, teachers, and office

---

27 Private email exchange with the Iveta Hajdakova, September 21, 2020. See also her text on the same issue: http://thisbloodyplace.com/ill-just-never-know/.

28 See Alan Liu's definition in *The Laws of Cool*: "Cool is information designed to resist information." We could update Lui's phrase "I work here, but I'm cool" to "I hang out here, but I'm cool."

29 https://news.stanford.edu/2021/02/23/four-causes-zoom-fatigue-solutions/. The research paper can be found here: https://tmb.apaopen.org/pub/nonverbal-overload/relea. A contextualization of the present can be found in this 1980s history of the "tired eyes" by Laine Noorey: https://www.vice.com/en/article/y3dda7/how-the-personal-computer-broke-the-human-body.

workers to work online has been rephrased as "prolonged video chats." The required presence of many hours and even days is presented as a choice: "Just because you *can* use video doesn't mean you *have* to." To take the pressure off the eyes, the researcher recommends exiting full-screen mode, reducing the size of the window, minimizing face size, and using an external keyboard. In addition, users should employ the "hide self-view" button, install an external camera, give themselves audio-only breaks, and turn their bodies away from the screen. Power relations, within education and beyond, have not been taken into account. In most cases, any form of "absence" from the screen will, intuitively or not, be read as disengagement and punished accordingly. Such tips tell Microsoft and Zoom how to improve their products and ultimately deliver more work to Stanford engineers. Instead of these "improvements," it is better to use a tool like Zoomscraper that "allows you to self-sabotage your audio stream, making your presence unbearable to others."

Six months into the pandemic, online conferences on spirituality and self-awareness began to offer counter-poison to their own endless sessions. They staged three-day Zoom events, twelve hours per day. They introduced Embodiment Circles, "a peer-led, free, online space to help us stay sane, healthy and connected in these uncertain and screen-filled times. The tried and tested one-hour formula combines some form of gentle movement, easy meditation and sharing with others."[30] The organizers promote "embodied self-care for online conferences. With such an amazing array of speakers and other offerings, the conference FOMO is real. Let's learn a few self-care practices that we can apply throughout the conference, so we arrive at the other end nourished, inspired, and well-worked ... rather than drained, overwhelmed, or with a vague sense of dread and insufficiency."[31] Given this context, should we be talking in terms of "harm reduction"? Online wellness is the craze of the day: our days on Zoom include breaks with live music performances, short yoga routines, or body scan sessions. It is Bernard Stiegler's *pharmakon* in a nutshell: technology that kills us will also save us.[32] According to this stance, if Zoom is the poison, online meditation is the antidote.

But our post-digital exodus needs no Zoom vaccine. Rather than medicalizing our working conditions, let's instead put forward some concrete demands. In late October 2020, students demonstrated at the Amsterdam Museumplein, demanding "physical education." We must now fight for the right to gather, debate, and learn in person. We need a strong collective commitment to reconvene "in real life"—and soon. For it is no longer self-evident that the promise to meet again will be fulfilled.

---

30 https://embodimentcircle.com/embodiment-circle-online/.

31 Quoted from communication related to https://icpr2020.net/.

32 *Pharmakon*: a Greek word meaning both poison and remedy. Bernard Stiegler argues that technics was a *pharmakon*, simultaneously curative and toxic.

Italian media theorist Donatella Della Ratta further opens up the debate by politicizing the online teaching situation. In her essay "Teaching into the Void," printed in this volume with a postscript, she reports about Zoom-specific face-lifts and the product hype of ring lights, face-upgrading technologies that made us all into influencers. In search of an exit, Della Ratta formulates a counter-politics "that finds and forms itself in the aural rather than the visual, one that is most present (and most potent) in the 'awkward moments' of lags, lapses, glitches, bandwidth failures, and frozen frames." Her focus is on subtle forms of refusal, such as students' ignoring their teachers' warnings and turning off their cameras during their Zoom lessons. What if you don't want to share your bedroom, kitchen, or living room with strangers? What if you look tired and bored and you're fed up with jolly backgrounds? Della Ratta's essay ends by praising awkwardness, that mental state that "thrives upon the clumsiness of your well-rehearsed professional performance that stumbles on impact with bad bandwidth, frozen frames, a kid screaming in the background, the family dog's impromptu barking."

Are there better precedents out there, better blueprints to build from? A media-archeological approach to Zoom might return to 1990s cyber fantasies of mass live castings, such as Castanet. The system was designed by dot-com start-up Marimba, a group described at the time as "a small group of Java Shakespeares" by *WIRED* magazine.[33] The idea was to make the web act more like TV by overthrowing the browser paradigm (a goal that the app would later partially achieve). Much like Zoom, Teams, and Skype, the Castanet application had to be downloaded and installed to maximize bandwidth capacity.

Two decades later, the basic choices are still more or less the same, and the players haven't even changed that much. Microsoft, for instance, which owns Skype and Teams, is still a key competitor. Each individual webcasting technology uses its own, proprietary mix of peer-to-peer and client-server technologies. Zoom, for instance, looks smooth because it compresses and stabilizes the signal of the webinar into one stream—instead of countless peer-to-peer ones that constantly need updating. It also pushes the user into a position of "interpassivity": a passive audience mutes its audio and shuts up, much like a pupil listening to a teacher in the classroom. This is in contrast to free software peer-to-peer architectures (such as Jitsi) that go back to the free music exchange platform Kazaa. Software like Jitsi, ironically enough, is also listed as one of the inspirations of Skype, which revolves around collaborative exchanges between equal partners. So, are we watching a spectacle as an audience or working together as a team? Are we permitted to vote, intervene, freely chat?

As the "hybrid event" future gets underway, we need to keep talking and thinking through Zoom fatigue rather than succumbing to fatalism. The era of "blended learning" that aims to merge the virtual and real has arrived. In the face of these

33 https://www.wired.com/1996/11/es-marimba/.

pressures, it is even more important to get organized, demanding a ban on the use of video conferencing in work both inside and outside the institution. Access to buildings will have to be a human right. Together, we should sabotage the real-estate mode of thinking and refuse online education as a cost-saving effort. Physical spaces are not "assets" but public goods.

Of course, that doesn't mean a technophobic retreat to some imagined utopia either. As always, mind the European off-line romanticism trap. Instead, let's make virtual meetings exceptional again. This starts by making virtual conferencing an issue of debate and global dialogue. In an age where the online population has passed the 5 billion users mark, other video conferencing platforms can become tools (among many) to overcome closed borders, reach out, organize, come together, and listen to those who have been excluded. The muted, top-down architectures of Teams and Zoom are the wrong start. It's time to go back to the drawing board, this time with an entirely different twenty-first-century cosmo-technical crew.

References

Abrahams, Annie, Daniel Pinheiro, Mauricio Carrasco, Daniel Zea, Tina La Porta, Alicia De Manuel, David Casacuberta, Pep Gatell, and Melissandre Varin. 2020. "Embodiment and Social Distancing: Projects." *Journal of Embodied Research* 3, no. 2 (October 9, 2020): 4 (video, 27:52). https://doi.org/10.16995/jer.67.

Bailenson, Jeremy N. 2021. "Nonverbal Overload: A Theoretical Argument for the Causes of Zoom Fatigue." *Technology, Mind, and Behavior* 2, no. 1 (February 23, 2021). https://doi.org/10.1037/tmb0000030.

Burleigh, Paula. 2020. "Square Roots." *Artforum* (blog). June 22, 2020. https://www.artforum.com/slant/paula-burleigh-on-the-zoom-grid-83272.

Diem, Cade, and Jaz Hee-jeong Choi. 2021. "Aesthetic Flattening." *The New Design Congress*, June 30, 2021. https://newdesigncongress.org/en/pub/aesthetic-flattening.

Embodiment Unlimited. 2022. "The Embodiment Studio." https://embodimentunlimited.com/studio.

Freund, Jesse. 1996. "Tuning in to Marimba." *Wired*. November 1, 1996. https://www.wired.com/1996/11/es-marimba.

Google Trends. 2022. Google Trends (online). Accessed April 4, 2022. https://trends.google.com/trends/explore?q=zoom%20fatigue.

Hajdakova, Iveta. 2020. "I'll Just Never Know." *This Bloody Place* (blog), July 10, 2020. http://thisbloodyplace.com/ill-just-never-know.

Han, Byung-Chul. 2015. *The Burnout Society*. Stanford, CA: Stanford University Press.

ICPR. 2020. Interdisciplinary Conference on Psychedelics Research. https://icpr2020.net.

Jiang, Manyu. 2020. "The reason Zoom calls drain your energy." *BBC Worklife*, April 22, 2020. https://www.bbc.com/worklife/article/20200421-why-zoom-video-chats-are-so-exhausting.

Kameir, Rawiya. 2014. "Is Internet Fatigue Ruining Your Life?" *Complex*, March 17, 2014. https://www.complex.com/pop-culture/2014/03/is-internet-fatigue-ruining-your-life.

Kasprzak, Michelle. 2022. "Anti-Video Chat Manifesto." *Dr. Kasprzak* (blog). https://michelle.kasprzak.ca/blog/writing-lecturing/anti-video-chat-manifesto.

Liu, Alan. 2004. *The Laws of Cool: Knowledge Work and the Culture of Information*. Chicago: University of Chicago Press.

Lorusso, Silvio. 2020. "Remote Work? Demand Dial-Up." *Platform BK*, June 12, 2020. https://www.platformbk.nl/en/remote-work-demand-dail-up.

Nooney, Laine. 2021. "How the Personal Computer Broke the Human Body." *Vice* (blog), May 12, 2021. https://www.vice.com/en/article/y3dda7/how-the-personal-computer-broke-the-human-body.

Ramachandran, Vignesh. 2021. "Stanford Researchers Identify Four Causes for 'Zoom Fatigue' and Their Simple Fixes." *Stanford News* (blog), February 23, 2021. https://news.stanford.edu/2021/02/23/four-causes-zoom-fatigue-solutions.

Sacasas, L. M. 2020. "A Theory of Zoom Fatigue." *The Convivial Society* (blog), April 21, 2020. https://theconvivialsociety.substack.com/p/a-theory-of-zoom-fatigue.

Shockley, Kristen M., Allison S. Gabriel, Daron Robertson, Christopher C. Rosen, Nitya Chawla, Mahira L. Ganster, and Maira E. Ezerins. 2021. "The Fatiguing Effects of Camera Use in Virtual Meetings: A Within-Person Field Experiment." *Journal of Applied Psychology* 106, no. 8 (August 2021): 1137–55. https://doi.org/10.1037/apl0000948.

Sklar, Julia. 2020. "'Zoom Fatigue' Is Taxing the Brain. Here's Why That Happens." *National Geographic*, April 24, 2020. https://www.nationalgeographic.com/science/article/coronavirus-zoom-fatigue-is-taxing-the-brain-here-is-why-that-happens.

Taparia, Siddarth. 2020. "This One Concept Will Transform the Future of Work Post-COVID-19." *Fast Company*, October 3, 2020. https://www.fastcompany.com/90558734/this-one-concept-will-transform-the-future-of-work-post-covid.

Wikipedia. 2022. "Zoombombing." Last modified March 28, 2022. https://en.wikipedia.org/w/index.php?title=Zoombombing&oldid=1079660583.

Wojnarowicz, David, and Olivia Laing. 2017. *Close to the Knives: A Memoir of Disintegration*. The Canons. Edinburgh: Canongate.

Xingyu, He. 2019. "Introduction: Posthuman Bodies—Judith Halberstam and Ira Livingston." *Making Difference: Architectures of Gender* (blog), May 9, 2019. https://blogs.ethz.ch/making-difference/2019/05/09/introduction-posthuman-bodies-judith-halberstam-and-ira-livingston.

The Need for Intentionally Equitable Hospitality in Video Conferencing

Maha Bali

As someone who has been living large chunks of my life online since before the COVID-19 pandemic, I experience Zoom fatigue (see also Lovink, this volume) as much as the next person, but I also actively resist making any meeting, presentation, workshop, or class session that I lead contribute to other people's Zoom fatigue. Before I share details, let's acknowledge that some elements of Zoom fatigue are just part and parcel of video conferencing: sitting for a long time in front of a screen, exhausting your eyes looking at it constantly, and trying to be social and compensate for the lack of physical togetherness that tends to give us (at least the extroverts among us!) a special kind of energy. Let's also acknowledge that some of the solutions to this are simple ones that we forget to do. For example, inviting participants to get up and move around as part of an activity in your workshop, or inviting people to write some stuff on paper and look away from the screen during a synchronous class meeting—we just forget to do these things when we are on a video conference. But there is no real reason to ignore the bodies behind the screens. Giving others breaks; giving ourselves breaks; and importantly, giving people grace with cameras (see also Della Ratta, this volume) when they want to keep their cameras off, whether it exhausts them to stare at themselves all day or be stared at by others. Enforcing the opening of cameras can be a kind of benevolent surveillance, and people should be able to opt-in or out. All of these things are easy to do and go a long way to avoiding exhaustion from long hours of video conferencing.

More than avoiding exhaustion, though, I believe that we can make video conferencing spaces welcoming spaces, ones that welcome diverse people with different needs and interests, and my colleagues and I call this Intentionally Equitable Hospitality (IEH) (Bali et al. 2019, Bali and Zamora forthcoming).

Intentionally Equitable Hospitality: Pre-design, Design, Facilitation and Beyond

Although we use technology to enact a lot of what we do during a video conference, all of the suggestions I bring here center the human and the connection between humans that video conferencing affords. It is really not about the technology at all, but about a kind of "entangled pedagogy" (Fawns 2021; Fawns 2022) where we are simultaneously hyper-aware of the affordances and limitations of technology and also deeply in touch with our pedagogy and our intentions. We use both of these skillsets to create learning experiences in specific contexts that allow us to do things we could not have done without marrying these two together. It is also essential as we meet people through screens not to ignore the embodied person behind the screen or the entire room surrounding them, for the environment and body are present and affect that person's engagement and presence, whether or not we can see or hear them. And it is also essential to recognize the interplay between power outside of and inside the video conferencing space and recognize our roles as facilitators to subvert inequity in spaces that we lead online.

IEH is "a facilitation praxis" (Bali & Zamora 2022), initially developed in the context of hybrid video conferencing by the co-directors of a grassroots movement called Virtually Connecting (Bali et al. 2019). Virtually Connecting (VC) has "challenged academic gatekeeping via rendering private hallway conversations that build social capital at face-to-face conferences into public hybrid conversations in which people who cannot attend conferences are able to participate" (Bali et al. 2019, para. 7). However, the key underpinning values can be applied to more formal educational contexts (Bali and Zamora, forthcoming).

The key foundation of IEH is that "the teacher or workshop facilitator is a 'host' of a space, responsible for hospitality, and welcoming others into that space" (Bali and Zamora, forthcoming). By "host" here, we evoke the analog meaning of the term: the person who invites others to a space, like a dinner party (not the "host" of a Zoom call, necessarily). The "host" is then responsible for intentionally making moves that promote and ensure equity every step of the way, and especially in these four phases (paraphrased from Bali and Zamora, forthcoming):

1. Pre-design: who is involved in designing the experience? Are the most marginalized groups of participants included, involved, and can they have power to adapt the design in different ways? Which platforms do we use and in what ways do they enable better economic access for people with different infrastructures and devices? In what ways do we account for differences in participants' cultures and how it might affect their participation? What kind of freedoms do we afford participants in the video conferencing platforms (e.g., freedom to use chat, share screen, rename themselves, choose breakout rooms if we use them)?

2. Design: in what ways can our designs enhance engagement in diverse ways that enable everyone's voice to be heard? What kind of guidelines or community practices can the host promote or build with participants to enhance equity? In what ways might the host use "generous authority" (Parker 2019) to ensure no one dominates at the expense of others, for example by suggesting and enforcing equal time at certain stages?
3. Facilitation: in the moment of facilitation, in what ways can hosts be fully present and use "intentional adaptation" (brown 2017) to modify plans in order to better respond to participants' needs, and in order to address unanticipated inequalities that emerge during a live session?
4. Beyond the moment: in what ways can we build community equitably before and after a session to help create a welcoming space for diverse participants? How do we use asynchronous spaces and semisynchronous, less formal spaces (like WhatsApp, Twitter, or Slack) to strengthen the relationships between participants and help enhance cohesion during live video conferencing sessions?

Less Prep, More Presence: Intentional Adaptation

I want to emphasize the importance of "less prep, more presence" (brown 2017), as a central praxis within IEH. I will give some examples of activities that are really great when you plan them, but need some quick thinking during a session to modify for the audience. One example is an activity called Wild/Mad Tea (watch a demo of the Liberating Structure in Development by Bali et al., undated). This is a speed-networking warm-up activity, where people answer questions quickly in pairs and then move on to answering another question in another pair, and so on. In a situation where people face connectivity issues, something that was planned for a pair to take two minutes may be quickly modified for a trio in four minutes, which gives people more time to move to rooms, and is less likely to result in one person ending up alone in a breakout room. An activity that is planned to ask participants to do something on video where people are unable or reluctant to open cameras can be modified to ask participants to share screens and share an image rather than their own faces. And so on. One can initially plan for an open dialogue where people unmute and speak, but in the moment, one can modify the activity and ask all participants to respond in the chat, making room for an initial response from all voices, before asking people to unmute and speak. When we find an audience that is quieter than we anticipate, we can convert a full-group activity into a small-group activity, and ask participants to work in groups of three or four before sharing back to the main group, or create a slide deck for them to edit while in their breakout rooms to give them something concrete to share back.

These are all moves that take a minute or two to set up, but can make a huge difference in shifting the energy and dynamic of a group. Having all these options in our arsenal to draw upon when we need them is what makes for a good facilitator who is more "present," rather than simply "prepared."

Harnessing the Power of the Platform for Equity

On video conferencing platforms there is a phenomenon we do not experience in our daily lives: when we join, we see other people's video, but we also see ourselves on video; "you are forced to see yourself being seen" (Caines 2020, para 5), which "routinizes a kind of self-surveillance" (Caines 2020, para 6). This can cause anxiety for some people, or distract them as they keep checking their own image. At the very least, individuals should be able to control whether they have their cameras on or off, and when. And yet, hosts need to be aware of the ways in which they might naturally focus more on the people with cameras on when others have their cameras off, taking visual cues from them on pacing and so on. Hosts also need to be aware that someone having their camera on is not really a proxy for engagement, as they can be looking at something else on their device or even behind their screen: a laugh or a frown can be related to the sound of a child screaming nearby, or an email they just saw.

By design, video conferencing platforms tend to give the "host" particular privileges (see also Distelmeyer, this volume) such as controlling who speaks and who is muted, who can chat with whom, who can come in and leave, whose video is "spotlighted" for others to focus on, who is allowed to share their screen or not. For a more democratic video conferencing experience, a host can enable others to do whatever they want here—opening up choices to chat, share screen, etc., but paradoxically, sometimes in order to protect participants, a host may limit some of these actions. For example, to avoid random "Zoom-bombing," hosts often disable screensharing and annotating for participants as a default. Occasionally, a host needs to "mute all" participants in order to avoid noise/feedback coming from an unknown source.

One of the biggest forms of control a host has is with breakout rooms on Zoom: a participant can find themselves in a small group conversation with others selected by the host (or randomly) for a certain amount of time, and they have little control over how long they stay or who they are with, and then suddenly the host can yank them back into the main room before they are ready to return. This makes managing time in classes and workshops much easier than in person, where you often cannot stop people from chatting in smaller groups. A host can flip their control here with breakout rooms, by giving participants control over which rooms to join and when to come back to the main room—but they would need to be willing to let go of a lot of control, and in many contexts, this is not the case.

Promoting equity in these spaces involves an awareness of the power the platform affords, and finding ways to promote participants' agency over how they use it, wherever possible. Sometimes a platform will directly allow participants to move themselves to breakout rooms, for example, or rename themselves, or blur their video background; other platforms will require the host to find ways around that, such as asking participants where they want to go before moving them, or inviting people to turn their cameras off if they choose.

Video Conferencing Is About More Than Video

This should be obvious, especially given the kinds of examples I've shared so far, but it bears repeating. Being together on video conferencing does not mean that all we do is share video of ourselves and "conference" (speak). The chat feature of video conferencing enables so much public backchanneling that can enable connection and create space for EVERYONE to participate in ways that are nearly impossible in a face-to-face environment! The opportunities for breakout rooms (especially well done by Zoom) enables intimate small group conversations that many of us miss dearly in the larger conversations that overload our cognition. The capacity to share a screen and annotate collaboratively—to color, to draw, to point out and highlight things while reading or viewing together—provides an experience more complex than paper and pen activities in a classroom context.

And we are never ever limited by a video conferencing platform once we are online, because we do not always need to be talking together in real time, all the time: sometimes we want to write quietly together, create visuals in parallel, or give our opinions anonymously. Sometimes we want to collaborate over a long period of time at our own convenience. We can open Google slides or Google docs or a Miro or Mural board and start editing and creating together (see also Michell, this volume, and Kaldrack et al., this volume). We can use polling tools to hear from everyone, create word clouds of our thoughts, and give everyone an opportunity to share—we bend time and space when we use asynchronous and semisynchronous tools during synchronous sessions.

Conclusion

Creating equitable video conferencing spaces starts with setting our intention and iterating toward equity in the pre-design and design of our sessions, as well as adapting intentionally during sessions. A successful and equitable video conferencing experience can also be enhanced by the communication between participants outside of the time and space of the session itself. Awareness of the ways in which platforms themselves can reproduce or exacerbate inequalities is essential, as well as awareness of how power inside of the video conferencing space is negotiated, and the ways in which our pedagogy can influence power dynamics.

References

Bali, M., & Zamora, M. 2022. "Intentionally Equitable Hospitality as Critical Instructional Design." *Designing for Care*, edited by IS. M. Morris, J. Quinn, S. Jhangiani, & M. Burtis, Hybrid Pedagogy Publishing. https://designingforcare.pressbooks.com/chapter/intentionally-equitable-hospitality-as-critical-instructional-design/.

Bali, M., Caines, A., DeWaard, H., Hogue R. J., & Friedrich. C. 2019. "Intentionally Equitable Hospitality in Hybrid Video Dialogue: The context of Virtually Connecting." *eLearn Magazine*. Retrieved 2019, from https://elearnmag.acm.org/archive.cfm?aid=3331173.

Bali et al. (undated). "Wild/Mad Tea." *Equity Unbound/OneHE*. https://onehe.org/equity-unbound.

brown, a. m. 2017. *Emergent Strategy: Shaping Change, Changing Worlds*. Chico: AK Press.

Caines, A. 2020. "The Zoom gaze." *Real Life Magazine*, 7.12.2020, https://reallifemag.com/the-zoom-gaze/.

Fawns, T. 2021. "An Entangled Pedagogy: Views of the relationship between technology and pedagogy." *OpenEd Edinburgh*. https://open.ed.ac.uk/an-entangled-pedagogy-views-of-the-relationship-between-technology-and-pedagogy/.

Fawns, T. 2022. "An Entangled Pedagogy: Looking Beyond the Pedagogy—Technology Dichotomy." *Postdigital Science and Education*. https://doi.org/10.1007/s42438-022-00302-7.

Parker, P. 2019. *The Art of Gathering: How We Meet and Why It Matters*. London: Pinguin.

Infrastructuring | Interfacing

Laws of Zoom

Kim Albrecht

In November 2019, the Zoom Video Communications platform had 10 million meeting participants per day worldwide. By April 2020, this number grew 30 times to 300 million daily participants (Vailshery 2022). The COVID 19 pandemic changed the role and scale of video conferencing in the daily lives of millions. But not only has the usage of Zoom expanded exponentially, so has the scope and breadth of what the application is technically capable of. This increase is not only internal but also externalized through a Zoom App Marketplace that allows third-party developers to build applications onto the Zoom infrastructure. While the marketplace already opened in 2018 (Zoom[developer], n.d. b), in February 2021, Zoom announced that the marketplace had grown to 1000 connected applications. One year earlier, the marketplace consisted of only 200 apps (Mullin 2021). The outreach of Zoom increased not only as a customer service but also as a networked meta-application containing a vast amount of additional software applications from outside developers. Zoom, in this sense, is not only an application, an independent code structure within the operating system, but rather a platform, a networked connector of supply and demand, regulating and influencing a broader networked system (Lovink 2021). The communication layer of this interdependent structure of relations is organized by the application programming interface, or API, which this article will examine. *This article investigates the Zoom API documentation by visualizing its organizing principles and structures. By doing so, one layer of the subface platform structure becomes a surface.* This investigation offers a perspective of the laws of Zoom from an outside developer. How is this perspective different than the graphical interface layer of its millions of users? And what mechanisms determine Zoom's relation to external companies developing apps within their platform?

Observing some of the external developers' apps for Zoom showcases the meaning behind Zoom's marketing claim of "keeping you connected." The applications on Zoom's Marketplace range from icebreaker games for teams (Playco, n.d.), to AI Note-taking transcriptions (Fathom, n.d.), to live Poll Apps, Quizzes, and Word Clouds (Mentimeter, n.d.). The categories by which Zoom separated its marketplace caught my attention—for example, analytics, customer relationship management, monitoring, productivity, or recordings. The question becomes, what is kept connected by Zoom? The "you" in keeping you connected suggests a human or user connection, but what is connected are infrastructures of networked hard- and software. *Behind the term "you" lies a global infrastructure of hard- and software of internal and external code structures.*

Zoom is more than a tiled canvas of talking heads. Some of the Apps within the marketplace are very open about this. For example, the App People.ai advertises to "automatically capturing all contact and customer activity data" (People.ai, n.d.). What is labeled "actionable intelligence" is a software-based surveillance system to turn video conferencing into profit-oriented exploitation of surveillance data. Within the example screenshots of the app, various visualization dashboards are displayed, ranking individuals by meeting minutes, number of calls, "engagement," and the number of scheduled vs. unscheduled meetings. Rather than "keeping you connected," as Zoom articulates it on their home screen, the app store and the capabilities of listed apps convey the possibilities of surveillance and control. This is possible due to a wide-ranging Application Programming Interface (API) that Zoom opened up to outside developers.

The scope of the Zoom API is vast, with capabilities, affordances, and restrictions that lie far beyond the possibilities of this article. But like the just-introduced People.ai app, many others use visualization methods to monitor individuals using Zoom. I want to counteract this approach by using visualization methods to instead map the capabilities of the Zoom API. I will call this method *infrastructure imaging*. But before doing so, I will begin with a short introduction to the API.

What Is the Zoom API?

API is an acronym for Application Programming Interface; it is understood as an intermediary between computer programs. The graphical user interface, or GUI, allows humans to interact with computational systems. The API does something similar but on the level of software to software interaction. The API is not intended for human usage apart from the computer programmer who uses it to write new software connected to already existing programs. *Thus the API is a human concept, made by humans for humans, but only a particular type: the developer.*

Through the GUI, a Zoom meeting participant learns how to chat during the meeting, invite others into the call, or leave the meeting. Similarly, a developer knows how to build new software connections toward the Zoom platform through the API description. The API is a socio-technical complex making certain sets of particular programmatic components available for developers to other programmatic components (see also Snodgrass and Soon 2019). That the API is a socio-technical complex centered around the developer is noted by Zoom explicitly in their API reference:

Figure 1: All parameters of the Zoom API

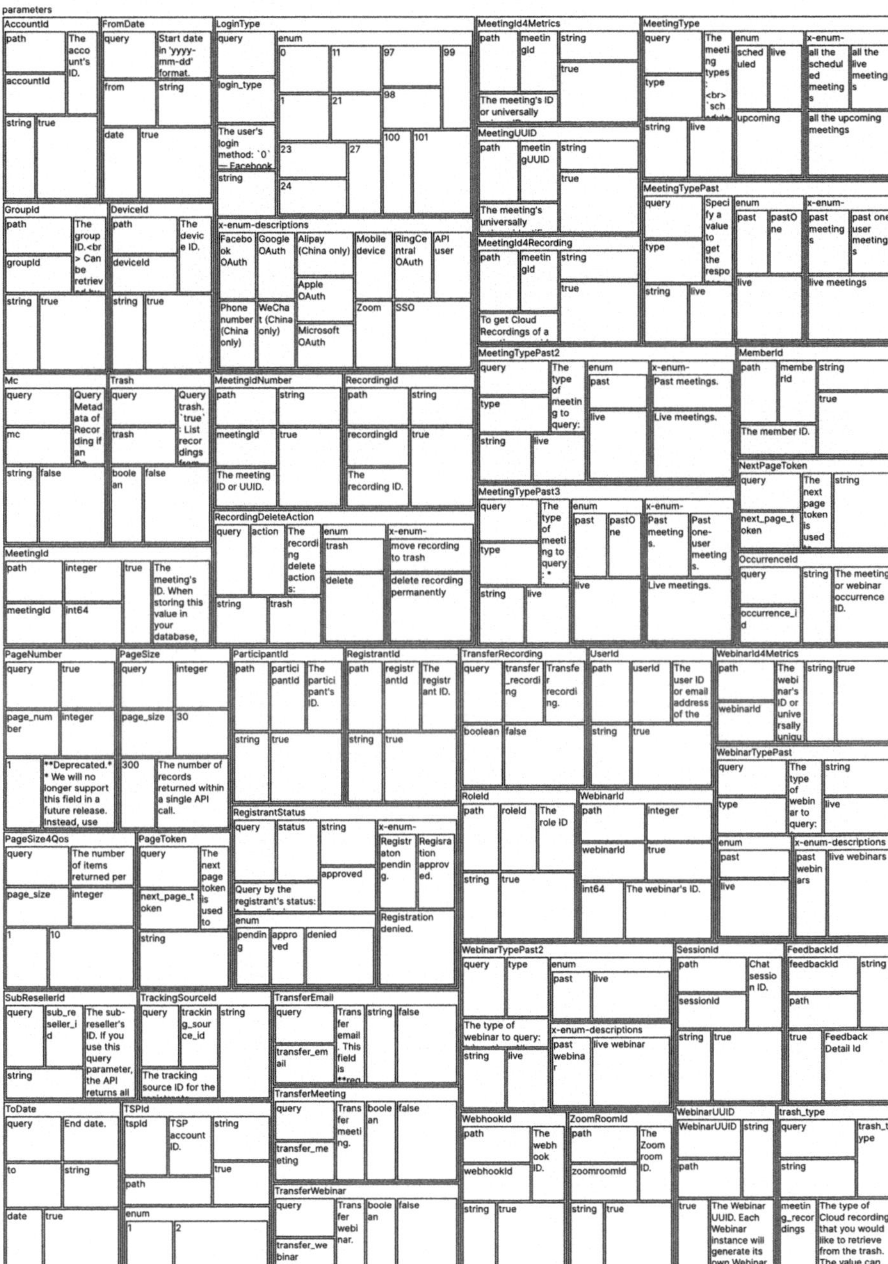

Figure 2: Within "paths" all setting for "rooms"

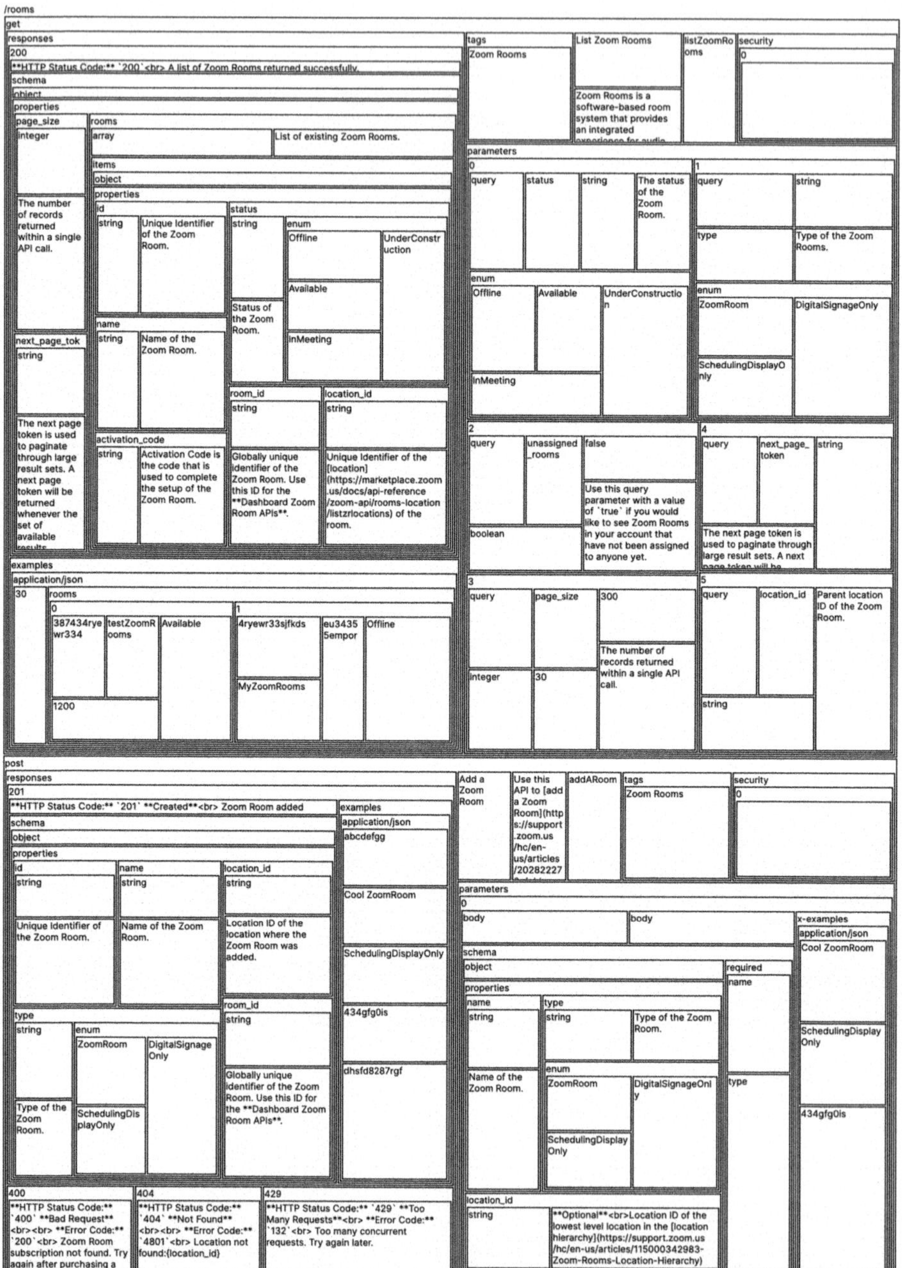

> The Zoom API is the primary means for developers to access a collection of resources from Zoom. Apps can read and write to the resources and mirror some of the most popular features available in Zoom Web Portal such as creating a new meeting, creating, adding and removing users, viewing reports and dashboards on various usage, and so on using the Zoom API. (Zoom[developer], n.d.a)

While the conception of the API is construed by Zoom as a move toward openness, toward giving access, sharing information, and allowing software to interact with one another, my argument would be that it is the opposite. The API is a designed system in which a cooperation, such as Zoom Video Communications, has complete control of what they share and what they keep hidden from the outside. The API hides how a system works and only shares specific access points by not sharing the entire code or working on an open platform. The API is open only under the conditions of the cooperation controlling the API, a modality of controlled sharing.

> Zoom APIs allow developers to request information from the Zoom, including (but not limited) to user details, meeting reports, dashboard data, as well as perform actions on the Zoom platform on a user's behalf. For example, creating a new user or deleting meeting recordings. (Zoom[developer], n.d. c)

In the case of Zoom, the accessibility of the API has two sides: the one keeps internal processes hidden from competitors, outside developers, and the public; the other opens their system toward free labor from developers creating applications for their app store. The API thus becomes a balancing act between sharing and hiding. The equation is simple: *Sharing power to retain power*. By inviting outside developers, Zoom withholds its power from the competition; by growing its platform, by doing much more than "keeping you connected," Zoom solidifies its power. Simultaneously, the API abstracts the underlying processes for outside usage, only exposing what is necessary.

Why Study the Zoom API?

The power relation of *sharing to retain power* makes the API an intriguing complex to investigate. The standard setting bar in the GUI of a Zoom conference consists of various buttons. From left to right, the mute button, the start and stop video button, the participant's list, the chat button, the share screen button, the record button, the reaction button, and furthest to the right, the red leave button. Depending on your role and settings, the bar changes and adjusts, restricting and providing access, it consists of five to ten possible operations. In contrast, the current (downloaded on 2021–12-09) Zoom API specification consists of 60.247 nested parameters. While the

API is an abstracted and restricted perspective on an application such as Zoom, one entirely under the company's control, the number of settings, as well as retrievable and postable data exchanges, is immense. Observing the API allows us to study an otherwise disguised system far beyond the graphical user interface. The API provides a perspective on processes hidden from human sight outside the developer's realm.

The Zoom API Specification Document

What I am investigating throughout the following pages is not the Zoom API itself; instead, what I will observe, visualize, and narrate is the API specification document. Every possible operation of receiving, changing, deleting, posting, or patching data from and to the Zoom API is listed within this file. What is not allowed is forbidden. *Anything outside the specified is unperformable.*

Within this context, Jan Distelmeyer refers to Wendy Chun in this publication. In "*Control and Freedom*," Chun argues that Lawrence Lessig's slogan "Code is Law" falls short. Code is not only law but, at the same time, execution (see Chun 2006, 66–67). As long as it is running, the code is simultaneously executing the laws. The separation of powers is not based on a *trias politica* model of a legislature, an executive, and a judiciary. The judiciary of the digital only contains two states: running or failing. There is a small margin of error within the system, that is, the glitch. Outside of that, the legislature "the developer" and the executive "the code" are the two folded systems of power in software infrastructures. The API specification document defines the possibilities for an outside developer to send and receive data with Zoom. *The Laws of Zoom from the perspective of an external developer are written within this documentation.*

The Zoom API reference is made available by Zoom through a web interface (Zoom[developer], n.d. d). In addition to the web documentation, Zoom provides the specification document in one JSON (JavaScript Object Notation) format, a machine-readable version of the document. The entire interaction with the Zoom API is listed within this document in a nested structure consisting of attribute-value pairs and arrays. As an example, 19 lines from the file:

```
"examples": {
    "application/json": {
      "page_size": 30,
      "rooms": [
      {
        "id": "387434ryewr334",
```

Figure 3: Within definitions all UserSettings

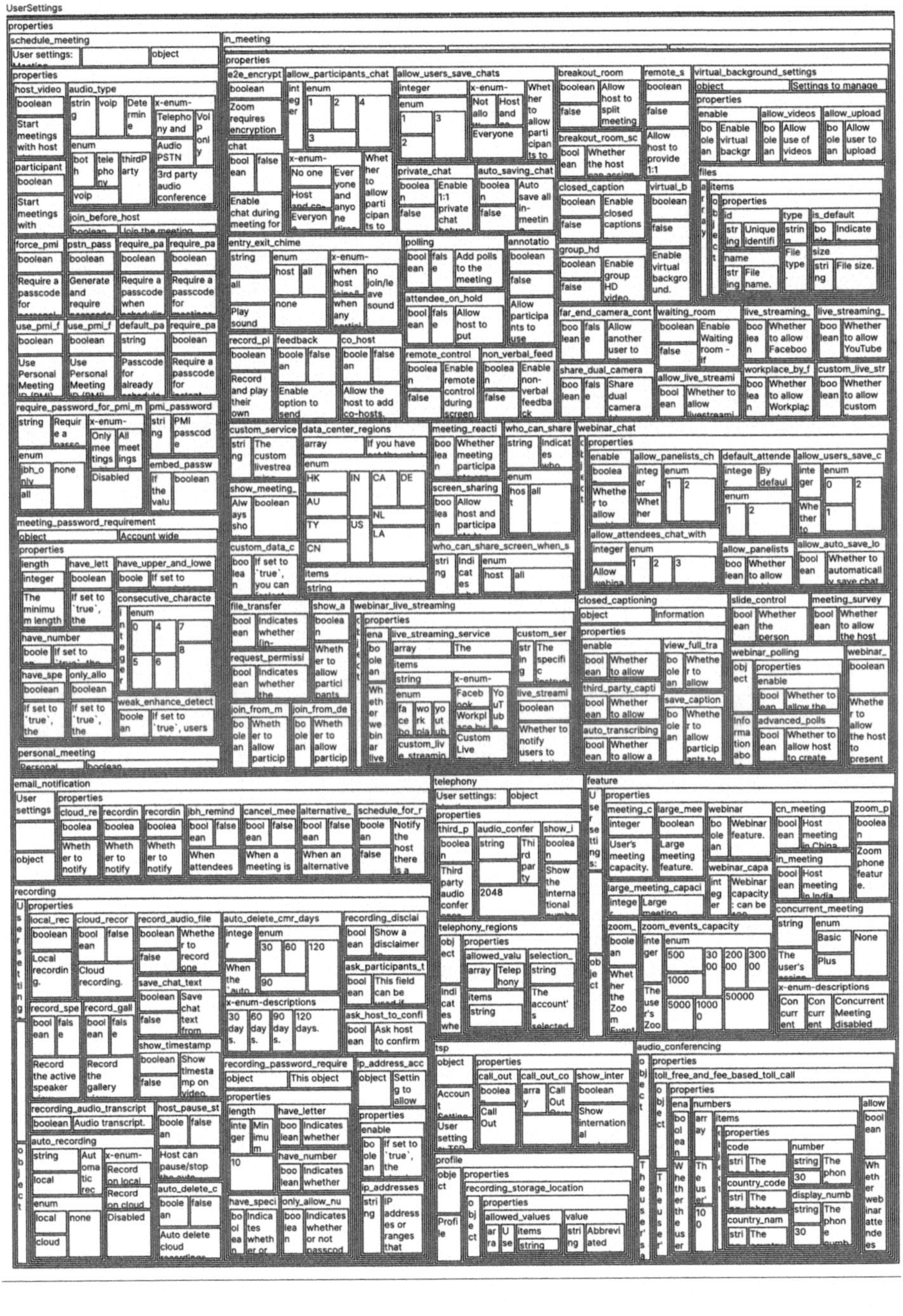

Figure 4: Within "paths" all setting for the messages of a specific User

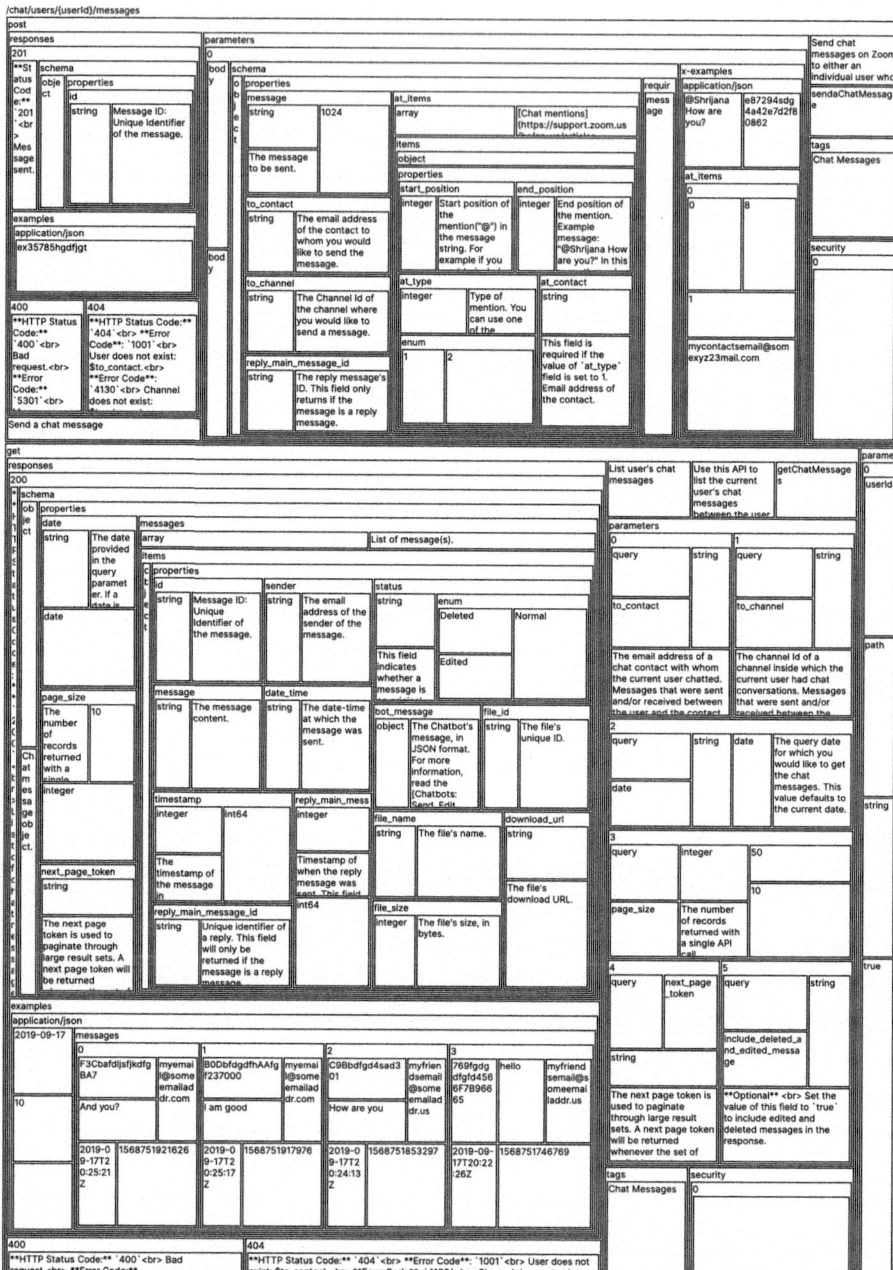

```
      "name": "testZoomRooms",
      "activation_code": "1200",
      "status": "Available"
    },
    {
      "id": "4ryewr33sjfkds",
      "name": "MyZoomRooms",
      "activation_code": "eu34355empor",
      "status": "Offline"
    }
    ]
  }
}
```

The JSON document consists of 80.392 nested lines of code; their basic building block is the so-called "key": "value" pairs. A colon and a value follow the key in double quotes. Commas separate each key/value pair; curly brackets distinguish objects from their nested relations. Square brackets, as in the above example after the key "rooms," create arrays, storages of multiple pieces of information on the same level of nestedness. Data is structured in hierarchical formations within the JSON format, nested buckets of distinctions. There are only two modes of structure: within and next to.

Infrastructure Imaging

Various methods can graphically represent hierarchical datasets, such as the Zoom specification document. The visualization for this publication is a rectangular area encoding in which size and position represent the relations of the nested commands. There are only two modes in which rectangles are positioned, placed next to one another and within one another. The artificial tiles of the talking heads of a Zoom interface are reflected and manifested within this data representation. The rectangles recursively subdivide the area; thus, the size of the rectangles represents the number of contained other rectangles. With 80.392 lines of code, the entire structure of the API specification document is representable with one symbol: the rectangle. The relations of the rectangles to one another are, like the data, twofold: within and next to. The four lines that constitute the rectangle define what is included and what is excluded. There are no overlaps; everything is perfectly expressed in stacked boxes, mirroring the definedness of the Zoom video tiles.

The first layer of stacked tiles within the API document consists of 13 objects: x-stoplight, swagger, info, host, basePath, schemes, consumes, produces, paths, parameters, definitions, tags, securityDefinitions, security. The quantities of nested structures within these terms vary tremendously. While "paths" contain 42,461 key/value pairs, others such as "x-stoplight," "swagger," "host," or "security" only contain one key/value pair.

Within the "parameters" key, 375 key/value pairs are nested (fig. 1). It is the third-largest set after "paths" (42,461 pairs) (fig. 2 & 4) and "definitions" (17,278 pairs) (fig. 3, 5, & 6). Within "parameters," for example, the "AccountId" and the "DeviceId" are stored. From the data, we can observe some of Zoom's constraints. For example, there are 12 different Login Types for Zoom. A user can log in to Zoom using the authorization protocol OAuth, using Facebook, Apple, Microsoft, and RingCentral, or by Mobile device, the API, a Zoom Work email, and Single Sign-On. In addition, Chinese Users can sign in by Phone number, WeChat, and Alipay. The entirety of possible login modes for users are listed within the document. What is not allowed is forbidden. But what is similarly intriguing is that the API documentation states all possibilities, even if they only apply to specific users, in this case, Chinese Users. Particular settings are only visible to specific users on the user interface level. *Within the API, all accessible operational potentialities must be defined.*

The API manual does not further define the logging in through Facebook; there are no details on the process or the data exchanged. Throughout the entire document, "Facebook" is mentioned 151 times. I am commenting on this as there have been some suspicious activities between the two services. An analysis published on March 26th 2020, by Vice's online magazine *Motherboard* (Cox 2020) found that Zoom sends data to Facebook's graph API upon using the iOS app. Zoom exchanged information about the used device, the phone carrier, location, and a unique advertising identifier, among others, with Facebook. Such an operation would have happened not through the public API but through Facebook's Software Development Kit for iOS. Six days after the Vice publication on April 1st, 2020, the CEO of Zoom, Eric S. Yuan, apologized for the data exchange and removed the possibility to log in on iOS through Facebook (Yuan 2020).

The Zoom API documentation is the law and executive manuscript for outside developers. Simultaneously as exemplified in the example with Facebook, outside of that law, Zoom can establish other exchanges. The Laws of the Zoom API documentation are far from universal. They are only defined for the specific use-cases of external developers building applications within the Zoom platform. That said, establishing an ethical commitment that all possible data exchanges are contained within an open API would be a tremendous step toward a more transparent relationship between users and their services. The Zoom Facebook exemplification shows how far we are from such a reality. The public Laws of Zoom only render visible what the company wants to be visible.

Figure 5: Within "definitions" all account settings

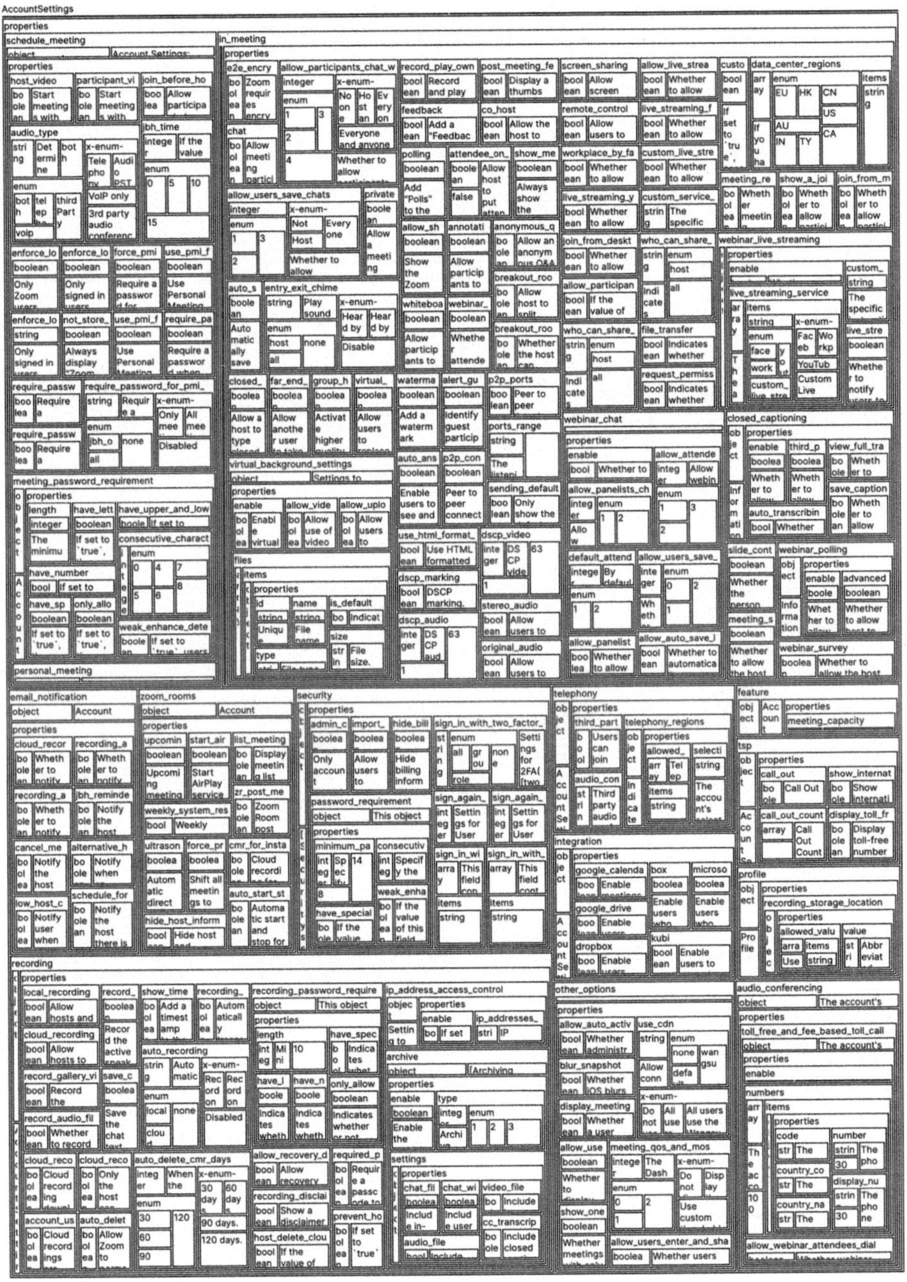

Figure 6: The meeting info within "definitions"

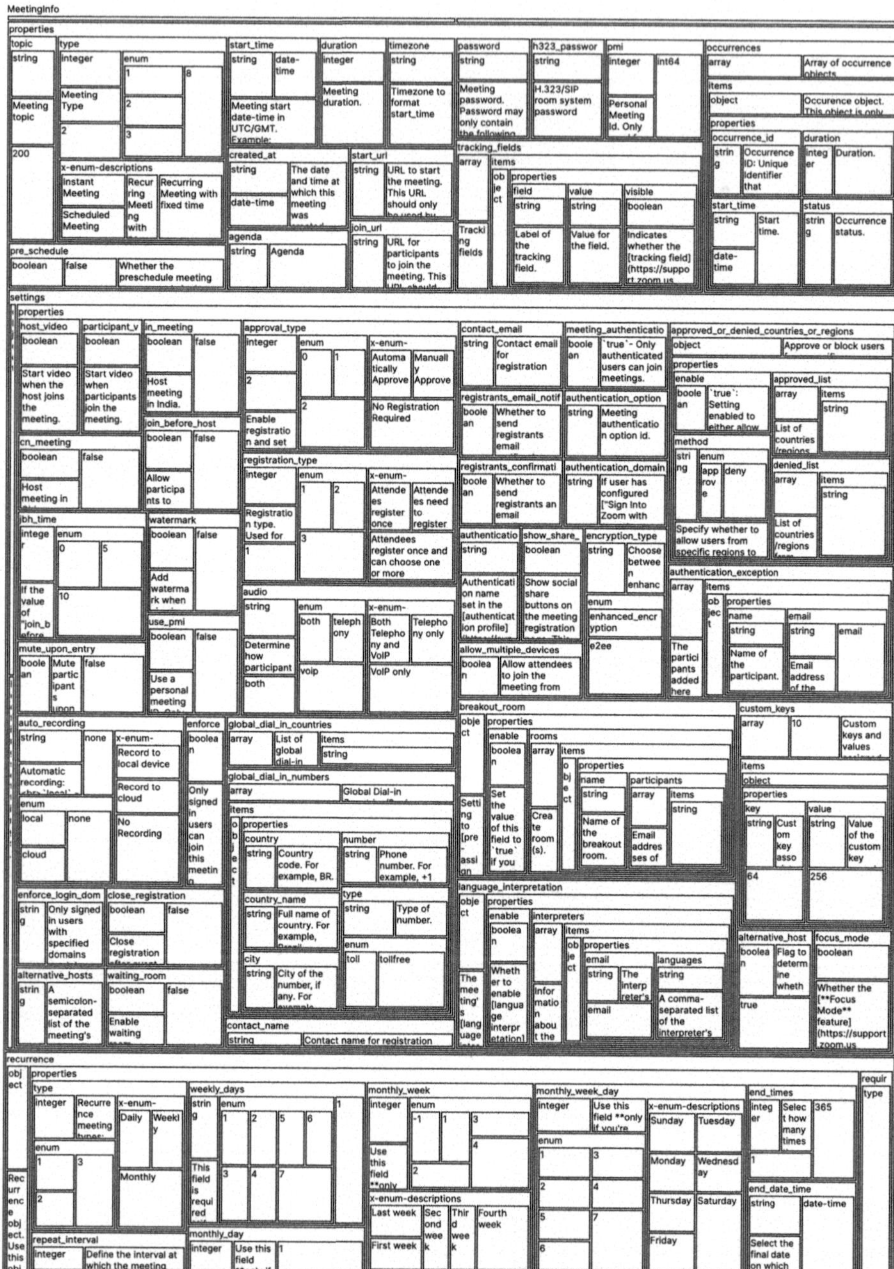

In researching computational systems, Frieder Nake suggested a differentiation between the surface, the visual graphic interface, and the subface, the underlying process, the algorithmic thing, making the surface visible (2015). The API is one layer of the subface—the processes hidden from view, the data exchanges occurring once you click the Zoom button without notice. The graphic representations of this paper turn parts of the subface into the surface—visualization as the alchemy of concealed operations. The subface of Zoom through the lens of the API specification document is at times strikingly similar to the interactions one has while using Zoom and at other times uncanny and distant. The specification document maps out the possibility space rather than actualities. It does not show what is but what is possible within the operational space for external developers.

Zoom's graphic user interface tiles, the talking heads, the arranged tiled structure, and the undercomplexity at which Zoom aestheticizes human interactions are also reflected in the orderings of Zoom's developer realities. The two-fold mechanism of next to and within reveals a facile worldview of a two-fold arrangement of reality. Simultaneously, the simple structure of the notation object overwhelms by scale. This observation divulges something fundamental about the complex, hidden behind the term "digital": Simple conceptions with few mechanics are scaled into feigned complexity. For a computer, it makes little difference whether there are 10, 100, 1,000, or 10,000 nestings. As long as the mechanics are constant, the quantity is secondary. Within the physical realities we inhabit, this is not the case. Repetition takes time and effort. The same is true about the observation of the graphic representations of the Zoom API. The number of rectangles is computed within milliseconds, but reading and comprehending these structures strongly depends on the amounts of represented dependencies. However, a discrepancy, an omission of a bracket, a change in the relationship between elements, or a missing comma would render the system incalculable. Quantities within hardware limits hardly make a difference in the system, but a conceptual difference breaks the system immediately. Remove one sign, and the file renders computationally illegible. Size and conceptual complexity are two dimensions that are diametrically opposed within the term digitality.

References

Chun, Wendy Hui Kyong. 2006. *Control and Freedom. Cambridge:* MIT Press.

Cox, Joseph. 2020. "Downloading Zoom for Work Raises Employee Privacy Concerns." March 26, 2020. https://mashable.com/article/zoom-conference-call-work-from-home-privacy-concerns.

Fathom. n.d. "Never Take Notes on a Zoom Call Again." Accessed April 10, 2022. https://fathom.video/.

Lovink, Geert. 2021. "Notes on the Platform Condition." Making and Breaking. 2021. https://makingandbreaking.org/article/notes-on-the-platform-condition/.

Mentimeter. n.d. "Mentimeter – Zoom App Marketplace." Accessed April 10, 2022. https://marketplace.zoom.us/apps/D6gsxPEXS8eNr9d6gNFCIg.

Mullin, Natalie. 2021. "We Now Have More Than 1,000 Apps on the Zoom App Marketplace to Help You Work Better." February 12, 2021. https://blog.zoom.us/we-now-have-more-than-1000-apps-on-the-zoom-app-marketplace/.

Nake, Frieder. 2015. "Surface, Interface, Subface. Three Cases of Interaction and One Concept." In *Paradoxes of Interactivity. Perspectives for Media Theory, Human-Computer Interaction, and Artistic Investigations*, edited by Uwe Seifert, Jin Hyun Kim, and Anthony Moore, 92–109. Bielefeld: transcript. https://doi.org/10.1515/9783839408421-005.

People.ai. n.d. "People.Ai." Accessed December 15, 2021. https://marketplace.zoom.us/apps/nea51Ug9Roaxh1J2qzofwA.

Playco. n.d. "#AskAway – Zoom App Marketplace." Accessed April 10, 2022. https://marketplace.zoom.us/apps/rVVDJnIqRgGEKcQ-SAdVJA.

Snodgrass, Eric, and Winnie Soon. 2019. "View of API Practices and Paradigms: Exploring the Protocological Parameters of APIs as Key Facilitators of Sociotechnical Forms of Exchange | First Monday." *First Monday*. https://doi.org/https://doi.org/10.5210/fm.v24i2.9553.

Vailshery, Lionel Sujay. 2022. "Zoom Daily Meeting Participants Worldwide 2019–2020." February 21, 2022. https://www.statista.com/statistics/1253972/zoom-daily-meeting-participants-global/.

Yuan, Eric S. 2020. "A Message to Our Users. Zoom Blog." April 1, 2020. https://blog.zoom.us/a-message-to-our-users/.

Zoom[developer]. n.d. "Introduction to Zoom API." Accessed April 30, 2022a. https://marketplace.zoom.us/docs/api-reference/introduction.

Zoom[developer]. n.d. "The Zoom Platform." Accessed December 15, 2021b. https://marketplace.zoom.us/docs/guides/getting-started/zoom-platform/.

Zoom[developer]. n.d. "Using Zoom APIs." Accessed December 15, 2021c. https://marketplace.zoom.us/docs/api-reference/using-zoom-apis/.

Zoom[developer]. n.d. "Zoom API." Accessed December 15, 2021d. https://marketplace.zoom.us/docs/api-reference/zoom-api/.

Video Conferencing as Programmatic Relations

Conditions, Consequences, and Mediality of Zoom & Co

Jan Distelmeyer

Introduction: Programmatic Comes to the Fore

This learning process was as quick as it was unprepared. Before the fight against the Covid 19 pandemic so radically restricted entry into public spaces, few people were aware of what video conferencing was supposed to be or be able to do. The switch to video conferencing in many areas of public and private life since 2020 has had numerous consequences, which were experienced and evaluated quite differently.

This included, for example, the advantage, easily overlooked by a majority, that some barriers now disappeared. This especially advantaged those who, due to different physical conditions, cannot easily enter standardized buildings and means of transport, but could now find other possibilities; those who were previously disabled from or had problems with attending, e.g., theaters, seminar rooms, or discussion events, but might now participate via video conferencing without leaving their own rooms. Architectural barriers disappeared—whereas, depending on equipment and dis/abilities, other barriers remained or emerged, as pointed out by Bieling et.al in this book.

Another and much more frequently publicly emphasized advantage concerns anthropogenic climate change and specifically the energy and CO_2 footprint. Numerous trips to business and/or private meetings could be replaced by video conferencing. This positive effect of reduced emissions has been emphasized by many studies, while—which I will come back to—the emissions caused by video conferencing have hardly been investigated (see Faber 2021). International meetings in particular thus became less complicated and cheaper.

The effect I want to address here leads less to a dichotomy of advantages and disadvantages than to questions of the specific mediality of video conferencing. For it was the very speed and emphasized urgency with which video conferencing became established as a new form of meeting in various contexts that provoked questions in this direction. What video conferencing is, how it works and should be practiced, what problems can arise and how they might be solved—these and other questions about the phenomenon of video conferencing were part of the everyday experience of many people who otherwise do not have to consider the conditions and effects of digital technology in this way (an unexpected contrast to the *it just works* promise).

As video conferencing became accepted as a new standard, the impact of this new normal also became more apparent, bringing conditions to the forefront that otherwise (should) remain in the operational background. Precisely because it was important to get used to new technical conditions as quickly as possible and also because problems in dealing with software and hardware ("my Internet seems to be unstable," "you're muted," "I cannot share my screen," "I'm not a cat,"[1] etc.) and their consequences (like "Zoom fatigue") were part of everyday experience and of public discourse, the connections between technology and aesthetics, between infrastructures and practices became pressing questions not only for researchers. Facing this ultimately calls, as I want to show, for more fundamental questions to be asked.

The experience with services and platforms such as BigBlueButton, Microsoft Teams, Jitsi, Webex, Skype, Google Meet, and especially Zoom enables a new approach to specific conditions that are hidden under buzzwords such as "digital transformation," "digital revolution," "digital life," or "digitality." The phenomenon of video conferencing opens up new opportunities in understanding this larger context of, to put it another way, computerization. The experience of and debates about video conferencing exemplify some of the conditions, processes, and effects that need to be considered when tasks are solved by and through networked computers and their programmable processes. Quite a few questions concerning the complex of digitality intensify here, become addressable and also—this seems to me to be the most remarkable consequence of the spread of video conferencing—perceptible.

It is particularly the (widespread and much-researched) exhaustion known as "Zoom fatigue" that plays a special role in this opportunity for understanding and questioning. Here, as I would like to show, becomes vivid and perceptible what otherwise is mostly assumed to be unnoticed and at best the subject of theoretical debates: the importance of interface processes that reach further and deeper than the presence of user interfaces.

Zoom is undoubtedly the most famous incarnation of this development since 2020, the eponym of the homonymous "fatigue" and the sparkling dummy for popular headlines like "We Live in Zoom Now" (*The New York Times*, 17.03.2020), "The Zoom Boom" (*The Guardian*, 21.05.2020; *BBC News*, 23.10.2020; *World Finance*, 09.02.2021), or "And It Made Zoom" (*die tageszeitung*, 29.04.2020, 13.06.2020, 24.10.2020). That is one reason why I discuss Zoom here as a central example, even though most of

---

1 This internet meme refers to a viral video in which a lawyer desperately tries to disable a video filter making him look like a talking cat: https://www.youtube.com/watch?v=lGOofzZOyl8.

the questions attached to it arise similarly in other systems. Including the aforementioned "fatigue" that can be observed in the same way with other commercial and open source systems—and therefore makes urgent the question as to why several hours of streaming with, e.g., Netflix or Mubi is considered entertainment but more than 50 minutes (see Gallo 2020) of a video conference seems to be a burden.

Furthermore, Zoom will also serve as a *pars pro toto* example because, in addition to the general questions about video conferencing, I would like to pay particular attention here to its widespread use at universities, which has led to the catchphrase "Zoom University" (*The New York Times*, 17.03.2020; *The Yale Review*, 20.04.2020; *Forbes*, 03.08.2020; *The Guardian*, 06.10.2020; *TechCrunch*, 18.02.2021). Since I am one of those who had to conceive and conduct university teaching via video conferencing (i.e., Zoom & Co) in 2020–22, this article is also partly an experience report: a participatory observation of processes that are described in this book by several authors such as Michell Kalani, Andreas Weich et al., Dontalla Della Ratta, and Maha Bali.

Against this background, I would like to discuss the form of connections and relations that are realized by services like Zoom & Co. as "programmatic." This accentuation underlines the centrality of computer technology along with its property of programmability and therefore seems to me to be particularly important here because the term video conferencing tends to obscure rather than emphasize this. The dependence of the interaction of image and sound (as video) on the interaction of networked computers that generate and distribute and present these signals (as hardware-software interaction) is elementary to the discussion of the phenomenon and mediality of video conferencing and its platformization. As a first step toward this discussion I want to look back at the year 2020.

Retrospection: "Digital Now Holds Us Together"

One of the early effects of the first "Covid-19 lockdown" was that suddenly two new classes appeared. Those considered working and whose work is thus seen and counted (as opposed to all unpaid forms of work) were divided: there were those who continued to "go to work" or would have gone because their jobs had nothing to do with office or information work; and those who were able to stay at home to work in the "home office."

The new class of home-based work, in which large pay gaps of course still exist, grew visibly. More and more were thus wondering what could be done from home, with this home of course being thought of as both a network node and "always on."

It was precisely this new and diverse class of home-based work that garnered increasing praise for "the digital" at the time in Germany. The Corona crisis also proved to previous "digital skeptics" that "digitalization is a gift for mankind" (von Gehlen

2020). Because: "Digital now holds us together." (Rosenfeld 2020). What was held up to the "cultural pessimists and skeptics of progress" (ibid.) in Germany was, unsurprisingly, the blessing of video conferencing.

The since the mid-2010s steadily increasing equation of the term "digital" with online procedures, with Internet-based services, did not refer solely to the understanding of video conferencing (see Distelmeyer 2022, 27–33). But it was particularly noticeable here, especially in the area of schools and universities. "Digital Classes" [Digitalunterricht], translated exemplarily by the *Süddeutsche Zeitung* "for people without school kids," means that children are "schooled over the Internet" (Rühle 2020).

Already in the first weeks of the pandemic, the "boom of videoconferencing as a quotidian media practice" (Volmar et al 2021) was repeatedly associated in this sense with the company called Zoom, which was hardly known before that time, such that "the rise of Zoom came to stand for a 'new normal' of networked, synchronous online sociality" (ibid.). At universities (not only in Germany), Zoom in particular became the popular solution to what has been called "digital teaching" with increasing emphasis since 2020 (see Kronmüller 2021). Just as this company and its "skyrocketing growth during the crisis" became a "household name" (Peng 2020, 3), the notorious tile aesthetic became a symbol of a new platform connectivity.

From "Zoom University" to "Zoom fatigue," as a sign of hope or dread, Zoom represents effects commonly associated with the international proliferation of online meetings. Even those disruptive maneuvers in which meetings are hacked (thanks to Twitter bots, among other things) and participants are subjected to sexist and racist attacks, in particular, have been associated, as "Zoom bombings," with the company (see Young 2021).

At the same time, it was also very specific data protection issues that brought Zoom into the headlines. Thus, the discovery of secret data transfer to Facebook at the end of March 2020, in which information about the users was exchanged through the Zoom app via the automatic connection to the Facebook's Graph API (application programming interface), caused a scandal to which the company reacted within a few days (see Cox 2000). Nevertheless, fundamental questions of data security do not only concern Zoom. In the summer of 2020, Berlin Commissioner for Data Protection and Information (BfDi) Maja Smoltczyk emphasized this with her criticism of leading video conferencing systems such as Microsoft Teams, Skype, Zoom, Google Meet, Cisco WebEx, and GoToMeeting: "Unfortunately, some of the vendors that provide technically mature solutions do not yet meet the data protection requirements" (DPA 2020), whereas open source solutions like Jitsi and BigBlueButton were rated positively.

Therefore, discussing Zoom means, for many reasons, discussing momentous processes of networking and shifting to platforms. In this context, Zoom offers itself not only as a dazzling and widespread example, but also as a symptomatic space of experience. In and with Zoom, essential basic structures of what might otherwise remain mythical and abstract as "the digital" can be experienced actively and physically. The new presence of video conferencing is also an emphatic confrontation with conditions of computerization. This was felt not only, but especially, by pupils and students. Many repeatedly reported—"I just looked at screens all day" (Himmelrath et al. 2021, 42)—frustration and loneliness. The study "Young Germans 2021" by social researchers Simon Schnetzer and Klaus Hurrelmann revealed that 53% of respondents between the ages of 14 and 29 experienced a "noticeable deterioration in mental health during the Corona crisis" (Schnetzer, Hurrelmann 2021, 10) and target group analyses show that they also suffer from the fact "that schools and universities have not offered suitable structures for digital teaching" (Schnetzer 2021, 42).

Everyday Experiences: Coming to Terms with "Zoom Fatigue"

The debate about the background of these experiences of frustration—which go far beyond this age group—quickly specified the problem of *looking at screens all day*. For the beginning of the historic heyday of video conferencing was also the beginning of reports of rapid-onset exhaustion, difficulty concentrating, and a strange apathy that reinforces the impression of not really being together. The lack of a sense of mood in a common space seems to make one's own mood even more present; experiences of fatigue, detachment, and isolation.

This phenomenon soon became so omnipresent and much discussed that "Zoom fatigue" became a household name, too—and a popular motif of cartoons about Zoom as a tool of torture and cause for support groups (see Tornoe 2020; Fishburne 2020; Margulies 2020). At the same time, new business ideas emerged with it: To alleviate the symptom that "video conferencing is so exhausting," for example, advertising for the app "mmhmm" initially diagnosed "that people just don't know how to come across as charismatic on video," so the app advertised aesthetic "fuseful" features—"funny and useful at the same time" (see Schwan 2021). Why countermeasures such as this or the presentation form "Immersive View" introduced by Zoom at the end of 2021, which supposes "to reduce fatigue during long class sessions, meetings, or other events" (Akolawala 2021), have little to do with the reasons for fatigue experiences identified so far, is related to the fact that different effects and causes are bundled in the term "Zoom fatigue."

Various contributions from media studies, communication studies, and social sciences, as well as medical and psychological research have dealt with the phenomenon. This includes Geert Lovink's contribution to this book, "Anatomy of Zoom Fatigue," as well as "Neuropsychological Exploration of Zoom Fatigue" (Lee 2020), "Understanding Zoom fatigue" (Nadler 2020), "Nonverbal Overload: A Theoretical Argument for the Causes of Zoom Fatigue" (Bailenson 2021), and "Zoom Exhaustion

& Fatigue Scale" (Fauville et al. 2021), in which four elements are highlighted in summary that "might be responsible for triggering Zoom fatigue" (ibid. 3). These include, according to Géraldine Fauville et al. (ibid.), unusual, sustained eye contact ("being stared at while speaking causes physiological arousal"), the complicated perception and classification of all forms of non-verbal communication ("During video conferences, the complex nature of nonverbal behavior remains while extra effort is needed to send and receive signals."), permanent self-examination ("video conferences participants often see a real time video feed that functions like a mirror"), as well as the lack of movement ("being forced to sit in view of the camera likely hinders movement, increases the amount of effort it takes to communicate").

While this outlines frequently mentioned causes, it is far from all of them listed in the articles available so far. One defining aspect, however, that recurs in all descriptions (and is listed under the second element in Fauville et al.) was summarized by student Gordon Kamer for *Harvard Political Review* like this: "[T]he reason why video calls in general are so pernicious is that the slight artifacts of video conferencing — the lag, the robot voices, et cetera — make us feel even more disconnected than if we never called at all" (Kamer 2020).

This coincides with the picture outlined by psychiatrist Gianpiero Petriglieri for the BBC under the title "The reason Zoom-calls drain your energy": "Our minds are together when our bodies feel we're not" (Jiang 2020). His diagnosis, "we need to work harder to process non-verbal cues like facial expressions, the tone and pitch of the voice, and body language" (ibid.), has been linked by Neta Alexander (2020, 29) to the problem of those slight artifacts: "Technical desynchronization between video and audio breeds a deeper sense of psychological and cognitive desynchronization." Fauville et al. (2020, 3) explain this with the fact that "additional cognitive resources are used to manage technological aspects of a video conference, such as image and audio latency." The desired and promised goal of "synchrony with others" is hard to do, as Randall Collins (2020, 491) points out, "with a screen full of faces, delayed real-time feedback, and lack of full body language."

So what is difficult to have in tile format—"trying to make out each other's smiles through the pixelation" (Kamer 2020)—is moreover conveyed under special conditions. The narrow limits of the user interface have a stronger effect the more disturbances appear in them. "The problem is," as summed up the *New York Times*, "the way the video images are digitally encoded and decoded, altered and adjusted, patched and synthesized introduces all kinds of artifacts: blocking, freezing, blurring, jerkiness and out-of-sync audio. These disruptions, some below our conscious awareness, confound perception and scramble subtle social cues. Our brains strain to fill in the gaps and make sense of the disorder, which makes us feel vaguely disturbed, uneasy and tired without quite knowing why" (Murphy 2020).

These aspects of "Zoom fatigue" open up something. They lead to the peculiarity of this media technology, to the conditions of computer interconnections and those processes that are usually not emphasized and noticed. They can be read like traces, which help—"in the course of using media the medium itself appears 'only' as a trace of its message" (Krämer 2015, 190)—to open up the mediality of video conferencing: What "Zoom fatigue" shows (and is therefore discussed in the debates) are those processes of networks and relations between software and hardware that are otherwise supposed to work imperceptibly and effectively. Their interplay is supposed to surface only in the form promised by the *Zoom Guide*: as the "consistent user interface" and "seamless, real-time interactive experience" (Zoom 2019, 3).

This is where the now widely established and more complex concept of interface becomes important, to which Christian Andersen and Søren Pold also refer in this book. In recent years, media studies have benefited more and more from the hint of software studies that the term interface encompasses much more than just user interfaces (see Fuller, Cramer 2008). Rather, the latter are only a small yet important part of the interface complex, which consists of various levels and processes of interfacing between hardware and hardware, software and hardware, software and software, and between computers and non-computers. Interfaces as such make computers work—on various interrelated levels. They create interconnections thanks to which computers function and operate, are networked with other computers, and are able to establish relationships with humans, other machines, and further parts of the world. In short: Interfaces perform mediation processes both to enable computer work and as part of it (see Distelmeyer 2022, 51–58).

For Zoom's promised "consistent user interface," this has clear consequences. That user interface, the stable tile world of faces and operational images, can only offer something like a "seamless, real-time interactive experience" because and insofar as specific infrastructures and numerous other interface levels mediate (lead and conduct) processes both between computers and between hardware and software.

This insight—the effective dependence of user interface operations on various other interface operations, which need not be equally perceptible—has previously been an insight of disciplines such as software studies, media studies, and app studies (e.g., Fuller, Cramer 2008; Andersen, Pold 2018; Distelmeyer 2018 + 2022; Weltevrede, Jansen 2019). Now this context is, to a certain extent, coming to the fore. It—or more precisely: something of it—becomes physically perceptible and thus, as the reports of experience and debates about the effects of video conferencing vividly demonstrate, turns from an object of reflection into an object of experience: brought up as "Zoom fatigue," "technical desynchronization," "slight artifacts," and "image and audio latency." For the problems described concern both the tile interface of the well-organized vis-à-vis encounters, manifesting effective and indeed ideological models of e.g. users, participation, and communication, as well as malfunctions and delays in the hidden interface processes of data processing and data transfer that run in and between the "protocological" (Galloway 2004, xviii) networked computers.

These disturbances are effects of interfaces between software and hardware, which otherwise do not have to interest us. For a long time they did not matter in the common understanding of the term "interface," which usually denotes our access to and interaction with computers. Humans first. Now—in the mode of disturbance with noticeable physical effects—what is otherwise supposed to work under the radar of human perception comes to the foreground: the dependence of surface effects on those interface processes that run computers and their networks. This is openly addressed as "mechanical malfunctions and networks struggling to handle increased traffic" (Wiederhold 2020, 1), resulting in one of the issues with video conferencing "that online communication, while extremely useful, is not completely synchronous" (ibid.).

Mediality and Platformization: Rules, Materilities, and Flow of Data

The experience and discourse of "Zoom fatigue" thus demonstrates, or at least gives a tangible indication, that online meetings are not just about video conversations. They are not just a (tele)presence through sound and image, but are characterized by a specific mediality. Since the 1990s, the concept of mediality has served media studies to describe both the general characteristics of media as an elementary dimension of life and culture and the specific qualities of concrete media and media constellations (see Hickethier 2003, 25–32, Krämer 2015 and Bergermann 2016, 434–436). "Mediality," as Sybille Krämer (1998, 15) has put it, "expresses the fact that our relation to the world, and thus all our activities and experiences with a world-opening—rather than simply world-constructing—function, are shaped by the possibilities for distinction that media open up and the limitations they impose in doing so." In video conferencing, this medial qualities prove to be deeply dependent on the protocological networking of computers and their running programs, realized through embedded models and by diverse interface processes and materialities.

In order for the tile interfaces to become productive and to provide the shared video streams, software-software interfaces, among other things, ensure that computers connect to each other at all and exchange data according to the rules of the Internet protocols. Software-software interfaces form the basis for every service we expect from the Internet; and at the same time they also open the doors for the criticized violations of data protection. Rules of engagement: Every human exchange here—and thus also with Zoom & Co—is always only possible thanks to and through data exchange, and software-software interfaces in the form of application programming interfaces (APIs) allow programs to interact, which makes the use of apps that correspond with each other so effective and "seamless" and is the reason behind the Zoom-Facebook data flow discovered in March 2020. Jean-Christophe Plantin, Carl Lagoze, Paul Edwards, and Christian Sandviget (2018, 303) illustrated this key role of APIs using the example of Facebook as follows: "APIs permit other programs to 'plug in,' in order to exchange data or perform other functions; unlike electrical sockets, however, APIs create a two-way flow of data. In the language of infrastructure studies, an API is a gateway, permitting other systems to interact with Facebook to form a seamlessly interactive network."

For this border-crossing flow of data, for this traffic, hardware interfaces are always needed, interfaces between those processing machines for which arm-thick submarine cables form the profoundly material connections of the Internet. Power consumption of the computers connected in this way, which has already been an issue in streaming (see Marks 2020), is therefore also becoming a topic here. As Grant Faber points out, many studies have shown "that emissions from virtual meetings are far lower than those generated by in-person meetings, but there can still be a considerable impact from conducting such conferences that ought to be measured and mitigated over time" (Faber 2021, 13).

According to a study published in 2021 by Purdue University, Yale University, and MIT, significant CO_2 emissions and water consumption can be reduced by 96% by not using the camera: "If one were to have 15 1-hour meetings a week, their monthly carbon footprint would be 9.4 kg CO_2e. Simply turning off the video, however, would reduce the monthly emissions to 377 g CO_2e" (Obringer et al. 2021, 3). And by proposing "a modifiable framework for systematically measuring the emissions attributable" to video conferencing (with the use of, among others, "participant computers, Internet energy intensity, network data transfer, server power ratings"), Grant Faber (2021, 1) simultaneously illuminates multiple programmatic conditions and dependencies that are part of the distinctiveness—the mediality—of video conferencing.

What the experience and discussion of "Zoom fatigue" as well as resource and energy consumption thus point to is anything but a purely technical, a purely practical, a purely social, or a purely energetic phenomenon. Rather, it is the combination of these factors (supplemented by political and aesthetic ones) that constitutes the mediality of video conferencing—that is, the peculiarity of the mediation taking place. This proves the dependence of video conferencing on functioning infrastructures, on "sociotechnical systems that are designed and configured to support the distribution of audiovisual signal traffic" (Parks, Starosielski 2015, 4). These dependencies include both deeply (and often latent) material formations, as well as running effective processes that are also ideological in that they act out and habitualize certain models concerning people, machines, and (their) interaction.

In order to further outline these dependencies, it seems important to me to remember that it is computers that make video conferencing possible—the technology that, by contrast with others, is distinguished by the fact that it is programmable. This is the reason for its status as a general-purpose machine: What computers accomplish, they accomplish both by virtue of their programmability and through the concrete execution of the specific programs. Computers become productive *because* they are programmable and *by* executing programs.

As a terminological strategy to keep this significant aspect in mind, I would like to propose the attribute *programmatic* here. In this sense, all forms and utilizations of computing can be understood as programmatic and thus all encounters via Zoom, Microsoft Teams, BigBlueButton, and similar services as programmatic relations. What becomes possible on Zoom & Co is only possible under the conditions of the respective software and the hardware processing it—a circumstance that connects the debates about "Zoom fatigue," emissions, data protection, and the responsibility of institutions (e.g., universities) and can also be experienced, as I would like to show in conclusion, in the concrete handling of Zoom by way of example. Thus *programmatic* here refers both to the significance of programmability as a characteristic of computer technology and to the trend-setting dimension of these relations of Zoom & Co, which bring fundamental questions of digitality before our eyes and ears.

The question of which software should actually be licensed (i.e., rented as a service) in order to meet the requirement to switch to "digital teaching," for example, is so sensitive precisely because each program has its own conditions. Binding regulations: As Wendy Chun has shown, Lawrence Lessig's famous slogan "Code is Law" falls short here. Because code—as long as it runs on computers and is not bypassed or reprogrammed—is both law and its uncontradicted enforcement. Programs are execution commands; they run. Code is "better than law," it is "an inhumanly perfect 'performative' uttered by no one" (Chun 2006, 66). That is why the computer scientist Seda Gürses asks one of the most important questions of higher education policies on this issue: What if universities invest in public infrastructure instead of software licenses? (see Gürses 2020)

In Germany, for example, where usually multi-million euro licensing solutions are the rule, this possibility is demonstrated by the exceptions of some universities that host free software on their own servers. For example, for a "data protection respecting" solution with BigBlueButton, the University of Applied Sciences Darmstadt "has even developed its own extension and documentation tailored for universities" (Kronmüller 2021). In summer 2021 the University of Marburg announced that it would offer BigBlueButton as a "federated service" of the German Research Network (DFN) and "thus also make it available to other universities" (Scheid 2021, 47).

However, the dependence on a contractor such as Zoom Video Communications Inc. for whose services German universities paid an estimated 6.4 million euros in 2020 (see Kronmüller 2021) is not limited to the licensed software alone. At the same time, this implies a much broader and, for platforms, symptomatic dependence—because Zoom provides its services with the help of subcontractors who ensure that the data transfer, which is controlled by the software, can be processed as smoothly as possible. In April 2020, Eric Yuan, CEO and founder of Zoom Video

Communications, himself explained the important role that server hosting and cloud service providers such as Oracle and especially Amazon Web Services (AWS) played in Zoom's service and the skyrocketing growth during the crisis already cited. "During this pandemic crisis," Yuan was quoted, "every day is a new record" and "our own existing data center[s] really cannot handle this traffic" (Judge 2020), which is why AWS brought thousands of new servers online for Zoom every day: "Amazon really offered great support to us. Andy and his team offered tons of server size, and every night added 5,000 to 6,000 servers ... a lot of servers to help us worldwide" (ibid.).

It is clearly stated here that the licensing and use of Zoom is by no means limited to Zoom's service and processes. Zoom is more than a software or self-contained service. Rather, Zoom exemplifies the ramified conditional structure of platforms and is an illuminating example of the elusive complexity that platform studies address. Since the mid-2010s, they have been focusing on various interdependencies, technical and social contexts, and new economic models, discussing key features of platforms such as "programmability, affordances and constraints, connection of heterogeneous actors, and accessibility of data and logic through application programming interfaces (APIs)" (Plantin et al. 2018, 294). What connects this concept of platform with the concept of infrastructure, which has also been receiving new attention since the mid-2010s, is the interest in processes and materialities, in effective concepts and ongoing procedures that are not identical with what becomes visible or perceptible in the results of these structures: "Both infrastructure and platform refer to structures that underlie or support something more salient" (ibid.).

For this reason of underlying structures and supporting processes, it is important to understand and discuss video conferencing both as (basically) programmatic relations as well as "platformized" (Plantin et al. 2018, 301). This platformization can be seen on the one hand in the abundance of apps enlisted in the "Zoom App Marketplace" that allows for compatibility with "the Zoom platform" to "leverage Zoom within your daily workflows" (Zoom 2021b). On the other hand, this programmatic connection of heterogeneous actors is also expressed in the business relation with AWS—the leading example of a cloud platform, the importance of which for data extraction is that "its rental model enables it to constantly collect data" (Srnicek 2017, 63). For the promise of video conferencing to be fulfilled on the surface of the user interfaces, the interaction of shared image and sound (as video) is not only based on the interaction of networked computers but furthermore on the programmatic interaction of heterogeneous actors who create a platformized effectivity and business model, as Kim Albrecht has shown in his reflections and visualization in this book. Against this background, it would actually be more appropriate to speak of *video/platform conferencing*.

This (infra-)structure of video conferencing as programmatic and platformized relations is therefore also very important for questions of politics. This becomes vivid in the area of data protection already mentioned: "When a university uses a video conferencing system," explains political scientist and information scientist Sven Hirsch, Data Protection Officer (DPO) of the Potsdam University of Applied Sciences, "the university is responsible for the contractor that is hired and also responsible for any subcontractors" (Hirsch 2021). However, how the flow and exchange of data can be monitored in accordance with legal requirements remains an open question: "As a data protection officer, I can know next to nothing about the path of the data, have virtually no way to control it. There are also no interfaces, no visualizations to see the data flow. It's legally provided and contracted that we can control the contractor, but practically unfeasible so far" (ibid.).

How to reconcile opportunities for access and privacy remains a central question that confronts institutions and individuals alike with problems hard to solve. A telling example for incompatibility was given by a video conference of the network "Dis-/Abilities and Digital Media" in spring 2021. Fitting with the theme of the conference, "Assistance—On the History of Assistive Ensembles," consideration was given to adding a closed caption feature to the session for accessibility. However, online services of transcription software such as Otter and Rev were rejected by the responsible universities. Such third-party applications use voice data that is transferred to the US; there, it is processed (saved) and then returned as closed captions to the video conference meeting. "The main problem," explains co-organizer Robert Stock, "was Terms of Service of the apps claiming exclusive rights to the processed user content,"[2] what was considered a violation of the EU's General Data Protection Regulation (GDRP). Thus, how the GDRP could conflict with the UN Convention on the Rights of Persons with Disabilities (to remove obstacles and barriers to accessibility) here, as Robert Stock emphasizes, makes the programmatic conditions of video conferencing only more apparent—video, sound and image, are here first and foremost (and also in the legal sense) data that are processed by software according to certain rules and thus link questions about, for example, human practices with those about the practices of platforms.

Programmatic Aesthetics: Tiles, Hosts, and Participants

On various levels that are difficult to oversee, the new presence of video conferencing massively confronts what is particularly crucial in digital technology and its diverse manifestations: the programmability of procedures and circumstances. On this basis of programmability runs what is bindingly given and at the same time can always become different.

2 Robert Stock, private email exchange with author after the workshop "Video Conferencing: Practices, Politics, Aesthetics," March 11, 2022.

Based on this, new defaults can be created and running systems can be interfered with, systems can be hacked, and new safeguards can be put in place. For example, in April 2020, Zoom Video Communications Inc. responded to the "Zoom bombing" issues by announcing new software defaults for password entry and automatic "waiting rooms." And on the same basis of programmatic flexibility, other forms of "interface mise-en-scène" (Distelmeyer 2018) can always be created and replaced. The presentation of user interfaces can easily change as Microsoft Teams has shown with "Together Mode"—"In a conceived scene setting [like a cinema, a curved outside amphitheater or a boardroom table], participants have seats with video streams" (Microsoft 2021)—followed by Zoom with its "Immersive View" of similarly staged seating arrangement, including a kitchen, an art gallery, a classroom, or a ski lift: "Immersive View allows hosts to arrange video participants and webinar panelists into a single virtual background, bringing people together into one scene to connect and collaborate in a cohesive virtual meeting space" (Zoom 2021a).

Hence, this special, programmatic characteristic of digital technology is nothing that inevitably eludes us in the sense of a "black box." Rather, it can also be observed and experienced concretely. The notorious tile aesthetic of Zoom is a good example of that.

Only what the program has provided can be done here. When it comes to active participation, I can, for example, activate my microphone and raise my voice, raise a virtual hand, set emojis, write a comment in the chat, or share my screen. Commanding and complying: As always in dealing with user interfaces, as always in programmatic circumstances, programming has provided and predefined what is possible and how (see Distelmeyer 2022, 62–70). But what is special here (and similarly with BigBlueButton) is that these interface options vary depending on the status of use.

The leading authority of a Zoom meeting, called "Host," has different and more possibilities than the category "Participant." Yet these interface options are not only different and thus an expression of programmatic flexibility. They can also limit and determine the possibilities as a "Participant." Depending on the presetting of modes like "Meeting" or "Webinar," a "Participant" can, for example, be muted by the "Host" (both individually and via the "Mute All" button) or could be "Put in Waiting Room." Other ways to act as a "Host" by clicking/commanding a "Participant" are specified with buttons like "Make Host," "Allow Record," "Lower Hand," "Rename," and "Remove." In turn, a "Participant" does not have these options in relation to the "Host" or other "Participants." It seems as if Jean Baudrillard's (2003, 281) old definition, "power belongs to the one who can give and cannot be repaid," is given a new embodiment here.

Even more, the possibilities of the "host" do not even appear as buttons in the user interface of the "participant," which emphasizes an aesthetic dimension of the mediality of video conferencing platforms: "The" aesthetics and "the" user interface of platforms like Zoom do not exist at all—rather, there are different variants that differ depending on status and default settings and are subject to change. In other words, "the" aesthetics and "the" user interface of platforms like Zoom are also programmatic, subject to the power and flexibility of programming that opens up certain spaces for action and invites negotiation processes. This difference leads to a particular, built-in imbalance and incidentally complicates "walkthrough" analyses of apps and their user interfaces (see Light, Burgess, and Duguay 2018).

The question of power that arises here online (and is answered via interface actions) naturally also concerns seminar rooms and lecture halls. As little as these (partly transfigured) locations were and are hierarchy-free spaces, the romantic "digital vs. analog" dichotomy is misleading. Rather, it is important to ask what specific conditions of power (and also of disruption) are actually at work here. This makes it all the more relevant what goes on and what goes missing in the interface mise-en-scène of Zoom & Co.

The classic frontal seating, for example, which in a lecture hall makes it clear right from the start who is to be in charge, does not have to be communicated in this way on Zoom. In the tile grid of the "Gallery" view, at least, we are all the same, no spacing, tile size, or the like emphasizes who is "Host" here. The same is true for those seats in the "Kitchen," in the "Art Gallery," or in the "Boardroom," which can be taken/ allocated in Zoom's "Immersive View." The (thus) invisible leading authority shows up differently. It is experienced by "Participants," e.g., when they try to use the "Share Screen" button, and stopped by a dialog box that only allows the option to be confirmed with "Ok": "Host disabled participant screen sharing."

The authority of the "Host," which no "Participant" can assume on his own initiative when storming the teacher's desk, proves itself in the process. It becomes apparent when interface options are denied, settings are undone, or a "Participant"—the host moves in mysterious ways—suddenly finds himself as a "Co-Host" or in the "waiting room" that Neta Alexander (2020, 25) has discussed as a "timely metaphor for corona-capitalism." Unlike seminar rooms and lecture halls (*code is better than law*), on Zoom any rule can become an unspoken condition of existence. Also, power acts programmatically here. What Søren Pold has called the "Zoomoptikon" (Pold 2021) also aims at this context of programmatic relations: it aims at the power imbalance that lies in the uncertainty of what exactly is being captured and thus part of a data transfer and economy whose concrete processes remain hidden.

These special power relations—that on Zoom & Co in the (tile) space of the user interfaces no hierarchy has to be expressed, while it always already works on the level of programmatic determination—have consequences. For precisely because the question of power is based on programmability, it is also to be decided at this level. Practices of hacking, data abuse, and also "Zoom bombing" tell about it. This increases the relevance of the political question of which institutions work with which software and which platform (in which jurisdiction).

Conclusion: People and Platforms Under Pressure

Video conferencing poses many familiar questions anew. The fact that these questions arise with noticeable urgency under the impression of the COVID19 pandemic marks a special historical situation at the beginning of the 2020s. The reason for this could be very simple: Perhaps the programmatic conditions of computerization and all its (interface) effects become so clear and perceptible here because of the obvious and nearly inevitable comparison with what video conferencing platforms are supposed to replace. The sudden compensation of all possible meeting spaces and forms by hardware, software, and platforms puts not only the people under pressure, but also the hardware, software, and platforms.

This creates attention for those traces that the medial conditions leave in video conferencing. Thus the truism becomes more tangible that such platformized and programmatic relations are first and foremost regulated relations in and among computers, before they thereby (thanks to camera, microphone, monitor, touch screen, etc.) bring people into relations. Spatial boundaries between people are overcome by using interface processes between computers whose formulaic nature, decentralized networking, and hard determination logic are also communicated—at least to a certain extent—at the level of human interaction.

At the same time, the experiences with Zoom & Co and the various contributions to this book show that the concrete forms of dealing with this technology are far from technically determined. Rather, it turns out to be important to look for possibilities (and their limits) to deal with it creatively. How our forms and practices of meetings and reconciling human/planetary needs and technical affordances will evolve remains open. It depends in no small part on what we learn from the beginning of the 2020s.

References

Akolawala, Tasneem. 2021. Zoom Rolls Out Immersive View to Bring Participants Together in a Virtual Room. *Gadgets*, 27.04.2021, https://gadgets.ndtv.com/laptops/news/zoom-immersive-view-rollout-for-meetings-reduce-fatigue-single-background-for-all-2422841.

Alexander, Neta. 2020. The Waiting Room: Rethinking Latency after COVID-19. *Pandemic Media: Preliminary Notes Toward an Inventory*, eds. Philipp Dominik Keidl et al., 25–31. Lüneburg: meson press.

Andersen, Christian Ulrik and Søren Pold. 2018. *The Metainterface. The Art of Platforms, Cities and Clouds*. Cambridge: MIT Press.

Bailenson, Jeremy N. 2021. Nonverbal Overload: A Theoretical Argument for the Causes of Zoom Fatigue. *Technology Mind and Behavior* 2(1). https://doi.org/10.1037/tmb0000030.

Baudrillard, Jean. 2003. Requiem for the Media (1972). *The New Media Reader*, eds. Noah Wardrip-Fruin and Nick Montfort, 277–288. Cambridge: MIT Press.

Bergermann, Ulrike. 2016. *Leere Fächer. Gründungsdiskurse in Kybernetik und Medienwissenschaft*. Münster: LIT.

Chun, Wendy Hui Kyong. 2006. *Control and Freedom. Power and Paranoia in the Age of Fiber Optics*. Cambridge: MIT Press.

Collins, Randall. 2020. Social distancing as a critical test of the micro-sociology of solidarity. *American Journal of Cultural Sociology* 8, 477–497.

Cox, Joseph. 2020. Zoom iOS App Sends Data to Facebook Even if You Don't Have a Facebook Account. *Motherboard*, 26.03.2020. https://www.vice.com/en/article/k7e599/zoom-ios-app-sends-data-to-facebook-even-if-you-dont-have-a-facebook-account.

Cramer, Florian and Matthew Fuller. 2008. Interface. In *Software studies: A Lexicon*, ed. Matthew Fuller, 149–152. Cambridge: MIT Press.

Distelmeyer, Jan. 2018. Drawing connections: How interfaces matter. *Interface Critique Journal* 1: 22–32.

Distelmeyer, Jan. 2022. *Critique of Digitality*. Wiesbaden: Palgrave Macmillan.

DPA. 2020. Datenschutzbeauftragte: Ampel auf rot. *Süddeutsche Zeitung*, 03.07.2020.

Faber, Grant. 2021. A framework to estimate emissions from virtual conferences. *International Journal of Environmental Studies*. https://doi.org/10.1080/00207233.2020.1864190.

Fauville, Géraldine et al. 2021. Zoom Exhaustion & Fatigue Scale. *Computers in Human Behavior Reports* 4: 1–10. https://doi.org/10.17605/OSF.IO/YAFRG.

Fishburne, Tom. 2020. Zoom Fatigue. *Marketoonist*, 14.05.2020. https://marketoonist.com/2020/06/zoom.html.

Gallo, Carmine. 2020. The 50-Minute Rule Makes Virtual Meetings More Productive, According To Neuroscience. *Forbes*, 29.09.2020. https://www.forbes.com/sites/carminegallo/2020/09/29/the-50-minute-rule-makes-virtual-meetings-more-productive-according-to-neuroscience/?sh=2a98295314ff.

Galloway, Alexander R. 2004. *Protocol: How Control Exists After Decentralization*. Cambridge: MIT Press.

Gürses, Seda. 2020. Rectangles-R-Us: What happened when the university went online? https://www.youtube.com/watch?v=fLuCDj3EOKM.

Hickethier, Knut. 2003. *Einführung in die Medienwissenschaft*. Stuttgart/Weimar: Metzler.

Himmelrath, Armin et al. 2021. Vor dem Bildschirm vergessen. *Der Spiegel* 49: 42–44.

Hirsch, Sven. 2021. Unpublished interview (with Jan Distelmeyer, recorded 08.11.2021).

Jiang, Manyu. 2020. The Reason Zoom Calls Drain your Energy. *BBC*, 22.04.2020. https://www.bbc.com/worklife/article/20200421-why-zoom-video-chats-are-so-exhausting.

Judge, Peter. 2020. Most of Zoom runs on AWS, not Oracle—says AWS. *Data Center Dynamics*, 20.04.2020, https://www.datacenterdynamics.com/en/news/most-zoom-runs-aws-not-oracle-says-aws/.

Kamer, Gordon. 2020. Why Zoom Ain't It, Chief. *Harvard Political Review*, 03.12.2020. https://harvardpolitics.com/why-zoom-aint-it-chief/.

Krämer, Sybille. 1998. *Medium, Computer, Realität. Wirklichkeitsvorstellungen und Neue Medien*, Frankfurt am Main: Suhrkamp, 1998.

Krämer, Sybille. 2015. *Medium, Messenger, Transmission. An Approach to Media Philosophy*. Amsterdam: Amsterdam University Press.

Kronmüller, Max. 2021. So viel bezahlen Hochschulen für Zoom. *netzpolitik.org*, 24.03.2021. https://netzpolitik.org/2021/private-infrastruktur-fuer-die-lehre-so-viel-bezahlen-hochschulen-fuer-zoom/.

Light, Ben, Jean Burgess and Stefanie Duguay. 2018. The walkthrough method: An approach to the study of apps. *New Media & Society* 20 (3): 881–90.

Margulies Jimmy. 2020 Thanksgiving 2020. *Cagle*, 12.11.2020. https://www.cagle.com/jimmy-margulies/2020/11/thanksgiving-2020.

Marks, Laura U. 2020. Streaming video, a link between pandemic and climate crisis. *Rosa Mercedes* 2, https://www.harun-farocki-institut.org/en/2020/04/16/streaming-video-a-link-between-pandemic-and-climate-crisis-journal-of-visual-culture-hafi-2/.

Microsoft Corp. 2021. Custom Together Mode scenes in Teams. 18.11.2021, https://docs.microsoft.com/en-us/microsoftteams/platform/apps-in-teams-meetings/teams-together-mode.

Murphy, Kate. 2020. Why Zoom Is Terrible. *The New York Times*, 29.04.2020. https://www.nytimes.com/2020/04/29/sunday-review/zoom-video-conference.html.

Obringer, Renee et al. 2021. The overlooked environmental footprint of increasing Internet use. *Resources, Conservation & Recycling* 167. https://doi.org/10.1016/j.resconrec.2020.105389.

Parks, Lisa and Nicole Starosielski. 2015. Introduction. In *Signal Traffic: Critical Studies of Media Infrastructures*, eds. Lisa Parls and Nicole Starosielski, 1–27. Illinois: University of Illinois Press.

Peng, Mike W. 2020. *Global Strategy* (Fifth Edition). Boston : Cengage.

Plantin, Jean-Christophe et al. 2018. Infrastructure studies meet platform studies in the age of Google and Facebook. *New Media & Society*, 20 (1): 293–310.

Pold, Søren Bro. 2021. Vi er blevet zoombier. *Akademikerbladet*, 11.02.2021, https://www.akademikerbladet.dk/debat/vi-er-blevet-zoombier.

Rosenfeld, Dagmar. 2020. Das Digitale hält uns jetzt zusammen. *Die Welt*, 17.03.2020.

Rühle, Axel. 2020. Es gibt ihn doch, den guten digitalen Unterricht. *Süddeutsche Zeitung*, 04.09.2020. https://www.sueddeutsche.de/kultur/unterricht-digital-corona-1.5019384.

Samara, Omar and Anthony Monzon. 2021. Zoom Burnout Amidst a Pandemic: Perspective from a Medical Student and Learner. *Therapeutic Advances in Infectious Disease* 8 (1–3). https://doi.org/10.1177/20499361211026717.

Scheid, Christoph. 2021. Das BigBlueButton-Cluster der Philipps-Universität Marburg. *DFN Mitteilungen* 99: 45–48.

Schnetzer, Simon and Klaus Hurrelmann. 2021. *Jugend und Corona in Deutschland—die junge Generation am Ende ihrer Geduld* (Sommer 2021). Kempten: Datajockey.

Schnetzer, Simon. 2021. Jugend und Corona in Deutschland. *JUPF Info* 122: 42.

Schwan, Ben. 2021. Videokonferenzen: "Video hat wahnsinnige Vorteile." *Heise online*, 01.02.2021. https://www.heise.de/hintergrund/Videokonferenzen-Video-hat-wahnsinnige-Vorteile-5040618.html.

Srnicek, Nick. 2017. *Platform Capitalism*. Cambridge, Polity Press.

Tornoe, Rob. 2020. Political cartoon: Feel exhausted after a Zoom meeting? *The Philadelphia Inquirer*, 15.05.2020. https://www.inquirer.com/opinion/cartoons/coronavirus-zoom-fatigue-tired-video-chat-exhausting-anxious-political-cartoon-20200515.html.

Volmar, Axel, Charline Kindervater, Sebastian Randerath, and Aikaterini Mniestri. 2023. "Mainstreaming Zoom: Covid-19, Social Distancing, and the Rise of Video-Mediated Remote Cooperation." In *Varieties of Cooperation: Mutually Making the Conditions of Mutual Making*, edited by Clemens Eisenmann, Kathrin Englert, Cornelius Schubert, and Ehler Voss, 99–133. Media of Cooperation. Wiesbaden: Springer.

von Gehlen, Dirk. 2020. Durch Corona wird das Internet zur Selbstverständlichkeit. *Süddeutsche Zeitung*, 22.03.2020, https://sz.de/1.4846552.

Weltevrede, Esther and Fieke Jansen. 2019. Infrastructures of intimate data: Mapping the inbound and outbound data flows of dating apps. *Computational Culture* 7. https://computationalculture.net/infrastructures-of-intimate-data-mapping-the-inbound-and-outbound-data-flows-of-dating-apps/.

Wiederhold, Brenda K. 2020. Connecting Through Technology During the Coronavirus Disease 2019 Pandemic: Avoiding "Zoom Fatigue." *Cyberpsychology, Behavior, and Social Networking*, 23 (7). https://doi.org/10.1089/cyber.2020.29188.bkw.

Young, Sarah. 2021. Zoombombing your toddler: user experience and the communication of Zoom's privacy crisis. *Journal of Business and Technical Communication* 35 (1): 147–153.

Zoom Video Communications, Inc. 2019. Design Guide Best Practices. https://explore.zoom.us/docs/doc/DesignGuide-BestPractices-Sep2019.pdf.

Zoom Video Communications, Inc. 2021a. Introducing Immersive View, A Fun New Way to Meet. 26.04.2021, https://blog.zoom.us/introducing-zoom-immersive-view/.

Zoom Video Communications, Inc. 2021b. We Now Have More Than 1,000 Apps on the Zoom App Marketplace to Help You Work Better. 12.02.2021, https://blog.zoom.us/we-now-have-more-than-1000-apps-on-the-zoom-app-marketplace/.

Techniques of the Face

The Art and Politics of Video Conferencing (Inter)Faces

Christian Ulrik Andersen and Søren Bro Pold

Due to the COVID pandemic, the 2021 edition of Electronic Literature Organization's 2021 conference and festival was held online, and most interaction between participants took place via the platform Zoom. At the final day of the event, the performance artist duo SALYER + SCHAAG (Katie Schaag and Andrew Salyer) posed as Kristin S. Wiley and Alfred S. Fox, CCOs ("Chief Corporeal Officials") of "Good Movement, Inc.," presenting *Perfect Movement Engineering for Better Everyday Zooming*. In short, the performance was a live instruction tailored to help the conference participants optimize their bodily behavior in front of the camera. The performance was a continuation of a former piece, *Perfect Movement Engineering for Better Everyday Living* from 2014, which they, in an alleged patent application, have described as a "gesture and movement-based system" that can "capture a user's motion and display a model that maps the user's motion, including gestures that are applicable for control." Based on the observation and capture of these gestures (such as holding a wine glass at an artist opening) they suggest a systematic analysis of the participants' gestures in order to "determine those cases where assistance to the user on performing the gesture is appropriate" (Fox and Wiley 2014).

With their performance SALYER + SCHAAG draw attention to how gestures and movements are intrinsically tied to social settings, including academic conferences, and by bringing their performance into Zoom, they explicitly highlight how Zoom, as a corporate software interface for conferencing, makes its users execute certain meeting-like gestures. Zoom is an interface design not only *for* use, but also *of* its users; or, one might say that Zoom (and similar software) does the same thing to our meetings, as Power Point has previously done to our presentations (even down to the fact that software, which is now used for all sorts of social interaction, is labeled "conferencing"—just as presentations have become "slide shows" that externalize some sort of truth, as pointed out by Kalani Michell in this book). One example is "the attentive nod," which they break down into specific facial mechanic features: the correct movement and tempo of the head ("continuously, and even," "not too much," "not too fast"), the position of the eyebrows ("not too much, you don't want to seem surprised"), the leaning forward toward the camera (again, "not too much"), etc.—all of which convey subtle differences (and failures) of the nod. As Kristin S. Wiley reminds the audience, "be careful to manage the micromovements of the face, your eyebrows and your cheekbones. People will notice what is happening across your entire face" (SALYER+SCHAAG 2021, [28:00-]).

As the performance demonstrates, any social context brings about a corporeal management, and with the wider proliferation of video conferencing interfaces, the micromanagement of *the face* becomes increasingly important—the faces we must learn to read and manage on camera, the faces that we "pin," the faces that leer at us and our homes, the faces that can be detected by the software and decorated with filters, the faces of colleagues that are there at our table top, and so on. The management of the video conferencing face has, in other words, become a familiar phenomenon. With this, and as pointed out by several authors in this book, the face has become a site for struggles over power and control: the subtle changes in our facial gestures and in our facial performances reflect a much larger politics of the face, rooted in the interface. This will be the subject of this article.

Within philosophy, the face has been debated by thinkers such as Emanuel Levinas (as a question of hospitality, identity, and the other) (Levinas 1969), Gilles Deleuze and Felix Guattari (as a question of the "faciality," or the social production of the face) (Deleuze and Guattari 1987), Frantz Fanon (as a question of black skin and race) (Fanon 1986 [1952]), and many more. Evidently, the aesthetics and politics of the face get further complicated by the proliferation of video conferencing, and also many other interfaces, such as Facebook, Tinder, Snapchat, and so on, which suggests that contemporary platforms are built on a reorganization of what one might call facial practices, and hence also of what a face is, means, and does. Our intention here, in this article, is by no means to provide a comprehensive overview of these philosophies of the face or the many and diverse facial practices of an interface culture, but to argue that the increased proliferation of interfaces for video conferencing makes the aesthetics and politics of the face ever-more present aspects of our everyday lives; and furthermore, to ask what are the conditions of this facial production? What does a face become in a video conferencing interface?

Studying video conferencing as an intrinsic part of interface culture inclines us to think of how the technology affects the way we see a face and how it presents itself. The face arguably plays a significant role in human communication, representing feelings or emotions, but also social significance (such as the color of the skin, the shape of a beard, the use of lipstick, and so on). Similarly, one might argue, as the art historian Hans Belting has, that the role of media is to "capture" the face in an image, and that this inevitably results in "masks" that do not represent the person as much as they point to a (more general) "depiction" (which has a long and rich cultural history, including portraiture as well as ritual masks) (Belting 2017). What is particular for our current situation (the interface) is that we find ourselves in what he calls a "digital masquerade" that not only includes faces that do not depict any person (such as faces produced by generative adversarial networks [GAN]), but also, as mentioned, an overwhelming number of faces on Facebook, Twitter, Tinder, Snapchat, and numerous other platforms that we swipe through every day. As noted by Tomáš Jirsa and Rebecca Rosenberg, Belting suggests a somewhat dystopian era where the face "rejects any traditional claim of 'true' resemblance and likeness of a real human being, marking a shift to a condition in which the relation between the face and the subject is more than ever before grounded in a radical disembodiment," a situation where the face has become an "(inter)face." Like Jirsa and Rosenberg, speculate about the possibilities for "less somber" perspectives on this (inter)face (2019, 3).

Our underlying argument is that a number of artistic video conferencing performances offers this, and that they do so, not by mourning the loss of bodily representational identity, but by exploring the body (and especially the face) as a technical object; that is, they explore the face from within the interface—from within the production of the face as a technical object.

The (Video) Art of Faces

Many of the contributions to this book highlight the relevance of artistic exploration of video conferencing (including Donatella Della Ratta's, Tilman Baumgärtel's, Martina Leeker's, and also Kim Albrecht's own interventions). In our search for possible answers, and also ways to even understand facial production and its contemporary conditions, we too will turn to a number of artistic experimentations with video conferencing interfaces. Within the arts, and predominantly video art, there is a long tradition of experimentation with video conferencing systems. Most famously, perhaps, is Sherrie Rabinowitz and Kit Galloway's 1980 work *Hole in Space*, a three-day performance where they connected the public space of the Lincoln Center for the Performing Arts in New York City, and a department store in Century City, Los Angeles, with two life-size, live television projections. Rabinowitz and Galloway defined themselves as "avantpreneurs": artists who are "alert to emerging trends in science and technology," and who "articulate the intrinsic qualities and dangers of unclaimed territory not yet targeted for total exploitation by the entrepreneurs" (1989). One might claim that *Hole in Space* is now superseded by video conferencing platforms instigating a huge variety of social encounters, and many more than Rabinowitz and Galloway probably imagined at the time. In light of this, we therefore stipulate that there is a societal need for artistic practices that explore this as a tendency.

According to Walter Benjamin, for art to explore a "tendency" it needs to explore its own conditions of production; hence, tendency is not to be understood merely as a general "trend" of an epoque (Andersen and Pold 2018, 24). Media technologies generally bring about new techniques of production—of making images, text, sound, and as suggested in this article, also faces. According to Benjamin, the artistic technique itself is a way to relate to this production—not necessarily how it is good or bad, right or wrong, but to position oneself *within* the production and in doing so investigate its wider conditions: "The technical revolutions are the fracture points of artistic development; it is there that the different political tendencies may be said to come to the surface" (1999, 2:17). In other words, we stipulate that the art works will reveal the fracture-points in the platformed production of faces, the making of faces, thus helping us better understand the video conferencing platform as a production apparatus of the face, including how it produces bodies, subjects, territories, and more.

In our analysis of this tendency of video conferencing we focus specifically on the production of the face; or, facial production. The art historian Rosalind Krauss has, in 1976, done an elaborate analysis of video art that in one way or the other features the body of the performer, and often also the face: "Unlike the other visual arts, video is capable of recording and transmitting at the same time—producing instant feedback. The body is therefore as it were centered between the two machines that are the opening and closing of a parenthesis. The first of these is the camera; the second is the monitor, which re-projects the performer's image with the immediacy of a mirror" (Krauss 1976, 52). Although depending on physical mechanisms, as an apparatus, video cannot, according to Krauss, be defined on technical terms. Krauss therefore turns to psychology, claiming that there is a certain narcissism drawing artists to the medium; or, as she also suggests, that video enacts narcissism. Video presents a mirror reflection of absolute feedback, which inclines us to "bracket out" the electronic equipment as a simple appurtenance: "video's real medium is a psychological situation" (Krauss 1976, 57).

Arguably, both artists and users engaging with video conferencing interfaces may recognize an "ego-libido" (in the words of Freud), but in pursuit of what Krauss later, inspired by Benjamin, coined as "The Optical Unconscious," it is our stipulation that one cannot exclude the technical object in this. In fact, we claim that the body, caught in the video feedback, is itself a technical object. In other words, we are particularly interested in art that, in the words of Krauss, "represent[s] a physical assault on the video mechanism in order to break out of its psychological hold" (1976, 59).

To frame a non-psychological understanding of the body (and the face of the video conferencing interface), we begin with the French sociologist Marcel Mauss who in 1934 gave a lecture entitled "Techniques of the Body." In the following, it is our ambition to, through a selection of artworks, provide an understanding of how video conferencing interfaces capture the body and the face as part of its apparatus; of the various techniques of the face in video conferencing systems—from the individual and collective techniques of users performing in front of each other (listening, acknowledging, etc.), to the techniques of users performing in front of the interface (navigating, adjusting camera angles, turning the camera on/off etc.), not to mention the bodily techniques performed by users with varying dis/abilities (as outlined by Bieling et al in this book). In this, we also want to argue that the aesthetics of the video conferencing interface is not so much about the construction of a (narcissist) self, as it is an example of how contemporary platform interfaces exercise power and control by way of subjectivation—a process through which one becomes the face of a producing subject in a platform economy. Video conferencing art may help us see this.

The Techniques of the Face

Figure 1: Kristin S. Wiley and Alfred S. Fox performing variations of "the attentive nod"

Source: Screen shot from *Movement Engineering for Better Everyday Zooming*.

To enter the discussion of the aesthetics and politics of the face, one may begin by considering, as SALYER + SCHAAG have, the micromovements of the face. The "attentive nod," along with other micromovements of the face, practiced in *Perfect Movement Engineering for Better Everyday Zooming*, belongs to what Marcel Mauss has labeled "techniques of the body." Mauss notes how different cultures, genders, and generations "move" in different ways. As an example, he explains how swimming techniques undergo changes in a generation's life-time. Once, children were taught how to dive closing their eyes and opening them under water. Later (in 1934), it is the other way around: children are taught to control their ocular reflexes as a way to familiarize with the water (Mauss 1973, 71). Likewise, Maori women, he claims, quoting the ethnographer Elsdon Best, acquired a "loose-jointed swinging of the hips that looks ungainly to us, but was admired by the Maori. Mothers drilled their daughters in this accomplishment, termed onioni" (Mauss 1973, 73). In Mauss' thinking lies the assumption that the technical cannot be entirely separated from the human bodily. In fact, "The body is man's (sic) first and most natural instrument. Or more accurately, not to speak of instruments, man's first and most natural technical object, and at the same time technical means, is his body" (Mauss 1973, 75).

Walking, swimming, and add to that also the micromovements of the face, are, as "techniques of the body," also habits. The habitual in the techniques of the body is here to be understood as a habitus (in Latin), rather than a habitude (in French): they are to be understood as an "acquired ability" or a "faculty," and not a "mysterious 'memory'" (Mauss 1973, 73). To Mauss, habits are first and foremost acquired through socio-cultural education. For instance, one can recognize a person raised in a convent if they walk with their fists closed: "I can still remember my third-form teacher shouting at me: 'Idiot! why do you walk around the whole time with your hands flapping wide open?'" (1973, 72). And yet, he claims, they cannot entirely be taught. Underneath the educational lies a game of imitation that some people, despite having the same education, master better than others (also bound to an element of social prestige): "The individual borrows the series of movements which constitute it from the action executed in front of him or with him by others" (Mauss 1973, 72). Evidently, the education and imitation of a habitus also compares well to *Perfect Movement Engineering for Better Everyday Zooming* and the staged game of mimicry that it engages the participants in, with sarcastic promises of increased social prestige.

Assuming that the micromovements of the face are an acquired ability also means—contrary to conventional assumptions—that the face is not simply a mirror of the soul or one's inner identity, feelings, or emotions. Rather, it is a mirror of a shared habitus. As expressed by Mauss: "'habits' do not just vary with individuals and their imitations, they vary especially between societies, educations, proprieties and fashions, prestiges. In them we should see the techniques and work of collective and individual practical reason rather than, in the ordinary way, merely the soul and its repetitive faculties" (1973, 73). One could also say that they are not just personal habits, but that we are inhabited by them, that they point to a culturally specific logic of sensemaking that also defines us, as human beings. As Wendy Chun has argued, media seem to matter the most not when they are new, but when they structure our lives (2016). Or, as Slavoj Žižek phrases it: "Belonging to a society involves a paradoxical point at which each of us is ordered to embrace freely, as the result of our choice, what is anyway imposed on us" (2008, 676). Habits are ideology and politics in action, embraced and yet also enforced.

In other words, Mauss proposes that in a face one might recognize a person and a person's mechanical aims (such as a yawn), chemical aims (such as sweating and feeling hot), or psychological aims (such as feeling sad), but one also recognizes a more generalized face, a more habitual face of a collective, expressing a mode of being, and of reasoning and acting in the world. If thinking of the facial implications of video conferencing interfaces begins by first considering the body and face itself as a "natural technical object" and how it acquires a habitus, then what is the role of the technical instrument, such as the interface itself? What are the relations between the techniques of the face (the user's habitus) and the techniques of the video conferencing interface?

The Face as a Mass Ornament

Mauss himself does not explicitly explain the role of media and technologies, but he interestingly notes that cinema seems to play a significant role on the bodily habitus: "A kind of revelation came to me in hospital. I was ill in New York. I wondered where previously I had seen girls walking as my nurses walked. I had the time to think about it. At last I realised that it was at the cinema. Returning to France, I noticed how common this gait was, especially in Paris; the girls were French and they too were walking in this way. In fact, American walking fashions had begun to arrive over here, thanks to the cinema" (Mauss 1973, 71).

To generalize a little from this observation, one might stipulate that the mediated gaze on the techniques of the body plays a significant role in the social acquisition and prestige of bodily techniques, such as those of walking, and also those of the face for everyday zooming. One answer to the above question (of the relation between the techniques of the user's face and the techniques of the interface) is hence that the video conferencing interface provides a particular gaze on the habitus and common techniques of the face. For instance, the "gallery view" of faces all performing the "attentive nod" or other acquired facial techniques for video conferencing meetings provides the user with a particular gaze on the habitus of a collective face.

But Mauss also points to other potential relations and speculates that corporeal movements more generally may relate to larger systems of industrial production. In his immaculate elaboration of the techniques of swimming, he notes: "In my day swimmers thought of themselves as a kind of steam-boat. It was stupid, but in fact I still do this: I cannot get rid of my technique" (1973, 71). In this small note, and also in his note on the role of cinema and elaboration of the body as a technical object, Mauss resembles his contemporary, Siegfried Kracauer. As an interesting historical fact, the new swimming and diving technique that Mauss describes as replacing his own "steam-boat" technique is in fact perfected in the contemporary invention of synchronized swimming (a term coined by Olympic Gold medalist Norman Ross the very same year, 1934) and the cultural proliferation of aquatic musicals and ballets (Sydnor 1998, 255). The aqua musicals easily compare to the proliferation of both gymnast stadium shows and dance shows, also discussed by Kracauer (1995).

To Kracauer, these spectacles are examples of "mass ornaments" and function as an aesthetic reflection of capitalism's production processes. The proliferation of capitalist production had produced a new organization of the masses in which one took part by performing the same movements (such as the synchronized movements of the assembly line), but without seeing the larger picture of the mass. The mass can therefore only be recognized by the individual as an indirect experience; or, staged aesthetically as a spectacular mass ornament—such as the stadium gymnast shows, the dance shows, and also the aquatic shows.

In mediating the techniques of the body, the aquatic show, the dance show, or the gymnast stadium show are modern cultural phenomena that provide a new gaze on the techniques of the body, and a new way of appropriating a bodily habitus in a growing mass society. Evidently, the *Perfect Movement Engineering for Better Everyday Zooming*, in a Maussian interpretation, also leaves an impression of the participants taking part in a contemporary ornamentation of faces, leaving its spectators (and participants) with a perspective of how one by "everyday Zooming" takes part in a new mass organization, where all our faces (together and apart) perform the production of the platform, and where what is otherwise supposed to work under the radar of human perception comes to the fore, as pointed out by Jan Distelmeyer in this book.

The Navigating Face

Although most faces appearing in a gallery view of a video conferencing interface have not undergone the kind of official training offered by Wiley and Fox, one usually recognizes a video conferencing face, not by its attentive nodding, but by its impartial look directed slightly off the camera lens. Winfried Gerling's article in this book demonstrates a much larger cultural history of staring at screens, and one might see this navigating face as a continuation of this history and also how it compares to, e.g., the synchronized movement of the conveyor belt in the platform production system. In the work *People Staring at Computers* from 2011, artist Kyle McDonald installed a custom application in a series of computers located in Apple's hardware stores in New York. Every minute, the application takes a picture with the computer's inbuilt camera, uploads it to a server (the blogging site Tumblr), and projects the collection of portraits back to the viewer; that is, the application takes a picture and slowly fades in that photo, and then begins to cycle through older photos of previous users. As it is stated in the project's presentation video: "maybe if we could see what our computers see we would stare back at them differently? [...] But most people just hit escape" (McDonald 2011).

Figure 2: People Staring at Computers by Kyle McDonald

Source: Screen shot from video.

As often as the video conferencing face is an attentive nodding listener, it is a seemingly impartial and motionless face (perhaps only the eyes are moving). Its habitus is not just a social quality, trained and caring for social appearance, perception, and prestige; its technique (the one we recognize across all users) is also that of the interface itself. Even when faced with its own mirror, it will inevitably assume a navigating attitude (and hit escape). In other words, we all, as users, look alike because we are navigating or watching the interface. It is probably safe to assume that many users are (whilst conferencing) occupied with checking emails, looking at documents, turning on or off a filter, configuring or navigating some other interface, and so on (and sometimes even playing games). Unless we share our screen, this is never quite visible to the other users; our face remains the same: mostly impartial and motionless.

The navigating face is a mediated face, but it is trained in a different manner than, say, the women of Paris in 1934 were trained to walk by Hollywood cinema (as Mauss claims). If the girls of Paris walk the same way as the women of New York, it may not only be that they imitate the walk, but also that the moving camera itself instigates a way of walking; and similarly, if all users carry the same face, it is not simply because we imitate each other, but because of the interface itself.

To explain a little further. The graphical user interface (GUI), including that of the video conferencing system, originates in a design ideology of user empowerment, with the potential of revolutionizing life in all its aspects. For instance, the marketing video for the first Macintosh computer and graphical operating system in 1984 (directed by Ridley Scott) shows an Orwellian society where Big Brother speaks through a screen to a community of users (or slaves of the machine) and ends with a young athlete smashing her sledgehammer through the screen. With voiceover and text, the advertisement reads: "On January 24th, Apple Computer will introduce Macintosh. And you'll see why 1984 won't be like '1984'" (Scott 1984). As noted by hypertext and literary scholar Gregory Ulmer, this vision compares to former cultural industries (such as Hollywood cinema), in that it expresses "the 'twin peaks' of American ideology"—realism (or media transparency) and individualism, now built into the computer as an apparatus of production (1991).

User empowerment comes about by making the medium disappear, not in the sense that it becomes invisible, but in the sense that it becomes a habitus (i.e., the acquired ability or a faculty of using an interface). Firstly, as highlighted by early scholars of "new media" (as it was once labeled around the turn of the Millennium), the GUI was built on recognizable procedures of former media and instruments (a page, a menu, a button) (Bolter and Grusin 1999; Manovich 2001). Secondly, as highlighted in the field of human-computer interaction, user behavior is not only directed by the objective of tasks (editing a page, selecting from a menu, etc.), but also by what, more broadly, makes sense to the user. One might seek to obtain a goal or task, but the actual activity follows a process of sense-making in which the relation between people and artifacts is highly situated: the meaning of actions originates in neither human nor artifact, but is distributed and dependent on the situation, where people in the everyday tend to use the opportunistic structures of the GUI (Rogers and Marshall 2017, 13, 17).

The navigating face is thus quite different than the face as a stamp of the user's identity, and it also produces a different kind of spectacle than that of the mass ornament (the stadium show, for instance). What one sees in *People Staring at Computers* is neither just the faces of the individual people, nor is it just an ornament of mediated faces in which one can mirror one's own belonging to a collective. In fact, when video conferencing, the navigating faces of other users often remain hidden behind other interfaces—the text document, the email, the spread sheet, and so on, that the user is navigating. This seems to suggest that the faciality of the navigating face is the product of a process of subjectivation—of a software design, the design of not only use, but also of an opportunistic and navigating user. What one sees in Kyle McDonald's work is a mirror of the opportunistic user, making sense of the situation, such as clicking "escape" when the interface behaves unexpectedly or in other ways trying to navigate the graphical user interface. In this sense, there is a certain irony in the (revolutionary) face of user empowerment and opportunism being stationary and motionless, and only distinguishable in its micro mechanics (the movements of the eyes and subtle muscular contractions that are almost invisible to human perception).

The rhetorical question hence is: does faciality and the techniques of the face *necessarily* have to do with the construction of an ego? Without entering into deeper philosophical discussions of faciality, the navigating face can also be understood differently, as a *milieu* or sur-face, resulting from a process of subjectivation. As noted by Michael Hardt in his reading of Gilles Deleuze and Felix Guattari's *A Thousand Plateaus*, "The face is … a field or a milieu on which signification or subjectification can take place […] It is constructed so as to make certain meanings and subjectivities appear" (n.d.). The interface, as a technological vision of a new cultural platform industry built on realism and individualism, conjures up signification with signal, meaning with sur-face. Subsequently, one might say that the navigating face as a technique (opportunistically navigating the interface) is the appearance of the revolution of the graphical interface and the promises of a tech industry.

The Captured Face

The bodily (technical) process of facialization is also, as Hardt notes, close to Guy Debord's understanding of a spectacle (and perhaps Kracauer's, too): "like the spectacle, the face corresponds to or determines a form of rule" (Hardt n.d.). However, unlike the media spectacle, which in Debord's line of thinking mediates social relations by way of spectacular images and representations, the navigating face is itself a media spectacle. In other words, there is a correspondence between the corporate nature of video conferencing software and the bodily, corporeal, facial technique as the surface of this. This dimension of the techniques of the body is also the subject of Alexandra Saum-Pascual's *Corporate Poetry* from 2020. The work is, as expressed by Saum-Pascual, "an exploration into how corporate language related to that other corpora that is our body" (2020a).

The work consists of a number of rooms that repurposes corporate software such as Google Forms, Survey Monkey, Qualtrics, and also Zoom, "in order to domesticate the neoliberal intent of these data gathering technologies." In several ways, Saum-Pascual's work bears resemblance to that of early netart of the 1990s. In 1997 the pioneering netartist Alexei Shulgin launched the seminal "Form Art Competition" at the Austrian festival for electronic art, Ars Electronica. The competition submissions featured a number of works that used drop down menus, frames, text fields, radio buttons, and more to interrogate the (then) new formal language of the web (Andersen and Pold 2018, 46). By focusing on software for gathering user information, Saum-Pascual seeks to bring attention not only to the formal language itself, but to how it (since then) has become a language of an embodied reality. The formal language is the surface of a digital infrastructure "that is unintentionally brought into our homes whenever we participate in an online survey or take a video conferencing call" (Saum-Pascual). In the works, Saum-Pascual combines the utilitarian goals of the software with the vulnerable situation of the user (especially during the Covid pandemic). For instance, "Room #1" is made using Google Forms. As a "room" it exemplifies an intimate space of motherhood, contrasted by a very formal language of software that seeks to inhabit the room. The work thereby draws our attention to how software and digital infrastructure's capture of personal data occupies "the domestic and personal space that poetry tends to inhabit."

Figure 3: Corporate Poetry, Room#1 by Alexandra Saum-Pascual, using Google Forms for poetry

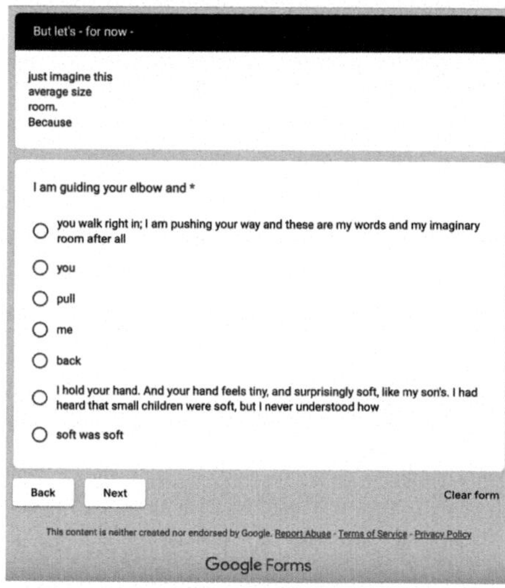

Source: Screen shot from poem.

The responsive text promises a sort of adaptation to one's intimate room (letting the reader's poetic choices create a poem about the intimacies of motherhood). By this, Saum-Pascual seems to disrupt how contemporary interfaces often build on the capture of intimate data, and how data feeds into a corporate network of interfaces. In this network, the textual inscription of the user, the corporeal and intimate, often implies an incomprehensible and invisible system where a dissolute calculation makes the interface appear smart and customized to the user's inner needs and desires. In this way, the formal structures and infrastructures of software interface come to inhabit the intimate and corporeal, but Saum-Pascual reverses this, and lets the corporeal inhabit the software interface.

In "Room #3," created in 2020 during the Covid pandemic, Saum-Pascual turns to how Zoom instigates a similar juxtaposition of the corporate (formal software infrastructures) and the corporeal (intimate space of the "room"). The work is an offline website where "webness is stripped from the global network to be rooted, deeply, at home" (Saum-Pascual 2020b). In other words, to actually witness the work means that one has to visit Saum-Pascual (in itself a paradox, as the pandemic prevents most people from doing this). The offline website presents a series of recordings from Zoom where Saum-Pascual appears in different versions of herself. In the first window she enacts the routine of forgetting to turn on the sound; in the second, the routine of asking the host of the meeting to unmute herself; in the third, the routine of notifying the other that her camera is off (pointing at her ears); and in the fourth, as the one appearing only by name, forgetting to turn on the camera. The final gallery view then runs continuously in a loop.

Figure 4: Corporate Poetry, Room#3 where Alexandra Saum-Pascual performs different techniques of capture

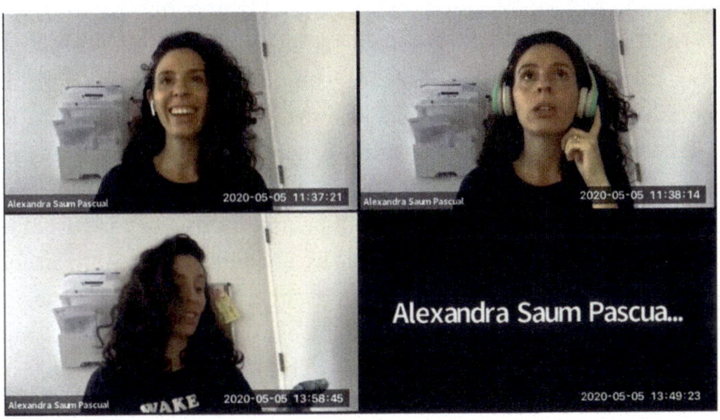

Source: Screen shot from video.

The Zoom interface can be seen as emblematic for a contemporary condition built on the formal capture of the user. Again, referring back to the turn in HCI in the mid/late 1990s toward the users' activities as processes of sense-making, the technique of capture may itself be seen as meaningful to the user. As discussed by Philip E. Agre, the capture of data is part of "a tradition of applied representational work that has informed organizational practice the world over" (2003, 745). And "as human activities become intertwined with the mechanisms of computerized tracking, the notion of human interactions with a 'computer'— understood as a discrete, physically localized entity—begins to lose its force; in its place we encounter activity-systems that are thoroughly integrated with distributed computational processes" (Agre 2003, 743). As exemplified in Saum-Pascual's work, the formal graphical user interface of the survey (as well as other corporate software) provides the user with a scheme for interaction, which is considered a meaningful activity—an extension of human activity through the interface.

As noted by Till A. Heilman, capture ("the systematic recording of activities and their 'grammarization' in data sets"), is today "the economic mechanism that drives the Internet in its current form" (2015, 40). Heilman compares this to a new kind of labor ("data labor"), built not on the exchange of labor for wage, but labor for "a 'space' of options for action opened up by media technology (which those affected consider useful, entertaining, or similar)" (2015, 43). It is the actions of this space, and increasingly more intimate actions, that are captured systematically in data sets, and which feeds into a corporate network of interfaces, providing the grounds for new intimate services. If "Room #1" stresses the capture of corporeal and intimate data, and urges us to consider the networked nature of this, and how corporeal data is used to generate customized and intelligent interfaces, "Room #3" further directs our attention toward the user herself. The four routines represent different common techniques of the Zoom face. As bodily techniques they are a mirror of how users act in front of a camera (smiling, gestures of "no sound" [pointing toward the headphones], surprised eyes, etc.), but they also expose the techniques of capture, vis-à-vis how software increasingly surfaces and *becomes* us, how it inhabits our bodies. The opportunism of the navigating user, trying to make sense, drives her not only toward the task of communicating with the other (which in the works seems reduced to nonsense), but toward an activity of capture, of being captured by the corporate interface, of making the face visible and audible as a bodily habitus. And, although this might seem as a mere process of capture by a camera and a microphone, it also involves a grammatization in data sets, used for, e.g., gaze-correction (as outlined by Rapoport and Tollman in this book) and more commonly filtering out backgrounds and adding facial filters (see also Andersen 2022).

In other words, the capture of the face is not only an extension of human activity through technology, but rather a structural condition for the use of the video conferencing interface, which again reflects a larger condition of corporate software that feeds on the inhabitation of the intimate and corporeal. As bodily habitus, the techniques of the face can, expressed in the line of Agre's thinking, be conceived as a surfacing grammar of action of the video conferencing system and a corporate business model of capture; and in itself, the technique embodies its own means of production.

Conclusion

"The face is politics," as put by Deleuze and Guattari (1987, 181), and we have proposed that the proliferation of video conferencing software exemplifies how software not only enables new functionalities in our lives, but also intersects with the construction of subjects. Or, put differently, we have proposed that subjectivation is a process of faciality, deeply entangled with the proliferation of the interface. First and foremost, it is the interface that sets the face at the fore; not just in a traditional sense, as a face of someone appearing on a screen, a stamp of an identity, but as a technical object which has become an intrinsic part of interface culture. Following Marcel Mauss, considering the body and the face as a technical instrument and object allows us to understand how the video conferencing face takes part in a contemporary spectacle. On the one hand, this is a mediated spectacle where our face, and all the other faces appearing at our screens, can be considered a mass ornament, or an aesthetic reflection of platform capitalism's production processes: all the faces, acting in similar and even synchronous ways (nodding attentively, glaring into the screen while navigating, pointing to the headphones, or in other ways drawing attention to the interface's capture of the user), display a new organization of the masses, and how we all, as users, perform the production of the platform. On the other hand, the faces of this spectacle are not just mediated images on a screen (a radical disembodiment), they *are* us; it is how we (the users) surface as software and perform the platform as producing subjects.

We have also, as did Mauss with the body, suggested that there are different techniques of the face, or different ways of (sur)facing. These can be read in a number of artworks attempting to deconstruct the platformed production of the face and in how contemporary video conferencing software as facial software takes part in a process of subjectivation. Put differently, if a face as a technical object corresponds to or determines a form of rule that becomes and inhabits us, then as Deleuze and Guattari also point out, there is no choice but to begin with our faces: "If the face is a politics, dismantling the face is also a politics involving real becomings ... know your faces; it is the only way you will be able to dismantle them and draw your lines of flight" (1987, 188). We believe that the artists presented in this article demonstrate this, by pointing out the production apparatus of the face—from how the corporate style conferencing interface moderates facial techniques in social interaction, to how its own techniques come to inhabit us, as techniques of navigation and capture.

Acknowledgements

This article is based on previous work presented to the Electronic Literature community, and published in the conference proceedings to the Electronic Literature Organization's (ELO) 2022 conference (see Andersen 2022).

References

Agre, Philip E. 2003. "Surveillance and Capture: Two Models of Privacy." In *The New Media Reader*, edited by Noah Wardrip-Fruin and Nick Montfort. Cambridge, MA: The MIT Press.

Andersen, Christian Ulrik. 2022. "Techniques of the face—on video conference art and politics." ELO 2022, Como, Italy.

Andersen, Christian Ulrik, and Søren Pold. 2018. *The Metainterface—the art of platforms, cities and clouds, Software Studies*. Cambridge, Massachusetts; London, England: MIT Press.

Belting, Hans. 2017. *Face and Mask: A Double History*. Princeton, NJ: Princeton University Press.

Benjamin, Walter. 1999. "Reply to Oscar A. H. Schmitz." In *Selected Writings*, edited by Howard Eiland Michael William Jennings, and Gary Smith Cambridge, MA:: Belknap Press of Harvard University Press.

Bolter, Jay David, and Richard Grusin. 1999. *Remediation. Understanding New Media*. Cambridge, Mass. London: MIT Press.

Chun, Wendy Hui Kyong. 2016. *Updating to Remain the Same—Habitual New Media*. Cambridge, Massachusetts; London, England: MIT Press.

Deleuze, Gilles, and Felix Guattari. 1987. *A thousand plateaus: capitalism and schizophrenia*. Translated by Brian Massumi. Minneapolis, MN: University of Minnesota Press.

Fanon, Frantz. 1986 [1952]. *Black skin, white masks*. Translated by Charles Lam Markman. London: Pluto Press.

Fox, Alfred S., and Kristen S. Wiley. 2014. *Better Movement Engineering for Perfect Everyday Living*. Edited by United States Patent Application: Better Movement, Inc. (Madison, WI, US).

Galloway, Kit, and Sherrie Rabinowitz. 1989. "Avantpreneur (The New Practitioner)." Accessed Dec. 8. http://www.ecafe.com/museum/avant.html.

Hardt, Michael. "Notes on *Milles Plateaux*, Chapter 7–9." Unknown. https://people.duke.edu/~hardt/mp3.htm.

Heilmann, Till A. 2015. "Datenarbeit im 'Capture'-Kapitalismus. Zur Ausweitung der Verwertungszone im Zeitalter informatischer Überwachung." *Zeitschrift für Medienwissenschaft. Heft 13: Überwachung und Kontrolle* 7 (2): 35–48. https://doi.org/10.25969/mediarep/1561.

Jirsa, Tomáš, and Rebecca Rosenberg. 2019. "(Inter)faces, or how to think faces in the era of cyberfaces." *World Literature Studies* 11 (4): 2–10.

Kracauer, Siegfried. 1995. *The Mass Ornament—Weimar Essays*. Translated by Thomas Y. Levin. Cambridge, MA: Harvard University Press.

Krauss, Rosalind. 1976. "Video: The Aesthetics of Narcissism." *October* 1 (Spring): 50–64.

Levinas, Emmanuel. 1969. *Totality and Infinity: An Essay On Exteriority*, Duquesne studies Philosophical series. Pittsburgh, PA: Duquesne University Press.

Manovich, Lev. 2001. *The Language of New Media*. Cambridge, Mass. London: MIT Press.

Mauss, Marcel. 1973. "Techniques of the body." *Economy and Society* 2 (1): 70–88. https://doi.org/10.1080/03085147300000003.

McDonald, Kyle. 2011. People Staring at Computers.

Rogers, Yvonne, and Paul Marshall. 2017. "Research in the Wild." Synthesis Lectures on Human-Centered Informatics.

SALYER+SCHAAG. 2021. *Perfect Movement Engineering for Better Everyday Zooming*: ELMCIP.

Saum-Pascual, Alexandra. "Rooms #1 and #2." *The New River* (Spring 2020).

Saum-Pascual, Alexandra. 2020a. "Corporate Poetry." https://www.alexsaum.com/corporate-poetry/.

Saum-Pascual, Alexandra. 2020b. "The Offline Website Project." http://www.alexsaum.com/the-offline-website-project/.

Scott, Ridley. 1984. *1984 (advdertisement)*: Apple Inc.

Sydnor, Synthia. 1998. "A History of Synchronized Swimming." *Journal of Sport History* 25 (2): 252–67.

Ulmer, Gregory L. 1991. "Grammatology Hypermedia." *Postmodern Culture* 1 (2).

Žižek, Slavoj. 2008. "Tolerance as an Ideological Category." *Critical Inquiry* 34 (4): 23.

Performing | Appearing

Sociospatiality between Agency and Fixation
Framing the Fixed View in Video Conferencing Arrangements

Laura Katharina Mücke

What Pandemic Zoom Pranks, Fails, and Glitches Can Tell

When I search the internet for keywords on "Zoom pranks," "fails," or "glitches" uploaded during the world's first pandemic-driven lockdown in March to June 2020, numerous results pop up. On the one hand, these derailments prove the technical and social insufficiency to cope with the exploding requests for video conferences (and their digital tools) since the pandemic's outbreak. On the other, they also point directly to the most basal construction principles and infrastructures of those new video environments: they mirror the way video conferences (should) work or do not. One well-known incident was the case of Will Reeve, a *Good Morning America* anchor, who was uncovered by his laptop's camera angle to be wearing only shorts under his jacket, tie, and shirt during a real-time and home-conducted interview on April 28 (fig. 1).

Figure 1: Will Reeve in shorts on Good Morning America on April 28, 2020

Source: https://www.youtube.com/watch?v=HlNKLiI9h3g (accessed: October 26, 2022).

He afterward dealt with his globally distributed fail in a relaxed and self-reflexive manner, posting a screenshot of it on his Twitter account, adding "I have ARRIVED*, *in the most hilariously mortifying way possible."[1] Multitudes of colleagues, friends, and strangers reacted supportively, nearly praising him through reenacting his habitude[2] and through comments like "You're a legend, man!" (@RandomWhig, April 28, 2020) or "This should be the new standard. Do not let the world shout u down, u are a hero and a prophet" (@Kwite, April 28, 2020). Will could have easily avoided his presumed faux pas, for example, by carefully following advice many online Zoom manuals published during the first weeks of video conferencing in 2020—ranging from suggestions to adjust what can be seen by tilting the laptop camera and covering up one's untidy background to the creation of an office-like desk arrangement.[3]

While Will's "framing carelessness" was just an unlucky coincidence, other transgressions of Zoom's spatial arrangements were intended ones: a little earlier, on April 12, 2020, TikTok influencer Samuel Grubbs had uploaded a video of a Zoom prank on his account, fueling an international hype during times of homeschooling.

His short clip displayed a school class "passing around" a pencil through different video tiles (fig. 2)—an arrangement that could technically only be managed through (1) different pencils and (2) manually arranged Zoom windows in a previously discussed spatial order. For someone not involved in the prank's construction, the project created the impression that an object can transgress formerly separate (and only digitally connected) spaces. Grubbs's video, which was watched over 40 million times and emulated on the same platform as well as on other ones,[4] kept subsequent Zoom users busy (re)arranging and trying out themselves in relation to the new digital constraints, redefining the margins of their spatial configurations.

---

1 For Will's Twitter post and related comments, see https://twitter.com/ReeveWill/status/1255141549450473473 (accessed: October 26, 2022).

2 The anchor's appearance on GMA and some reenactments can be found in a short clip compiled by *Entertainment Tonight*, which was already uploaded on April 29, 2020. See https://www.youtube.com/watch?v=HINKLiI9h3g (accessed: October 26, 2022).

3 For examples of online manuals and supposed dos and don'ts related to Zoom, which were published during the first lockdown, see https://blacknight.blog/12-tips-for-having-a-successful-video-conference-call-zoom-office-365-teams-meeting.html and https://www.thewrap.com/fox-news-dana-perinos-tips-for-working-from-home-keep-a-schedule-step-away-when-you-can (accessed: October 26, 2022).

4 A video compilation uploaded on YouTube on October 31, 2020, shows many of the TikTok-Zoom pranks done during online school classes. The video has (until now) been watched over seven million times. See https://www.youtube.com/watch?v=ACkJivTdmgA (accessed: October 26, 2022).

Figure 2: One of the first TikToks on Zoom pranks that influencer Samuel Grubbs (@samuel-grubbs) uploaded: a school class "passing around" a pencil

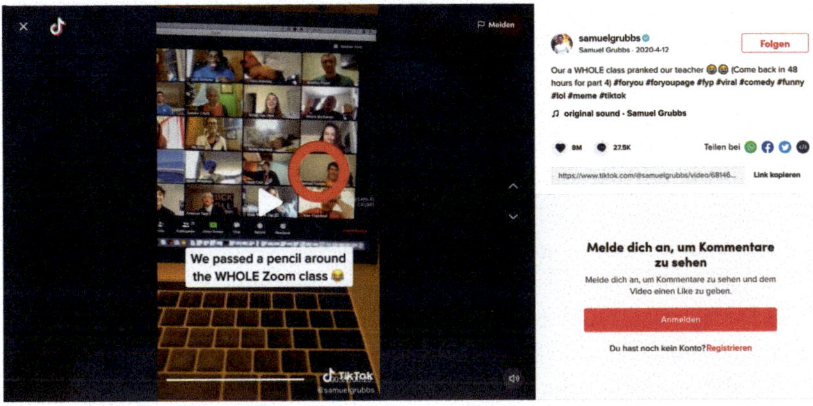

Source: https://www.tiktok.com/@samuelgrubbs/video/6814652084529466630?is_copy_url=1&is_from_webapp=v1 (accessed: October 26, 2022).

Both Grubbs's TikTok challenge and Reeve's appearance without trousers refer to various spatio-ontological, social, technological, and infrastructural layers of what the pandemic's instant move to a fully mediated conversation culture in 2020 (and beyond) has demanded: new kinds of digitally connected solidarity. Even if video conferencing had technically been possible since the late 1960s and commonly used since the late 1990s, during the pandemic video calls replaced most social interactions worldwide. Face-to-face meetings were restaged in both working and private contexts and, despite the complications for technically not well-equipped or well-educated people, video conferencing became *the* standard mode and needed expertise to audiovisually stay in touch with others—to gather in shared virtual spaces and experience the urgent feeling of copresence as telepresence during the hardest times of isolation.

In media studies, the immediate need to cope with the fundamental social changes has, as Jan Distelmeyer recently (2021) remarked, accelerated discussions about how new spaces of digitalization and the internet could be integrated in our thinking about "living environments"[5] in general. Video conferencing as a scientific topic had until now been assessed mainly from the fields of social science and communication studies (see, e.g., Dickson and Bowers 1974 and, for a

5 I use the term *Lebenswelt* in the meaning Jürgen Habermas has borrowed it from Alfred Schütz and Edmund Husserl: as the socioculturally driven conglomerate of evidence, which emerges out of specific and unspecific interrelations between subject's agency and its discursive system. See Habermas (1982, 209).

quick historical overview, Held 2020). But their often descriptive and quantitative-empiric framing is not helpful for speaking about the sociospatial transformation that daily, exhausting video meetings demand today. Moreover, the corona crisis seems to have brought back some fundamental onto- and epistemological thoughts that had already been diversified during the past decades—especially nostalgic longings for the analog world as well as thoughts about face-to-face closeness and virtual distance, which seemingly went missing in "the pandemic world." The political implications within those well-known topics point to much broader current discussions about mediated spheres, such as on fake news, the power of images and technology, platform politics, and privacy settings—making questions about spatial rearrangements and orders of dispositives pressing again.

In the following, understanding the pandemic predominantly as a media-based social and/as *spatial crisis*, I will focus on sociospatial arrangements of video conferencing by primarily using genuine film studies concepts instead of referring to social sciences. I am interested in the politics and aesthetics of video conferences' spatial interrelations and their new framings of usership and spectatorship, social relations, platforms, infrastructures, and technologies as well as the way they affect and interact with our current ways of communicating and thinking. Based on my introductory examples, I will describe the interlocking of public and private spaces in video conferences through Adrian Martin's (2014) concept of *social mise-en-scène* in its paradoxical relation to Susanne Lummerding's (2005) political term *suture*. Because while *screening and being screened*, *seeing and being seen*, we are confronted with an inevitability of the fixed frontal camera perspective as both aesthetic concept and political corporeality. Accordingly, I will examine the new powerful scopic regimes (see Jay 1999) of visibility and invisibility, movement and fixation, and foreground and background. They bring with them the necessity to refocus on the precarious ephemeral and corporeal position of the person in front of the screen/camera, as well as on social and political negotiations of spaces themselves. I treat these (con)figurations as techniques of governance that accompany every video conferencing meeting we conduct. To have a specific video conferencing arrangement in mind, I will develop my argument referring to Zoom, as Zoom's metaphorical *spatial wording* seems the most fitting equivalent of my argument. Additionally, I will shortly reflect on the more play-like platform Gather.town, which offers some presumably different spatial conditions.

Distant Socializing and Personal as Public Spaces

As sociologists Hubert Knoblauch and Martina Löw have recently claimed, the pandemic has shown how fluid territorial demarcations in "networked spaces" had become:

> Territorial spaces follow logics of positioning and arrangement, which means to have clear-cut borders to external spaces and accept the restriction of the diversity in its inside. Usually, they are perceived as static constructions. In contrast, networked spaces follow logics of relationality and heterogeneity. In networked spaces, distant elements can be put in relations and their diversity is one of their main characteristics.[6] (Knoblauch and Löw 2020, trans. LKM)

According to their view, the pandemic has first and foremost entailed ruptures in spatial experiences. During the pandemic, "the physically delimited territories of the people, who are forced to retract themselves in their close private spaces, meet the boundless and body-less network of communication,"[7] which entails a contradictory logic. Considering this perspective and conceptualizing video conferencing spaces therefore with theories on pictorial spaces and visual experience, Shane Denson's reminder of Stanley Cavell's bifurcating the meaning of *screen* as "protective shield" *and* "world-accessing window" seems reasonable (2020, 317). When we enter a video session, our (visual and auditive) depiction becomes, paradoxically, an access to other windows of people accessing our window. At the same time, we can feel protected as we do not need to worry about real metaleptic entrances of others into our own window. That this protection can even become a severe obstacle has unfortunately been proved through several sad stories about Zoom participants helplessly watching someone die during a meeting.[8] Despite those sad happenings, the dialectic between Cavell's shield and window serves as tipping point for a complex fact: viewed from a phenomenological perspective, our witnessing of others is unavoidably linked with displaying ourselves within the bigger spatial arrangement of the tiles. This uncomfortableness of not primarily wanting but needing to *become a picture* is a condition that has immediately been written about—for example, in Kerim Dogruel's (2020) observation that most students switch off their cameras during class. This mere fact becomes even more paradoxical as many people wished to gain back analog encounters during the first lockdown: situations that depend on exactly those configurations of simultaneously seeing and being seen. Especially, the essays of Bill Ayers (2020) and Emmanuel Alloa (2020), which were published among

---

6 German original: "Territorialräume folgen einer Logik des Platzierens und Arrangierens, der zufolge klare Grenzen nach außen gezogen werden und eine Beschränkung der Diversität nach innen akzeptiert wird. Sie werden in der Regel als statisch wahrgenommen. Dagegen folgen Netzwerkräume einer Logik der Relationierung des Heterogenen. In Netzwerkräumen können distante Elemente in Beziehung gesetzt werden und die Differenz der Elemente ist ein wesentliches Kennzeichen der Netzwerkräume."

7 German original: "[Da] treffen die physikalisch begrenzten Territorien des Selbst von Menschen, die sich mehr und mehr auf den Nahraum zurückziehen und zurückziehen müssen, nun auf die entgrenzte und körperfreie Vernetzung der Kommunikation."

8 https://www.youtube.com/watch?v=OKJ3n8YNC_g (accessed: October 26, 2022).

many other pandemic-related texts in *Critical Inquiry* between March and June 2020, bemoaned the loss of analog encounters. Alloa proved his argument by complaining that the "disappearance of shared public space also corresponds to a disappearance of surprise"—as if it were not surprising and sharable enough to be confronted with a forced self-view while encountering others.

Against such proclaimed feelings of unsafety in becoming a constant picture, German sociologist Sascha Dickel, who coedited one of the first quickly published, Covid-related edited collections in social science in Germany in 2020, emphasizes the outdatedness of this bemoaned "society of presence" in evenly criticizing the oft-used term *social distancing*:

> [The] term social distancing is misleading. It is the outcome of a society, which is still misperceiving itself as a society of presence—and which interprets presence as co-presence of bodies in a physical space. (Dickel 2020, 80, trans. LKM)[9]

Dickel instead proposes to name the social condition during pandemic lockdowns *distant socializing*—a term that also gets new dimensions in thinking about shielding and screening as digital social categories. In describing closely what happens during video conferences from a spatioaesthetic perspective, in this article, I also aim to overcome those nostalgic separations of analog from digital spaces mentioned above. Focusing instead on the complex intersections of shielding and displaying, hiding and looking, private and public, and close and distant spaces, I want to grasp the tension that operates between the seemingly different spatial arrangements of one's own (felt) physical space and the other's (perceived) transmitted space. It is the Italian philosopher Chiara Cappelletto (2020) who adds in her contribution to the *Critical Inquiry* collection the call for a "new aesthetics of presence." To think about collective copresence as evenly assembling telepresence becomes here a new way to conceptualize societies as disperse but simultaneously connected groups. Via Zoom, spectators are now copresent in two ways at the same time: as spectators and as persons being screened. Presence thus becomes a term to elaborate on how we can relate to each other even digitally. Cappelletto reminds us of the consequences of those "new" sociospatial constellations. For example, she queries why even our home space, which should work as comfy private space and steady orientation, "suddenly becomes the cutoff point where our experience of near and far, neighbor and stranger, collapses."

---

9 German original: "[Der] Begriff des Social Distancing ist irreführend. Er ist das Produkt einer Gesellschaft, die sich immer noch weitgehend als Anwesenheitsgesellschaft selbstmissversteht—und dabei Anwesenheit als Kopräsenz von Körpern an einem physischen Ort interpretiert."

What Is Visible within the Video Conferencing Picture?

It is thus the spatial arrangement of the screen before us and the display of others that reframes our sociospatial perceptions. If able to switch my camera and mic on, I principally agree to the conditions of *being visible as being captured* within the window of a horizontal (medium) rectangular close-up frontal camera perspective. At the same time, I need to be prepared to display everything behind me, which is technically ensured by my laptop camera's depth of field. In adjusting the space appearing in my tile—in becoming "zoomable" (and Zoom can zoom!) I need to "customize" my *being-with as being-viewed* to the usually laptop-associated integrated webcam.[10] To change the visible part of my picture, I can only tilt my laptop's monitor on a vertical axis, for example, to hide what is in front of me[11] or right above my head. In the meantime, I am mostly forced to accept that anything visible can become witnessed. This implies that what appears in the space depictured behind me can be (or become) even more important than myself—a fact that offers a theoretical rebuttal of image theories emphasizing the importance of objects in the foreground over objects in the background of an image.[12] Also, my Zoom's background reveals much information about the situation and terrain I currently occupy. The social determination and individual side effects (and affects) of such uncalculated backgrounds are self-evident, as can be proven by another widely distributed Zoom fail, in which a girl becomes ashamed for her supposed boyfriend as he stumbles half-naked through the background of her official video meeting, bangs against a glass door while escaping from the frame, and bounces back into it.[13] But Zoom's

---

10 The evidence that many people used additional hardware to create a different spatial recording setting does not override my argument. Leaving aside the fact that this different setting had been affordable only for better-institutionalized people and became rarer with the increasing prices of external webcams (see, e.g., https://www.theverge.com/2020/4/9/21199521/webcam-shortage-price-raise-logitech-razer-amazon-best-buy-ebay [accessed: October 26, 2022]), I take the fixed view of the integrated camera as standard constellation because most calls (even the ones during train rides and the ones conducted through smartphones, etc.) are held in this fixed setting.

11 For example, I often witnessed mothers tilting the laptops to a higher angle while breastfeeding their babies. I consider this as a political act.

12 This perspective adds a new dimension on speaking about the close-up as an intimacy-inducing and highlighting perspective because it opposes the argument that what is in front of the screen counts for viewers always as most important. See Persson (2003, 130).

13 The video became very popular. See https://www.thesun.co.uk/news/11249319/coronavirus-lockdown-man-underpants-conference-call (accessed: October 26, 2022). The auditive signals in Zoom's surroundings are equally important to account for during video conferences. This is quickly explained by another Zoom fail: the story of Scott Connell, chief meteorologist for KSDK St. Louis, who tried to record the actual weather forecast multiple times because each take was disturbed by his distantly barking dog. See https://www.youtube.com/watch?v

technical dealing with those unwanted intruders in letting better-equipped laptops insert a so-called virtual (or blurred) background is equally prone to disturbance. First, virtual backgrounds create a nearly planimetric and weird appearance of dislocation of the body. Second, they make the integration of objects other than human bodies impossible because their technology is based on the AI principle of predictive "portrait segmentation"—which had been developed as a one human body–centered tool. The predictive blurring of all potential background objects then leads, for example, to the scenario that even other objects like books in front of the camera (to show them to others) just become rendered invisible.

As a reminder of this need to constantly and consciously (re)arrange and design one's spatial visibility to a rectangular window while video conferencing, Zoom comes with a prearranged quasi-*spatial structure*: Zoom meetings are metaphorically named "rooms," their "entrance" is modulated through "waiting areas" (see also Alexander 2020), and "leaving" the main session to small group spaces is called a "breakout." The entire vocabulary of video conferencing thus simulates sociospatial setups, even though those "spaces" are visually constructed only through the multi-split-screen and two-dimensional, side-by-side-arrangement of the tiles (see Hagener 2020). How little social liberty or *agency* those layouts offer can be shown in a brief approach to an alternative video conferencing tool that is considerably less known: Gather.town.

Besides its visual and ludic resemblance to a *Super Mario* world, Gather.town provides pixelated colorful spaces (fig. 3) instead of parallel tiles. Designed in particular for bigger groups of 10 to 50 people, Gather.town seems at a first view to serve a different concept of sociospatiality. Its social interactions are navigated by each participant's creation of an avatar and their walking through the computer game–like world each meeting is bound to. The usual video tiles of other participants, then, only pop up when one approaches other avatars spatially (and vanish when one walks away), mimicking the same adjustment of spatial closeness and distance through the sound layers as talking avatars become louder or softer in relation to their "spatial distance." Thus, Gather.town offers more flexible parameters for sociospatial orientation than Zoom. On the first view, its spatial pre-structures let users play with their locatability during meetings. Thus, it opens up navigable spaces of liberation and domestication.[14]

=Xg9lxTAE8ql. The incident became popular as well—so popular that even Ellen DeGeneres invited the meteorologist and his dog onto her show. See https://www.youtube.com/watch?v=0uTv8Aw9WD4 (accessed: October 26, 2022).

14 Discourses on the domestication of the screen become important again in talking about the intersection of home and working spaces in video conferences, as three nearly simultaneously conducted studies during 2020 in Singapur, Rotterdam and Sydney proved. See Lim and Wang (2021); Harteveld (2020); Watson, Lupton, and Michael (2021).

Figure 3: Gather.town advertising picture on the company's website

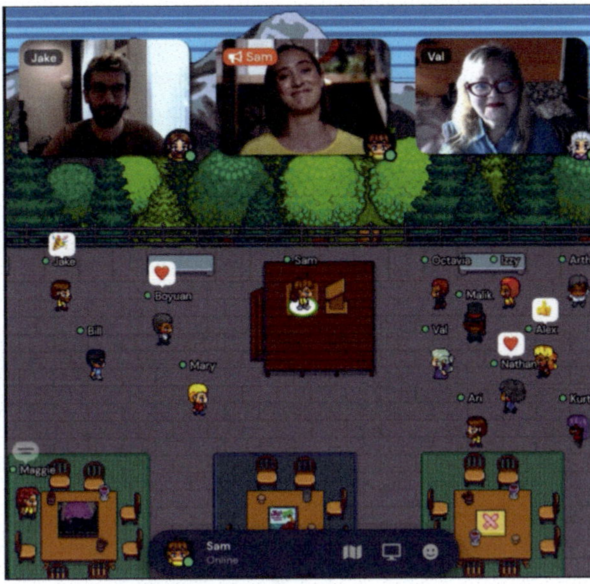

Source: https://www.gather.town/about (accessed: October 26, 2022).

On a second view, even Gather.town is tied to an interface of video tiles as *fixed frontal views* because what remains is that users still need to sit in front of their computers (and cameras) to operate their avatars and talk to others. The fixed frontal view of the camera is inevitable—even older spatial configurations of video conferences had them, as manuals on video conferences' technical arrangements from the 1980s can tell (fig. 4 and 5).

Also, video conferencing platforms in general already tend to *calculate* with this artificial modeling of a social space: Gather.town aims at a reconstruction of the displayed space in adjusting each user to a twofold ego, as avatar and as frontally viewed camera image. And Zoom has recently figured out a remodification of its spatial arrangement: the new "immersive view" function repositions viewers without a background into a drawn or technically layered environment. How nefarious this supposed freedom to collaborate can be is the topic Naomi Klein called attention to in focusing on the capitalist interests that "tech companies" hide behind these modelling of (infra)structure, calling it the "Screen New Deal":

> Far more hi-tech than anything we have seen during previous disasters, the future that is being rushed into being as the bodies still *pile up* treats our past weeks

> of physical isolation not as a painful necessity to save lives, but as a *living laboratory* for a permanent—and highly profitable—no-touch future. (Klein 2020, emph. LKM)

That those new sociospatial freedoms are primarily accessible by privileged groups of people sharply contrasts with the advertising slogans that Zoom Video Communications—the company behind Zoom as program—uses.

Figures 4 and 5: Even early manuals to analog video conferences depict the inevitability of a fixed frontal view

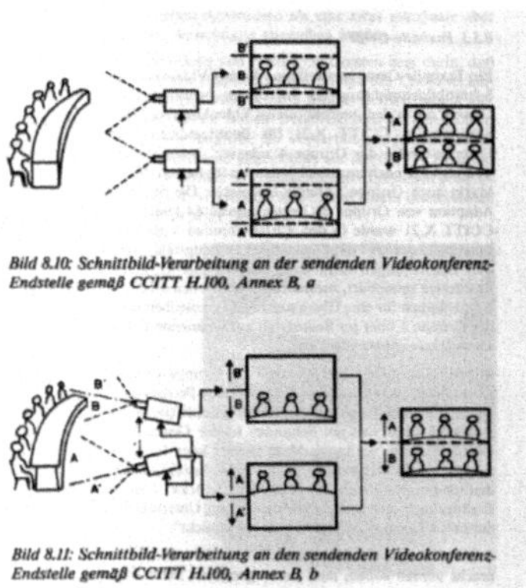

Source: Gerfen (1986, 73, 157).

Fixed Frontal View versus Spatial Liberty

It is necessary to point to those techno-ideologies and capitalist privileges as Zoom Video Communications is currently working on upgrading "our future" into the next level of sociospatial experience. Their big annual Zoomtopia event, a Virtual Zoom User Conference, mainly gives the impression that their users hold democratic agency because the company promise them especially more *spatial* liberty during a Zoom meeting.[15]

15 Zoom's style to give their users the feeling to be their protagonists appears as their main neoliberal selling strategy. On Zoomtopia's website, one finds written the following: "Zoomtopia 2021, the annual celebration of our customers, takes place virtually September 13–14, and we'd love to 'see you' there! This year's theme—The Imaginarium—highlights all the ways you've used Zoom to embrace change, enable hybrid workforces, and continue to grow

Figures 6 and 7: Two animated examples shown during Zoomtopia 2021 for future spaces in Zoom meetings via Oculus and VR

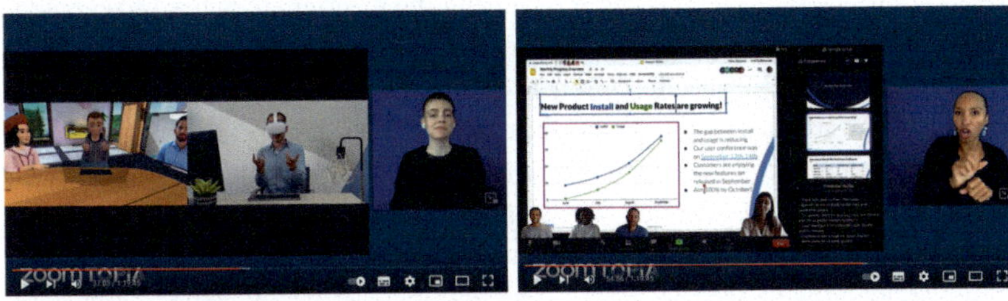

Source: https://www.youtube.com/watch?v=s2SBVnafMSI (accessed: October 26, 2022).

At their 2021 event, Zoom announced among other things its future cooperation with Oculus, planning to invent VR-Zoom meetings as the standard mode for digital encounters. They presented how those spatial enhancements would liberate users' decisions to move quasi-physically through the meeting space as the new hardware tools would enable them to virtually gesticulate and walk to the front of the virtual room. This is told without mentioning that special hardware requirements would exclude everyone who cannot afford them. And it thwarts Zoom's alleged wish to be "more inclusive," to "have room for everyone here on Zoom," or to "bring more capabilities to you"[16]—to control social collaboration by providing spatial flexibility. But even in their simulation of these future plans, one characteristic of video conferencing remains the same: the fixed view of the frontal camera (figs. 5 and 6).

I call this characteristic a fixed view because the frontal camera perspective can and will probably never be deleted from a video conference setting. Its basic condition is the already mentioned coercion to appear within the picture while operating the laptop. Because people are mostly unable to operate their computer from elsewhere, they become automatically screened as a rectangular, medium-close-up extract of their surrounding reality. Therefore, the fixed view serves as a specific *dispositive of watching while being watched*. In practice, it forces Zoomers to adapt to the Cavellian window magically and abruptly popping up in the middle of private spaces. In theory, this concrete requirement offers potential to also approach very basic film theories like apparatus theory from a new perspective: from both, a production *and* a reception-oriented point of view, as the question of the governance of the screen becomes conditional while coproducing the image.

---

your business in the face of immense challenges." See https://zoomtopia.com/zoomtopia-2021-spinifex.

16 These and similar phrases can be heard in the recorded introduction to the 2021 Zoomtopia conference on YouTube. See https://www.youtube.com/watch?v=s2SBVnafMSI (accessed: October 26, 2022).

Zoom forces their operators into specific bodily situations,[17] which has social consequences: users are kind of trapped within their computer's camera window. If they want to move around and be seen during the meeting, they need to carry the laptop and at the same time move in a still screenable way. In this manner, users also need to deal with a specific closeness of the camera. To look as good as possible, it suddenly seems to become important to worry about pimples, eye bags, and the smallest (even unintended) facial reaction, none of which would be visible if users could handle their keyboard from farther away. In this situation, switching the camera off, tilting the laptop's screen high up or down, and deciding not to operate the computer while being seen become the only options (even if they are not real options) for circumventing this unscrupulous fixed frontal view.

All these configurations become seriously problematic for new digital framings of visible social spaces, for example, since a 2021 Stanford study has already "proved" that the widely known phenomenon of "Zoom fatigue" can be explained by spatial reasons: by the "excessive amounts of close-up eye contact," by "seeing yourself during video chats constantly in real-time," and because "video chats dramatically reduce our usual mobility" (Bailenson 2021; see further Lovink 2020).

The Frontal View as Picture: Personal Spaces, Social Mise-en-Scène, and the Close-Up

From the perspective of a film scholar, these spatial ambivalences between seeing and being seen, simultaneity and non-simultaneity, and mobility and fixation are well known. That is why in calling attention to previously researched perspectives on enabling and impeding *agency* through camera angles and framings, I want to shed light on some overlooked aspects of the visual *appearance* of this inevitable fixed frontal view. My argument is that the collapsing of personal and private space can, through such a discursive approach, be explained—as it also brings with it, for example, new social conditions for spaces of fear, criminality, violence, protest, or party (see Sayman 2020). Before I proceed to this, it needs to be emphasized that, theoretically, the mediality of video conferencing can also be approached through many more media (and not only film aesthetic) genealogies—for instance, its close connection to such televisual characteristics as liveness and transmission as well as its recordability, which relates Zoom to tape recorders.[18] Still, film studies' visual

17 Anyway, other—and not just digital or filmic—dispositives are meanwhile at work since the most comfortable body position while zooming seems to be "sitting" (which also works here as dispositive), see the submission by Winfried Gerling in this volume.

18 I thank the participants of the author's workshop of this edited collection for pointing to this fact.

concepts, such as composition, shot length, and allocation of depicted spaces, better serve my concern for sociospatiality and a socioaesthetic analysis of the video conferencing image as *a screened image* in the first place. Medialities are anyhow characterized by their general hybridity, which invites us to view them beyond their essentialist ontologies (Creeber 2013, 3).

Strangely enough, such an approach to technical visibilities and their social consequences is relatively new even to film studies, as Adrian Martin has shown in his book chapter on *social mise-en-scène*. In calling attention to the fact that our social world is "already relatively strictly organized, codified, subject to a multitude of rule-sets that govern (or at least regulate) behavior, posture, gesture, level of emotion" (2014, 131), Martin states that we always need to relate questions of technical arrangements to social constructivism.[19] Kyle Stevens (2020) stated in another *Critical Inquiry* essay that the pandemic has led him to view movie scenes containing many closely interacting people now through the lens of his own fears of contagion. Also for Martin, social mise-en-scène "engineers a specific shift in critical/analytical perspective":

> With social mise en scène, rather than going directly or primarily to the unique, idiosyncratic sensibility or world-view of the maker, we attend to the newly grasped raw material of social codes, their constant exposure and deformation in the work of how a film articulates itself. In particular, it allows us to zero in on something specific: known rituals that are recreated, marked, inscribed in the flow of the film. (2014, 134)

Referring further to the felt "awkward arrangement of bodies that are positioned too closely" during the scene in the listening booth in *Before Sunrise* (1995), Martin highlights the "micro gestures" taking place during film experiences, which are directly connected to our felt sociospatial staging while watching someone so close via Zoom. Before Martin, it was mainly Per Persson in his 2003 book on "personal space" who had connected social with aesthetic theories in defining "variable framing" in film by pointing out its ability to also change the profilmic space, and as a powerful device "to create hierarchies" (2003, 101). Even if Persson has described his ideas without any connection to a theory of the "really felt closeness" of the viewers, as Guido Kirsten (2019, 51) has lamented, *or* to video conferencing tools *or* to the question of experiences of digital copresence, he spends much time describing especially the close-up as dense and "isolated" situation of "social communication" (123).

19 Martin states that even if big-name film and art scholars have mentioned this connection of form and sociality—such as Michel Mourlet ("Mise en Scene comme langage," 1987), Umberto Eco ("Articulation of the Cinematic Code," 1967), Pier Paolo Pasolini (2007), and Jean-Louis Comolli (1980)—it has never found its way into the main discourses of film analysis. See Martin (2014, 131–133).

He borrows his social perspective from Edward T. Hall's book *The Hidden Dimension* (1966), which is interested in the nonverbally expressed but culturally framed (and for Hall: clearly racially used) spatial conditions managing intimate communication situations. In describing the film's potential to "simulate personal-space behaviour," he equips the close-up with clear social consequences:

> the cut-in gives guidance and provides time for thinking, enabling spectators to attribute mental states, establish causal relations, and speculate about future events and their relevance to this character. (Persson 2003, 123)

The sociotechnical dimension of the close-up has already been linked to the pandemic by Guilherme da Silva Machado in his contribution to the edited collection *Pandemic Media: Preliminary Notes toward an Inventory* (2020). There, he writes about the visual proximity of the close-up as a recognition pattern and semantic field of the facial expression, leading to a probable self-voyeuristic experience of "seeing myself at work" (201) during Zoom meetings. But, phenomenologically speaking, the close-up is more than a question of the semantic field: it is an embodied experience.

The Frontal Perspective as Viewing Position and the Politics of Suture

The focus on the *embodiment* inscribed in social mise-en-scènes and close-ups at work during video conferences is neither the main goal of Persson or Martin nor of de Machado. But as Zoomers view close-ups while *experiencing* themselves as a *screened close-up*, questions about the experience of this strange double-framing-POV rise to the forefront. Embodiment can especially be elaborated in relation to film phenomenological positions that had thought about camera's inability to simulate a "perfect" viewer experience with a prevalent use of a subjective camera perspective, such as *Lady in the Lake* (1947) or *Le scaphandre et le papillon* (2007). Several times, this discussion has pointed to the inability of the viewers to relate empathically to those film's protagonists because of the inadequacy between the camera's work and *all* bodily perceptions (see Hanich 2016). Even if this discussion on POV shots seems different from my paper's interest, the elaboration on a "double inability" gives important hints for the (in)congruency between the doubled viewing-as-being-viewed position of the viewer on Zoom. The viewer's position, then, seems to be akin to Matthias Wittmann's (2020) outline of the ghostly standpoint users are bound to in 360-degree and VR settings. Wittmann describes the VR helmet–wearing user as seeing but having nothing unseen behind their back while simultaneously being unable to see themselves when looking down at their body. For Wittmann, the new technically fixed positions urge us to reconsider theoretical questions about the spectator, the imaginary, and the power of filmic spaces (as diegetic spaces) in relation to what Jean-Pierre Oudart has called *suture*—the subject's way of stitching itself and being stitched into the film's space. His description helps to deal with the volatile and precarious position that *viewers as users* occupy during video conferencing.

I want to follow Wittmann in stating that it is especially the concept of suture that "bridges" this double positioning of a Zoomer's body as seeing and being seen. I consider suture helpful, even if I am, here, not interested in the term's psychoanalytic background at all. I think, I can refer to the tension between different forms of (in)visibility, or off-screen spaces, that co-agitate during video conferencing in only remaining with suture's subject theoretical background. On Zoom, especially the *negotiations* of visible sociospatialities become blatant. In knowing now how socially perceived pictorial spaces are constructed and how fixed as well as connected the subject on Zoom is, we should further consider which forms of communication and social relationality emerge from it.

Hence, I prefer the conception of suture which had been developed by Susanne Lummerding (2005) in her book on agency with its political dimensions of technological visions. Lummerding evaluates Oudart's film-theoretical conceptualization of the relation between on- and off-screen spaces as a genuinely *discursive* term. Drawing now on Chantal Mouffe, Ernest Laclau, and Jacques-Alain Miller, Lummerding understands suture as the relation of a subject to the network of a discourse, letting questions about visibility as well as positionality appear as political *negotiations*. According to Lummerding, suture as theoretical concept always becomes crucial again when new technologies redefine our perception and experience. Thus, its politicalness is marked by its possibility to negotiate meaning, *through* new configurations of present and absent spaces, along the media's promise to capture reality (161). This understanding of suture as a process negotiating the political calls into question how strong tools of fixation and variability, visibility and invisibility, participate in framing us as socially located subjects.

In this perspective, in the last section of my article, I want to go back to the Zoom pranks and fails I described in the beginning as empowering examples of a (self-reflexive) negotiation of video conferencing's fixed framings. I therefore stick to Sascha Dickel again who points to the genuine advantages that virtual conferences bring, such as the maintenance of political "zones of informality" in all those parts of the picture that remain invisible (Dickel 2020, 83). The consideration of those "zones of informality" is inspiring for turning my thoughts about the aesthetics of the fixed view and the spectator as user to a more political reading of video conferencing infrastructures—a political reading that is based on all the presets I have discussed until now.

Politics of (In)Visibility as Zones of Informality

Also according to Katharina Block and Michael Ernst-Heidenreich (2020), the corona crisis is characterized particularly by the corresponding *in*visibility, *non*availability, and *in*accessibility of social spaces and the world as a whole. In this respect, it can be assumed that knowledge practices need to be reconfigured through new relata of the spatial in sociodigital environments. Thinking about suture in video conference dispositives, therefore, can mean to consider empowering zones of informality.

And maybe this is exactly what happens discoursively in my two starting examples, in Grubbs's idea and Reeve's fail. Both violated or declined some spatial "rules" of video conferencing programs: Reeve applied clothing etiquette only to his visible frame—and nowhere else. With this, he made it thinkable for others to also keep up zones of informality that they as private persons would "deserve" even during publicly screened private zones. The comments he got prove how welcoming and liberating his presumed faux pas had become. Samuel Grubbs's group took hierarchically focused attention away from the teacher, the break of the frame here seemed to create a feeling of collectivity lacking students in pandemic times. Important is that both cases became popular examples of "playing around" with new and old spatial structures. In their behavior, both Grubbs and Reeve not only emancipated themselves from the space provided by the screen and digital tool but also collected many encouraging and sympathizing reactions from others—as if they fulfilled wishes people enjoyed testing out while public spaces were locked down and amid a lack of face-to-face contact. Technically put, they seem to have done their fail and prank only because *they were (sociospatially) able to.*

In an environment of the private as a merely public space, Grubbs's and Reeve's derailment have gained back the twofold function Cavell assigned the screen: its abilities to connect people and to shield them indeed from each other by maintaining undefined/invisible areas of contingency and uncertainty. Both ways of dealing with the fixed view of video conferencing formats means acquiring agency as well as literally turning around the original concept of suture as imaginary placeholder of the subconscious. Rather, Grubbs's class and Reeve face the tightness of Zoom's tile arrangement with a very visible habitus of ambiguity, taking responsibility for the single and connected new *shared* spatiality they insert in "official" video conference meeting—even though, in Will's case, the incident was not caused on purpose. It is right there between visibility and invisibility, on-screen and off-screen, private and public space that this formerly *fixed agency* is negotiated again. Or finally, as Chiara Cappelletto (2020) puts it,

> We need to queer the dominant narrative and finally abandon the regime of "natural iconicity," with its divide between presence and absence, which has unfortu-

nately withstood decades of academic studies; we need to think about how embodied and gendered minds perform freely in spaces where the affective values of near and far have taken on enormous political relevance.

To start approaching questions of "near and far," of closeness and distance, and of "right" and "wrong" in digital public spaces, it is worth reflecting on destabilized categories and framings of the (in)visible and on sociospatiality in video conferencing in general. At the same time, we should keep in mind Naomi Klein's abovementioned impulse to be conscious that staying home is as much of a privilege as appearing in a Zoom tile. Because everyone who does not possess (anymore) the soft and hard skills that digital environments require or whose jobs bind them to reality and to an enhanced risk of getting sick from Covid is excluded from doing so. So, also: negotiating sociospatiality is a privilege. But in facing the fact that media environments will definitely not leave us alone in the future, complaining about their inability to replace analog worlds and their withdrawal from face-to-face spaces seems anachronistic and pointless. Instead, this article has proposed some theoretical linkages to aesthetic concepts that connect technological and aesthetic forms to their discursive role of communication and society. Thinking about the negotiations of those new sociospatialities can offer a starting point also for looking beyond our familiar onto- and epistemological categories.

References

Alexander, Neta. 2020. "The Waiting Room: Rethinking Latency after COVID-19." In *Pandemic Media: Preliminary Notes toward an Inventory*, edited by Philipp Dominik Keidl, Laliv Melamed, Vinzenz Hediger, and Antonio Somaini, 25–32. Lüneburg: Meson Press.

Alloa, Emmanuel. 2020. "Coronavirus: A Contingency That Eliminates Contingency." *Critical Inquiry* 47 (2), April 20, 2020. https://critinq.wordpress.com/2020/04/20/coronavirus-a-contingency-that-eliminates-contingency (accessed: October 26, 2022).

Ayers, Bill. 2020. "OK, Zoomer!" *Critical Inquiry* 47 (2), April 27, 2020. https://critinq.wordpress.com/2020/04/27/ok-zoomer (accessed: October 26, 2022).

Bailenson, Jeremy N. 2021. "Nonverbal Overload: A Theoretical Argument for the Causes of Zoom Fatigue." *Technology, Mind, and Behavior* 2 (1). https://doi.org/10.1037/tmb0000030.

Block, Katharina, and Michael Ernst-Heidenreich. 2020. "Das pandemische Unverfügbarwerden von Welt: Zeitdiagnostische Überlegungen zum sozialtheoretischen Denken in ökologischen Zusammenhängen." *Soziologie und Nachhaltigkeit*, 73–83. https://doi.org/10.17879/sun-2020-2941.

Cappelletto, Chiara. 2020. "Arguments for a New Aesthetic of Presence." *Critical Inquiry* 47 (2), May 13, 2020. https://critinq.wordpress.com/2020/05/13/arguments-for-a-new-aesthetic-of-presence (accessed: October 26, 2022).

Creeber, Glen. 2013. *Small Screen Aesthetics: From Television to the Internet*. Bloomsbury: Bloomsbury Publishing.

De Silva Machado, Guilherme. 2020. "Zoom in on the Face: The Close-Up at Work." In *Pandemic Media: Preliminary Notes toward an Inventory*, edited by Philipp Dominik Keidl, Laliv Melamed, Vinzenz Hediger, and Antonio Somaini, 195–204. Lüneburg: Meson Press.

Denson, Shane. 2020. "'Thus Isolation Is a Project.' Notes toward a Phenomenology of Screen-Mediated Life." In *Pandemic Media: Preliminary Notes toward an Inventory*, edited by Philipp Dominik Keidl, Laliv Melamed, Vinzenz Hediger, and Antonio Somaini, 315–24. Lüneburg: Meson Press.

Dickel, Sascha, 2020. "Gesellschaft funktioniert auch ohne anwesende Körper: Die Krise der Interaktion und die Routinen mediatisierter Sozialität." In *Die Corona-Gesellschaft: Analysen zur Lage und Perspektiven für die Zukunft*, edited by Michael Volkmer and Karin Werner, 79–88. Bielefeld: transcript.

Dickson, Edward M., and Raymond Bowers. 1974. *The Video Telephone: Impact of a New Era in Telecommunications: A Preliminary Technology Assessment*. New York: Praeger Publishers.

Distelmeyer, Jan. 2021. "Programmatische Verhältnisse." *Cargo* 49 (March): 28–34. https://www.cargo-film.de/heft/49/essay/programmatische-verhaltnisse (accessed: October 26, 2022).

Dogruel, Kerim. 2020. "Face Off." In *Pandemic Media: Preliminary Notes toward an Inventory*, edited by Philipp Dominik Keidl, Laliv Melamed, Vinzenz Hediger, and Antonio Somaini, 279–86. Lüneburg: Meson Press.

Gerfen, Wilfried. 1986. *Videokonferenz Alternative für weltweite geschäftliche Kommunikation—ein Leitfaden für Anwender*. Heidelberg: R. V. Decker.

Habermas, Jürgen. 1982. *Theorie des kommunikativen Handelns*, Band 2: Zur Kritik der funktionalistischen Vernunft. Frankfurt am Main: Suhrkamp.

Hagener, Malte. 2020. "Wir sind beobachtete Beobachter." Interview by Marco Kasang. *Spiegel Kultur*, April 21, 2020. https://www.spiegel.de/kultur/kino/splitscreen-realitaet-wir-sind-beobachtete-beobachter-a-3cc0d866-14fe-4315-a9f2-f1649a509c45 (accessed: October 26, 2022).

Hall, Edward T. (1966) 1990. *The Hidden Dimension*. New York: Anchor Books.

Hanich, Julian. 2016. "Experiencing Extended Point-of-View Shots: A Film-Phenomenological Perspective on Extreme Character Subjectivity." In *Subjectivity across Media: Interdisciplinary and Transmedial Perspectives*, edited by Maike Sarah Reinerth and Jan-Noel Thon, 127–144. London: Routledge.

Harteveld, Maurice. 2020. "Domestication Will Shape Future Public Spaces: A Report from Rotterdam." *The Journal of Public Space* 5 (3): 53–66. https://doi.org/10.32891/jps.v5i3.1379.

Held, Tobias. 2020. *Face-to-Interface: Eine Kultur- und Technikgeschichte der Videotelefonie*. Marburg: Büchner.

Jay, Martin. 1999: "Scopic Regimes of Modernity." In *The Visual Culture Reader*, edited by Nicholas Mirzoeff, 66–69. London: Routledge.

Kirsten, Guido. 2019. "Distanz wahren Proxemik und politische Semantik in Bad Day at Black Rock." *Montage AV* 28, 43–62.

Klein, Naomi. 2020. "How Big Tech Plans to Profit from the Pandemic." *The Guardian*, May 13, 2020. https://www.theguardian.com/news/2020/may/13/naomi-klein-how-big-tech-plans-to-profit-from-coronavirus-pandemic.

Knoblauch, Hubert, and Martina Löw. 2020. "Dichotopie. Refiguration von Räumen in Zeiten der Pandemie." *SFB 1265*, March 30, 2020. https://sfb1265.de/blog/dichotopie-refiguration-von-raeumen-in-zeiten-der-pandemie (accessed: October 26, 2022).

Lim, Sun, and Yang Wang. 2021. "Lessons from Our Living Rooms: Illuminating Lockdowns with Technology Domestication Insights." *Journal of Children and Media* 15 (1): 17–20. https://doi.org/10.1080/17482798.2020.1858909.

Lovink, Geert. 2020. "The Anatomy of Zoom Fatigue." *Networkedcultures*, November 2, 2020. https://networkcultures.org/geert/2020/11/21/anatomy-of-zoom-fatigue/?pdf=2166 (accessed: October 26, 2022).

Lummerding, Susanne. 2005. *Agency@? Cyber-Diskurse, Subjektkonstituierung und Handlungsfähigkeit im Feld des Politischen*. Wien, Köln, and Weimar: Böhlau.

Martin, Adrian. 2014. *Mise en Scène and Film Style: From Classical Hollywood to New Media Art*. Basingstoke: Palgrave Macmillan.

Persson, Per. 2003. *Understanding Cinema: A Psychological Theory of Moving Imagery*. Cambridge: Cambridge University Press.

Sayman, Volkan. 2020. "Corona und der öffentliche Raum." *SFB 1265*, April 17, 2020. https://sfb1265.de/blog/corona-und-der-oeffentliche-raum (accessed: October 26, 2022).

Stevens, Kyle. 2020. "When Movies Get Sick." *Critical Inquiry* 47 (2), March 25, 2020. https://critinq.wordpress.com/2020/03/25/when-movies-get-sick (accessed: October 26, 2022).

Watson, Ash, Deborah Lupton, and Mike Michael. 2021. "The COVID Digital Home Assemblage: Transforming the Home into a Work Space During the Crisis." *Convergence: The International Journal of Research into New Media Technologies* 27 (5): 1207–1221. https://doi.org/10.1177/13548565211030848.

Wittmann, Matthias. 2020. "Vom Ende der 'Suture.' Zeit- und Raumformen im 360° Film." *360storytelling*, January 2020. https://www.360storytelling.org/de/zeit-und-raumformen-im-360-film.

Eye Contact with the Machine

Gaze Correction in Video Conferencing

Robert Rapoport and Vera Tollmann

During video-mediated communication it is not possible to look the person on screen in the eyes. While the problem is simple, the solution is not. Attempted remedies have variously been called *attention correction*, *gaze redirection*, or *gaze correction*. Gaze correction, on a basic level, seeks to fix the problem of a lens acting as a proxy for the eye. During a video call, one cannot simultaneously look at the screen and (the laptop/smartphone) lens, creating a dissonance: *Will we produce eye contact for ourselves or our conversation partner?* Since the pandemic's beginning, many have experienced the "uncorrected gaze" of video-mediated communication, metabolized by some as "Zoom fatigue" (see Distelmeyer and Lovink in this volume). The goal of gaze correction, succinctly stated, is "to digitally alter the appearance of eyes in a way that changes the apparent direction of the gaze" (Ganin et al. 2016). It could be argued that gaze correction, narrowly applied, simply fixes a mechanical problem: we have to look at a screen instead of a lens when in a video call. But eye contact is so foundational to communication, its meaning so tacit, that unintended consequences need to be considered.

Social cognition, our ability to make sense of the behavior of others, relies on both linguistic and nonlinguistic cues. The consensual imperative of unmediated eye contact stimulates the release of the neurotransmitter oxytocin. Studies of video conferencing have found that this phenomenon does not work as well on screen (Auyeung et al. 2015). Moreover, neuroscience has recently identified concrete physiological effects of live rather than prerecorded screen-mediated eye contact (Noah et al. 2020). Near-infrared spectroscopy found increased coherence in brain activity during *live* eye-to-eye contact. Additional studies have described the reciprocal prelinguistic processing of information as the "rapid and reciprocal exchange of salient information in which each send and receive 'volley' is altered in response to the previous [signal]" (Hirsch et al. 2017, 314). Thus, if eye contact is essentially *relational*, gaze correction in its current form removes this relationality, delegating the task of eye contact to a platform.[1] To build on the metaphor of the volley: gaze correction could be akin to watching a game of tennis in which every shot is "corrected" so as to be returnable. Such a game, while graceful for a time, would quickly become lifeless for viewer and spectator alike.

Gaze correction will present us with a choice: Do we perform the essential emotional labor of maintaining eye contact, or do we delegate it? If two people do consent to have a platform mediate eye contact for them, what metric will be used by the platform to decide when contact should be stopped and started? On a social level, will this produce a space in which we are constantly second-guessing eye contact? Might gaze correction create affect-specific platforms in which users can choose between styles of eye contact and thus attention (e.g., engaged, skeptical, or rapt)? This would continue the logic of "style transfer" (the practice of transposing qualitative metrics between images) that already exists on various social platforms (see figs. 7 and 8). When one is able to simply select different styles of eye contact, gaze *optimization* rather than *correction* would seem more apt.

Computational optimization involves increasing the efficiency of a system while decreasing its resource use (Sedgewick 1984, 84). The questions for machine learning-based gaze correction is what the primary object of optimization will be? If Zoom optimized for attention rather than fidelity to the user's intent, what kind of social space would result?

The state of the art in gaze correction involves using a generative adversarial network (GAN) to make real-time modifications to the user's eyes. A GAN consists of two neural networks: a classifier and a generator (fig. 1). The generator's role is to render a constant stream of training images because "the classification constraint does not work without the adversarial learning, or in other words, the adversarial learning helps to avoid adversarial examples" (He et al. 2020). The best-known use of GANs is the deepfake paper from 2016 (Thies et al. 2016). The epistemic threat of this technology has been widely discussed (Fallis 2020). To extrapolate: gaze correction is a real-time deepfake intervening at the most fundamental level of social communication, eye contact. While not as obviously threatening for public discussion as deepfakes, gaze correction nonetheless poses a similar epistemic threat (fig. 8). Jonathan Crary has written that the "reductiveness" of machine learning causes "splintering of the interhuman basis of a shared social reality" (Crary 2022, 92). Widespread gaze correction adoption would render it particularly difficult to track instances of such splintering precisely because we lack the vocabulary to talk about

1 At the time of writing, the authors were unable to find research on how this relational quality of eye contact would be encoded in gaze correction. We speculate that this is due to the lack of a training dataset. Such a dataset would only be produced if/when gaze correction were widely adopted. Widespread adoption of a gaze correction program without some mechanism to account for the consensuality of gaze could produce unintended effects.

such unprecedented mediation. General terms like "Zoom fatigue" are symptomatic of this lack of vocabulary. Bailenson (2021) has argued that video conferencing is tiring in part because our social brain expects a level of feedback that is not forthcoming. While this disconnect is a problem for our social cognition, a platform might see it as a way to build dependence. It is precisely those patterns of social behavior that *precede* language that are most attractive to those who seek to make us click before becoming aware of it. If we are engaging in conversations where the "person" looking at us is in fact a machine learning model, a whole set of ethical and epistemological questions follow. Again, the opposition between correction (social cognition) and optimization (machine learning) provides a shorthand to think through this shift. There is no ruleset for when to maintain eye contact, though if gaze correction is widely adopted there functionally will be, even if known only to the implementing platform.

Gaze correction has been researched since the 1990s in both industry and academia. The most familiar attempt to roll out gaze correction was Apple's Eye Contact setting in iOS 13 (Bohannon 2013, 177). GPU producer Nvidia has attempted to take the technology a step further by redrawing not only eyes but also the interlocutor's head (Wang et. Al 2021). In what follows, we refer primarily to Zhang et al. (2020) as representative of the broad trend to correct gaze computationally rather than mechanically. The paper's GazeGAN model takes the initial live video feed and redraws the participant's pupils in real time (see fig. 4). This technique is called "Image Inpainting ... an important task in computer vision and graphics aims to fill the missing pixels of an image with plausibly synthesized contents" (Zhang et al. 2019).

Figure 1: Overview of GazeGAN network architecture

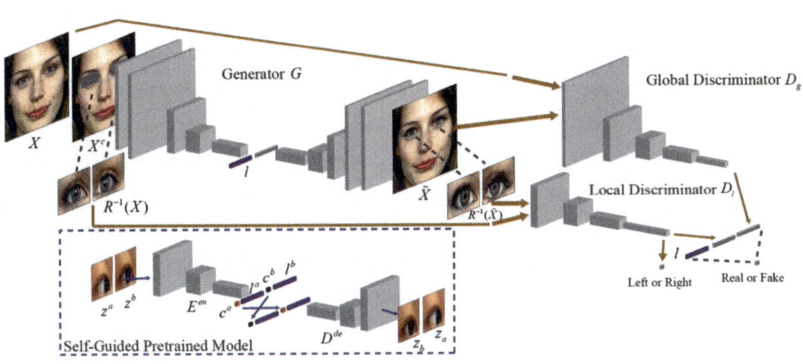

Source: Zhang et al. 2019.

Figure 2: Architecture of GazeGAN "in-painting" model

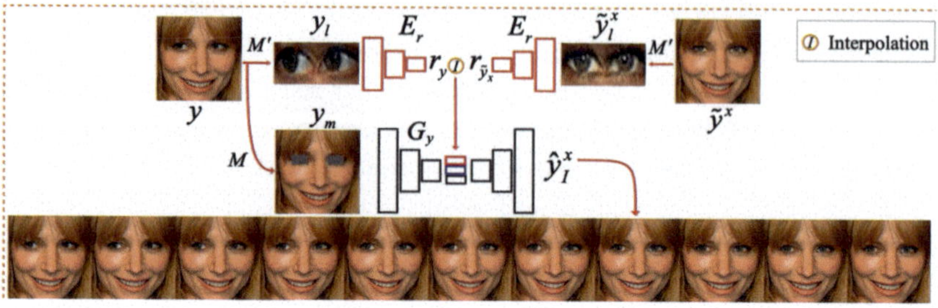

Source: Zhang et al. 2020.

Zhang et al.'s focus is primarily on how realistic the redrawn ("in-painted") pupils and eyes are (fig. 2). This concern for realism, however, does not extend to the social realm. The GazeGAN model cannot recreate the cadence of mutual eye contact, which includes involuntary looking away and other subtleties of joint attention (Bohannon et al. 2012). Such relational interpersonal behavior is difficult to encode. Our decision to maintain eye contact depends on tacit knowledge of the other that we are often unaware of. What would such a training set look like? To date, there exists no training the authors could find. This is not to say that there could not be several diverse ones. Such training sets would have to index qualitative values like empathy, which produce believable, authentic eye contact in lived experience. In a way, video conferencing sessions with machine learning interventions could prevent harmful, unsettling behavior, such as contemptuous looks and eyes full of doubt.

Hardware Antecedent to the *Split Gaze*

To understand the possible effects of gaze correction, we can look to one of its antecedents: the teleprompter, which shifts the eyes mechanically as opposed to computationally. In the early years of broadcasting, the teleprompter was developed as an aid for presenters of all types in the industry. In the postwar years, only professional film studios could afford the necessary hardware. Initially, the problems of the teleprompter were similar to those of twentieth-century video conferencing: *to align the display with the optical path of the camera*. The first "television prompting apparatus," patented in 1953, replaced a human prompter (who would read the text and give vocal cues when needed) (fig. 3).[2] The first generation of devices produced by the

2 Barkau, Fred H. 1953. Television Prompting Apparatus, US Patent 2,635,373A, filed April 21, 1949, and issued April 21, 1953.

TelePrompTer Corporation were attached to the side of the camera lens, making the presenter's gaze oblique to the viewers. Audiences at home could sense the presence of the mediating apparatus. After several iterations of the device, the shifted gaze of the presenter was effectively "corrected." In 1959, Jess Oppenheimer patented the first in-camera teleprompter, which used mirrors to project the script in front of the lens.[3] Below is a diagram from the initial patent.

Figure 3: Original patent diagram from Fred H. Barkau, 1953

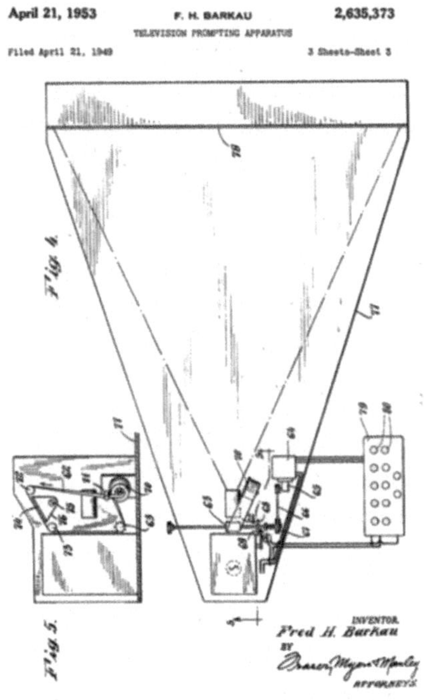

Source: Fred H. Barkau, 1953. Television Prompting Apparatus, US Patent 2,635,373A.

The mechanical gaze correction of teleprompters thus allowed a presenter to appear to look the individual viewer in the eye, laying out the tension more completely realized in computational gaze correction: it simulates eye contact while allowing the speaker to read from a script. The news broadcast, for example, gained a new

3 Oppenheimer, J. 1959. Prompting Apparatus US Patent 2,883,902, filed Oct. 14, 1954, and issued April 28, 1959

intimacy of address through this technical trick. And yet their mediation simultaneously created a power imbalance: the corrected gaze of the broadcaster meant the news was internalized using, to some degree, social cognition. The teleprompter thus effaced a one-to-many flow of information. Might gaze correction create a similar dynamic for users of video conferencing applications? What happens in a situation in which every user can feign a direct gaze? What subtler prompts might this assemblage be capable of?

Neural Nets, Datasets, and Gaze Correction

The foundations for today's developments in gaze correction were laid in 2009 with the publication of ImageNet—the largest ever training data set for machine vision (Deng et al. 2009). Because of its size and granularity ImageNet greatly accelerated research in the field. In 2012, the ImageNet Large Scale Visual Recognition Challenge (ILSVRC) was won by a team from the University of Toronto using a neural network (Hinton et al. 2012). This watershed moment is considered the start of the widespread use of neural nets in machine learning research. So many papers used ImageNet for training models that inherent biases in how it labeled images created ripple effects throughout areas of society (see Crawford and Paglen 2019).

The other lynchpin in the development of contemporary gaze correction was the publishing of the Columbia Gaze Data Set in 2013 (see fig. 6). The data set provides a systematic collection of 5,880 images of fifty-six people over five head poses and twenty-one gaze directions. It marked the most comprehensive and diverse data set related to gaze correction to date, even though it was initially created at Columbia University for training purposes in human-object interaction, specifically, "to train a detector to sense eye contact in an image" (Smith et al. 2013, 271). This data set, coupled with the learning neural nets, has served to accelerate research on gaze correction. Indeed, the research by Zhang et al. (2020) discussed above relies on both. Prior to the Columbia Gaze Data Set, data sets were generally scraped from the web rather than purpose built. These data sets attempt to model a mode of communication that is not situated. In a sense, the corrected gaze appears to employ the long-criticized "view from nowhere." Dan Kotliar has pointed out that the process of dataset compilation often glosses over the fact that users exist between "very distant localities" (2020, 922). This same tendency is evident in the training sets for gaze correction.

Apart from the cultural and geographical context, on a data level, as Taina Bucher (2020) elaborates with Jean-Luc Nancy's concept of "being as always already a being-with," everyone is already existing in a connected state. By applying Nancy's concept to the social networks, she argues that besides active users, offline actors are also connected with others (i.e., as contacts). This way, even those who are not logged in leave data traces, since machine learning algorithms classify, recommend, and match according to data input. In a different publication, Bucher (2021, 100) refers to these passive data doubles as "non-users." In comparison to the algorithmic layers relevant for gaze correction, training data sets of eyes and faces, such as the Columbia Gaze Data Set, are the ground truth that connects any one pair of eyes with another. While participating in video conferences or casual group calls with eye contact enabled, users might eventually experience the state of "being singular plural" in a rarely visible way.

Social Cognition, Gaze Direction, and Video Conferencing

Eye contact is the infant's first step toward building social cognition. While following the gaze of our parents, we begin to understand how the world appears to the other (Frith 2010; Lamm et al. 2016). "The ability to process its [gaze] direction starts very early in life and is fundamental for the development of normal social cognition" (Itier and Batty 2009, 845). Some researchers have even argued that our theory of mind begins with gaze tracking (Farroni et al. 2002). Gaze direction—and the attention it engenders—is so tacit in our making of social meaning that it is difficult to reflect on, creating a "non-verbal interpersonal communication including salient and emotional information" (Hirsch et al. 2017). And yet in video conferencing, the uncorrected gaze appears without the cues we expect. "Zoom fatigue" has been attributed in part to the fact that the mediated gaze is "perceptually realistic, but not socially realistic" (Farroni et al. 2002). As this dissonance is prelinguistic, how will we metabolize its predictive "correction" by machine learning? Put differently, is there a deep patterning to eye contact that machine learning will be able to "see" in a way we cannot? The "cooperative eye hypothesis" argues that the particular morphology of the human eye—compared to other primates—facilitates higher levels of joint attention. The whites of human eyes enable us to read more intention from the eyes of others, facilitating cooperation. Having a gaze legible to others is thus intrinsic to our humanity (Tomasello et al. 2007).

Eye contact is rarely discussed in procedural terms, but machine learning-based approaches to gaze correction will require as much. At the front end, or "surface," these systems learn to redirect pupils in real time. "The screen is the surface, the display buffer is the subface of the algorithmic thing that the two of us—we ourselves and the program—are engaged in. The algorithmic thing comes as a visible appearance for us" (Nake 2015, 106). Gradually, turning on one's webcam may signal tacit acceptance of a new intimacy with machine learning. The delicacy of this moment is evident in the fits and starts with which the technology has been approached by companies as formidable as Apple. Apple's Eye Correction feature brought gaze correction to FaceTime in iOS 14 (Peterson 2020). First iterations were tested under the label of "attention correction." The feature's name was later changed to the more benign Eye Contact, perhaps in a bid to sound less disciplinary. A cursory search for this feature will surface many tutorials about disabling this "creepy" new feature. If Eye Contact returns to iOS—and it likely will—the questions in this chapter will become much more immediate.

To apply the language of computation to the social world, we might say that humans are evolutionarily optimized to read the gazes of others. Each culture has differing tacit norms around eye contact, all optimizing for different outcomes, stochastically. Would an iOS-based gaze model account for cultural specificity? As ever, this would be a hard system to reverse engineer. Thus, video conferencing applications equipped with eye tracking and gaze correction would become obvious places to study the connections between gaze and more complex behaviors that have been intuited but not modeled. Once a model is developed, it would be hard for platforms to resist manipulating eye contact for specific outcomes.[4] The human infant recruits caretaking behavior from its parents using gaze. Could a platform do the same, using gaze correction as a means of inducing bonding between users, and by extension with the platform itself? While our question is motivated by social concerns, companies in platform capitalism follow market rules and advances in AI. (For example, Zoom started using user content to train their AI in July 2023. Zoom offers a virtual assistant and AI generated content such as conference summaries).

Indeed, biometric signals are already creating new approaches in behavioral health as they can help predict illness before it appears clinically (Bednarik et al. 2005). This medical approach to eye tracking overlaps with the larger field in behavioral medicine called digital phenotyping. The technique argues that the data traces left on the touch screen (and the other sensors on the smartphone) can be correlated to neurological states (Insel 2017). Insel's article does not list eye tracking, but its capture could follow the same logic. One can imagine that a society built on video-mediated communication would quickly produce a granular model of gaze direction that could be correlated to any number of other biometrics—as digital phenotyping attempts to do with thumbs, voice, and speech. The long-term population-level effects of such a data set would be profound. A data set of eye movements during video conferences could yield many insights for digital phenotyping (Nag et al. 2020).

4 For example, in a video conference about contractual negotiations, a platform could "optimize" gaze correction to nudge both parties toward agreement. This is a fairly heavy-handed example, but the subtler behaviors that gaze correction could model are abundant.

Theories of the Gaze

In some cultures, the gaze is perceived as an agent in its own right, particularly in the archetype of the "evil eye" (Breuer 2015). The term, widespread since antiquity, describes the belief that a gaze can inflict harm. One can "cast an eye" with the intent to punish someone. De-potentialization and fluidity of "negative" and "positive" looks characterize the modern discussion of the evil eye; where tradition is abstracted, which is one of the prerequisites for its circulation (van Loyen 2015). In this sense, eye contact with the machine could literally mean being subject to the agency of machine vision, adding a new case to the splitting of eye and gaze, the matter of making visible and being seen in public, studied by Sartre as well as Goffman. At different times and with regard todifferent media, Barthes and, later, Silverman investigated the mediated gaze.

In a memorable scene, Sartre depicts momentarily forgetting one's own body in the immersion of the clandestine gaze (Sartre 1956, 277). A person squats in front of a keyhole to watch others inside a room. The sound of footsteps in the hallway makes him feel "suddenly struck in his being." At this moment, the observer sees himself all at once from the outside, as a voyeur. Sartre suggests: the one who is looked at becomes acutely aware of having a body. Here the gaze has an activating function. In video conferencing, by contrast, each participant is constantly self-aware due to the mirror-like self-video.

When gazed upon, we are objectified, aware that we are for the other an object of consciousness; this triggers a new, reflexive relationship to ourselves. For Sartre, the gaze is existential. The existence of those looked at manifests itself from another point of view. What does it mean, then, when in video conferencing this reciprocal balance is delegated to machines?

There is a crucial difference between the perception of Sartre's individual gaze as an agent in an analog space and the unified space of a corrected gaze that glosses over individual differences. We may become more cognizant of our individual habits once gaze correction throws them into relief. Following Sartre's train of thought, then, we can interpret the GAN as a proxy for the real-world situation described by his phenomenological approach. In the particular case of machine learning–based eye contact, the structures of perception—the spectator's, the camera's, and the GAN's gazes—are all being shifted into a kind of "proxy politics" (Steyerl 2014). Communicating using GAN-augmented eyes means inhabiting the proxy.

Erving Goffman compared social life in public space to a theater play. People present themselves on the basis of roles. In "On Face-Work: An Analysis of Ritual Elements of Social Interaction" (1955), Goffman studied how communication routines were shaped by the media. He defined face-work as "the positive social value a person effectively claims" (Goffman 1967, 5). Goffman continues: "One's own face and the face of others are constructs of the same order; it is the rules of the group and the definition of the situation which determine how much feeling one is to have for a face and how this feeling is to be distributed among the faces involved" (Goffman 1967, 6). What does video conferencing do to our ability to perform face-work on the social level? Goffman might draw a conclusion similar to Sartre's: gaze correction does not restore this order, which video-mediated communication disrupts.

Whereas both Sartre and Goffman were reading unmediated gazes *en plein air*, Barthes focused on the representation of gaze in photography. In Barthes's view, what defines the gaze is its excessiveness. In "Right in the Eyes" ([1977] 1991), he suggests that science interprets the gaze in three combinable ways: as information, as relationships, and finally as possession. Accordingly, Barthes differentiates three functions: an optical, a linguistic, and a haptic role of the gaze, making it an unsteady sign. Barthes describes a person who appears to be looking straight at the spectator or photographer. He elaborates: "In reality, the portrait looks at no one, and I know it; it looks only into the lens, that is, into another, enigmatic eye: the eye of truth" ([1977] 1991, 240). The camera, like the GAN, seeks to self-efface. The GAN performs elaborate artifice in the name of making the medium disappear. We do not look into a lens, into an apparatus, but at the screen displaying a conversation partner.

In *The Threshold of the Visible World* (1996), Kaja Silverman develops the concept of the "productive look," which can see through the political motivations in images. She takes Farocki's film *Images of the World and the Inscription of War* (1989) as a starting point to consider the camera's departure from human vision. Aerial photographs intensified the move toward quantifiable images: "From this vantage point, the invention of the camera represents less a moment of rupture with earlier visual technologies than the moment at which their implicit disjuncture from the eye becomes manifest" (Silverman 1996, 143). The quantifiable image thus brought us closer to a paradigm in which machines not only model but act on increasingly unstructured data. Kotliar describes "algorithms' ability to characterize, conceptualize, and affect users" (2020, 919). The algorithmic gaze is thus "a multifocal one—as a gaze that stems from a complex combination of diverse types of lenses" (934). And here, lenses are employed in the metaphorical sense as a new instance of the *split gaze*: looking at the screen and the camera at the same time. Returning to Barthes, the gaze here becomes more about automation than relationship; eye contact is reduced to the level of information, becoming a steady sign in the process. Machine learning functionally separates the eye from the gaze, thereby splitting bodies from behavior while producing a "surveillant assemblage" (Haggerty and Ericson 2000). While Deleuze and Guattari criticize the "cementing" of the human face through social coding, gaze correction appears as an automated update. The space opened up by the intensities that build up in the interpersonal sphere short-circuit.

More recently, researchers at Strelka Institute have coined the term "peak face" to describe the declining reliability of the face as a source of information (Abbott et al. 2021). In the case of video-mediated communication, what might move into the void left after peak face? Might vocal intonation or chat come to the fore?[5]

In *The Right to Look*, Mirzoeff claims that "the right to look is not about seeing. It begins at a personal level with the look into someone else's eyes to express friendship, solidarity, or love. That look must be mutual, each person inventing the other, or it fails. As such it is *unrepresentable*" (2011, 1). Mirzoeff treats visuality as a tool of power and "a discursive practice that has material effects" (2011, 3).[6] With gaze correction however, we witness a new take on visuality and power, in which algorithms become disciplinary apparatuses[7]—exercising cultural practices such as sorting and classifying and therefore changing human knowledge and social experience. Building his argument on a decolonial discourse, Mirzoeff concludes that "classifying, separating, and aestheticizing together form … a 'complex of visuality'" (Mirzoeff 2011, 3–4). He continues: "The *right to look* claims autonomy from this authority, refuses to be segregated and spontaneously invents new forms" (2011, 3–4). The representation of the gaze in video conferencing could be read as a result of digital authority since it might determine the nonverbal communication of two participants or several participants in a group chat. In the physical world, eyelines establish a real space that cannot be recreated on a flat screen. In a different text, Mirzoeff sees the increased surveillance in the digital sphere as a state response to "the anxiety that imperial subjects might start to think and act in common. Because prior to all law, there is a relation between people. We move from the social to the individual" (2014, 228). Mirzoeff points out that in the digital sphere, the common is repressed in favor of control. Therefore, he argues based on neuroscientific findings about "mirror neurons" that relations between people cannot be represented, since *live* eye-contact is a condition.

The concept of machine vision doing things in the world instead of seeing is further elaborated by Luciana Parisi with the idea of "negative optics." In "Negative Optics in Vision Machines," Parisi explores how the automation of vision seems to "re-

5 Secondary signifiers such as voices—female voices in particular (Siegert and Niebuhr 2021)—are perceptually affected by speech signal compression, and the chat might be the last component where machine learning does not yet discriminate—as long as chatbots are not involved in the conversation (Nvidia, for example, integrates chatbots in their Conversational AI program). While, on the other hand, the industry argues for those modifications in terms of accessibility for disabled participants, it becomes evident once again that technology genders, surveils, and modulates behavior.

6 Relating to this approach, video conferencing could be considered as a democratizing tool, particularly in times of social distancing, since it enables mediated face-to-face contact (if not limited by digital divide).

7 Several media studies researchers have expressed their concern that algorithms exercise too influence over social realities (Beer 2009; Bucher 2012; Gillespie 2014; Kitchin and Dodge 2011).

define vision in terms of a mediatic function that does not rely on light" (2020, 1281). Furthermore, she speaks directly to the production process of GANs: "the invisible image of machines is part of alien epistemologies," training a model on generative artifice (1283). This moves us away from the binary of visible and invisible such that "image feedback is no longer assured by the interaction with the world" (1281). Thus, Parisi leaves the machinic eye tied to non-generative media behind, which was at its time of conceptualization bound to vision.[8] In Parisi's view, reflecting on GANs means reflecting on a new kind of "vision machine" that "generates an avalanche of inputs" divorced from the real world. A video conference dependent to some degree on the alien epistemology of the GAN may cast "eye contact" in a different light—beyond Western ocularcentrism.

An Augmented Social Reality?

Gaze correction is a subtle form of augmented reality, creating an interpretive ambiguity for the user. If eye contact to date could include only two people, how will we metabolize a third presence? In what is commonly called augmented reality, virtual objects are usually clearly distinguishable from the physical environment onto which they are projected. However, an inherent risk of AR technology is that "there may be people interested in misleading us by creating virtual objects which are disguised in the world scene as if they were real objects" (Ariso 2017, 7). Insofar as gaze correction requires the technique to efface itself, it creates an interpretive grey area around eye contact in general.

The augmented gaze is different from augmented reality applications we are familiar with. Augmented reality systems such as the ones studied by Tolentino (2019) and Wegenstein (2010, 29) in the context of TikTok and Snapchat face filters are self-referentially worn on the human body. Google Glasses also function in this self-referential mode with displays responding to the individual user's head movements (Mainzer 2017, 27). Gaze correction changes the reality a video conference participant looks at according to the movement of their partner. Therefore, this machine learning intervention is not interactive for the individual user and distinct from cosmetic augmentations (see fig. 5). The participant cannot parse whether the eye contact is mutual or machine based. Here, we could speak of a covert augmented reality.

8 In the context of drone operation, filmmaker Harun Farocki saw a transition from images representing the material world to what he called the "operational image"—one no longer representing an object but rather serving as a cue for action (Farocki 2004). Artist and writer Trevor Paglen builds on Harun Farocki's operational image for contemporary machine vision with the term "invisible images"—neural net operations based on image data never seen by humans, but nonetheless exercising agency (2016). For Paglen (2016), operations inside the black box are an "invisible world of machine-machine visual culture."

On both ends, this equivocal gaze itself begins to carry less information if we cannot trust the movement of the eyes. Elsewhere in this volume, Lovink reminds us that Zoom's centralized protocol is not the only one available to us. The reciprocity inherent in a peer-to-peer model might metabolize gaze correction without the optimization. In either case, the drive of automated eye contact is to make itself indistinguishable from reality.

Conclusion: Gaze Correction versus Optimization

We have yet to see widespread adoption of gaze correction. If and when this happens, we will each have to decide if we turn it off.[9] What type of user might willingly delegate the maintenance of eye contact to machine learning? Or will the technology insinuate itself as a default? In the emerging discourse on the effects of online learning, a deficit in social learning is frequently cited (Strouse and Samson 2021). The interruption of social cognition created by video-mediated communication will, one can extrapolate, encompass gaze correction. But once the technology is implemented, it would be a matter of turning the dial on a parameter to socialize increasingly reward-based behavior. In the race down the brain stem, we know that some signals in eye contact are functionally preconscious. Thus, to engage in speculation: if the technology is socially assimilated on a given platform, it would be very tempting for that platform to begin optimizing for user retention over the social accuracy of the technique. In the language of industry, this would mean "favoring engagement" over other factors. What happens when two people are having a conference and have fundamentally different experiences of each other than face-to-face because of neural nets optimizing their gaze style? What would be the larger atomizing effects on society?

Were it to be narrowly applied, gaze correction could be a net benefit. Given the power of gaze to alter behavior there are clear multiple profit incentives for platforms to optimize interactions. A user would not have to do anything to appear social but accept the *correction*. However, as we have argued, true correction is a product of social cognition. GANs, by contrast, are systems composed of deterministic functions to favor an output. If that output is eye contact, how much is too much?

Mutual eye contact creates an inherently affective space. The tone of this space has to date resisted capture or encoding. During eye contact there are patterns at play that we *cannot* know and that a platform will nonetheless approximate. This behavioral data is functionally preconscious, placing the results outside of social discourse. The capture of eye contact patterns would thus appear a further step in

9 At the time of writing a cursory search for "ios, facetime, eye contact" reveals many want to disable it.

the black-boxing of the social world by proprietary code. Apple's Eye Contact will not be the last attempt to introduce gaze correction as a default. The question moving forward will be whether we will learn to recognize the change before it becomes functionally invisible. Recall the initial reaction to gaze correction: users googled the iOS 13's Eye Contact in order to learn how to disable it. Returning to Mirzoeff's reflection on the gradual loss of the commons, gaze correction can be seen as one such incursion. In this liminal space we should remember to ask who or what we are making eye contact with.

Figure 4: Qualitative comparison of gaze correction models

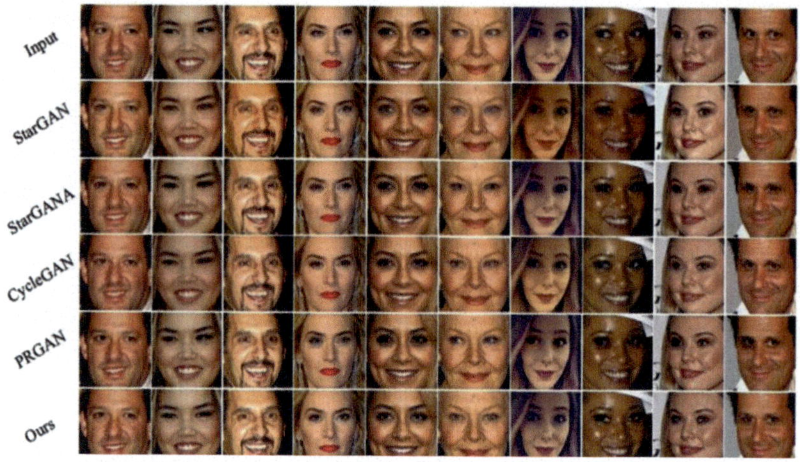

Source: Zhang et al. 2020.

Figure 5: An artistic response to machine vision and the automated gaze

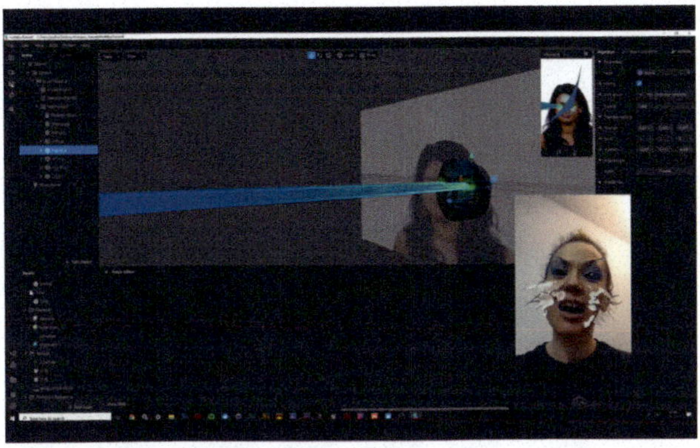

Source: Screenshot from Popp, Alla. 2021. "#alieneffect Face-filter Workshop for Beginners. Spy on Me #2 Online Program." HAU—Hebbel am Ufer, Berlin, accessed August 11, 2022. https://www.youtube.com/watch?v=lkp6zkyYh8Y.

Figure 6: Columbia Gaze Data Set

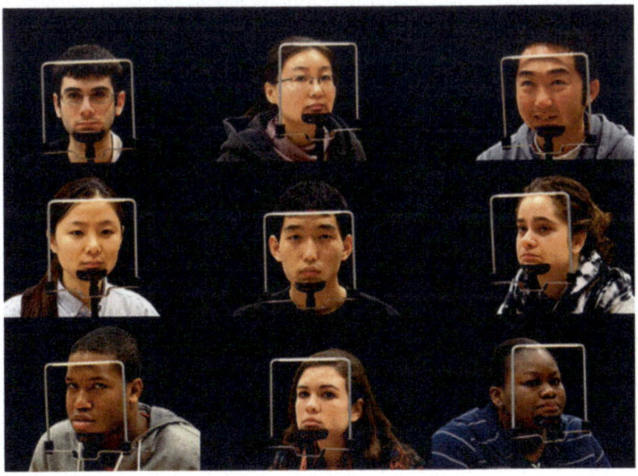

Source: Smith et al. 2013.

Figure 7: Generative adversarial style transfer networks for face aging

Source: Palsson et al. 2018.

Figure 8: An example of a deepfake synthesis procedure

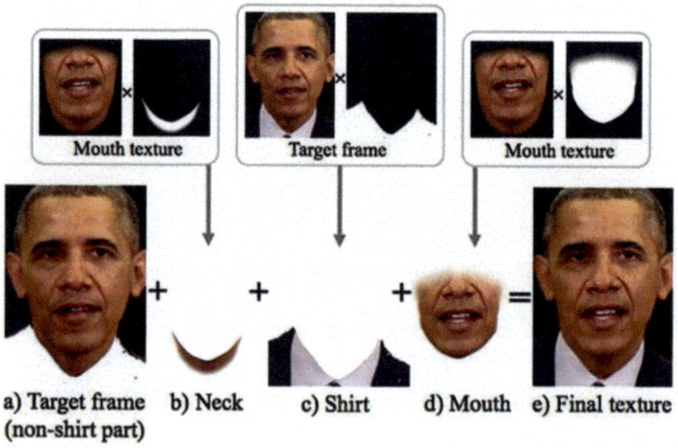

Source: Suwajanakorn et al. 2017.

References

Abbott, M. C., Maria Buey González, and Carl Olsson (Strelka Institute). 2021. "Peak Face." YouTube video, 11:09, August 23, 2021. https://www.youtube.com/watch?v=rcfp25LaWjI.

Ahuja, Abhimanyu S., Vineet Pasam Reddy, and Oge Marques. 2020. "Artificial Intelligence and COVID-19: A Multidisciplinary Approach." *Integrative Medicine Research* 9 (3): 100434. https://doi.org/10.1016/j.imr.2020.100434.

Akechi, Hironori, Atsushi Senju, Helen Uibo, Yukiko Kikuchi, Toshikazu Hasegawa, and Jari K. Hietanen. 2013. "Attention to Eye Contact in the West and East: Autonomic Responses and Evaluative Ratings." *PLoS ONE* 8 (3): e59312. https://doi.org/10.1371/journal.pone.0059312.

Althusser, Louis. 2012. "Ideology and Ideological State Apparatuses." In *Media and Cultural Studies: Keyworks*, edited by Meenakshi Gigi Durham and Douglas M. Kellner, 200–4. Malden, MA: Blackwell.

Ariso, José María. 2017. "Is Critical Thinking Particularly Necessary when Using Augmented Reality in Knowledge Society? An Introductory Paradox." In *Augmented Reality*, edited by José María Ariso, 3–22. Berlin: De Gruyter. http://doi.org/10.26530/oapen_628401

Ashwin, Chris, and Paola Ricciardelli. 2019. "Editorial: Typical and Atypical Processing of Gaze." *Frontiers in Psychology* 10. https://doi.org/10.3389/fpsyg.2019.02576.

Bailenson, Jeremy N. 2021. "Nonverbal Overload: A Theoretical Argument for the Causes of Zoom Fatigue." *Technology, Mind, and Behavior* 2 (1). https://doi.org/10.1037/tmb0000030.

Barthes, Roland. 1991. "Right in the Eyes." In *The Responsibility of Forms: Critical Essays on Music, Art, and Representation*, translated by Richard Howard, 237–242. Berkeley: University of California Press.

Barthes, Roland. 2015. "Auge in Auge." In *Roland Barthes: Auge in Auge*, edited by Peter Geimer and Bernd Stiegler, translated by Dieter Horning, 200–4. Berlin: Suhrkamp.

Bednarik, Roman, Tomi Kinnunen, Andrei Mihaila, and Pasi Fränti. 2005. "Eye-Movements as a Biometric." In *Image Analysis*, edited by Heikki Kalviainen, Jussi Parkkinen, and Arto Kaarna, 780–89. Berlin: Springer. https://doi.org/10.1007/11499145_79.

Bohannon, Leanne S., Andrew M. Herbert, Jeff B. Pelz, and Esa Rantanen. 2013. "Eye Contact and Video-Mediated Communication: A Review." Displays 34 (2): 177–85.

Breuer, Christian. 2015. *"Und wo blicke können tödten ...": Der Malocchio als begehrlicher Augen/Blick*. Berlin: Kadmos.

Bucher, Taina. 2020. "Nothing to Disconnect from? Being Singular Plural in an Age of Machine Learning." *Media, Culture & Society* 42 (4): 610–17. https://doi.org/10.1177/0163443720914028.

Bucher, Taina. 2021. *Facebook*. Cambridge, UK: Polity.

Crary, Jonathan. 2022. *Scorched Earth: Beyond the Digital Age to a Post-Capitalist World*. New York: Verso.

Deng, Jia, Wei Dong, Richard Socher, Li-Jia Li, Kai Li, and Li Fei-Fei. 2009. "Imagenet: A Large-Scale Hierarchical Image Database." *2009 IEEE Conference on Computer Vision and Pattern Recognition (CVPR)*: 248–55.

Ewing, Louise, Gillian Rhodes, and Elizabeth Pellicano. 2010. "Have You Got the Look? Gaze Direction Affects Judgements of Facial Attractiveness." *Visual Cognition* 18 (3): 321–30. https://doi.org/10.1080/13506280902965599.

Fallis, Don. 2020. "The Epistemic Threat of Deepfakes." *Philosophy & Technology* 1 (21): 623–43.

Farocki, Harun. 2004. "Phantom Images." *Public* 2 9: 12–22. http s://public.journals.yorku.ca/index.php/public/article/view/30354.

Farroni, Teresa, Gergely Csibra, Francesca Simion, and Mark H. Johnson. 2002. "Eye Contact Detection in Humans from Birth." *Proceedings of the National Academy of Sciences* 99 (14): 9602–5. https://www.pnas.org/doi/full/10.1073/pnas.152159999.

Fiske, Susan T., and Shelley E. Taylor. 2008. *Social Cognition: From Brains to Culture*. New York: McGraw-Hill.

Frith, Uta, and Chris Frith. 2010. "The Social Brain: Allowing Humans to Boldly Go Where No Other Species Has Been." *Philosophical Transactions of the Royal Society B: Biological Sciences* 365 (1537): 165–76.

Ganin, Yaroslav, Daniil Kononenko, Diana Sungatullina, Victor Lempitsky. 2016. "DeepWarp: Photorealistic Image Resynthesis for Gaze Manipulation." In *Computer Vision—ECCV 2016*, edited by Bastian Leibe, Jiri Matas, Nicu Sebe, Max Welling, 311–26. Cham, Switzerland: Springer. https://doi.org/10.1007/978-3-319-46475-6_20.

Glissant, Édouard. 1997. *Poetics of Relation*. Ann Arbor: University of Michigan Press.

Goffman, Erving. 1955. "On Face-Work: An Analysis of Ritual Elements." *Psychiatry* 18 (3): 213–31. https://doi.org/10.1080/00332747.1955.11023008.

Goodfellow, Ian, Jean Pouget-Abadie, Mehdi Mirza, Bing Xu, David Warde-Farley, Sherjil Ozair, Aaron Courville, and Yoshua Bengio. 2014. "Generative Adversarial Nets." *Advances in Neural Information Processing Systems* 27 (NIPS 2014): 2672–80.

Haensel, Jennifer X., Tim J. Smith, and Atsushi Senju. 2021. "Cultural Differences in Mutual Gaze during Face-to-Face Interactions: A Dual Head-Mounted Eye-Tracking Study." *Visual Cognition* 30 (1–2): 100–15. https://doi.org/10.1080/13506285.2021.1928354.

Haggerty, Kevin D., Ericson, Richard V. 2000. "The Surveillant Assemblage." *The British Journal of Sociology* 51: 605–22.

He, Zhenliang, Wangmeng Zuo, Meina Kan, Shiguang Shan, and Xilin Chen. 2019. "AttGAN: Facial Attribute Editing by Only Changing What You Want." *IEEE Transactions on Image Processing* 28 (11): 5464–78.

He, Zhenyi, Keru Wang, Brandon Yushan Feng, Ruofei Du, and Ken Perlin. 2021. "GazeChat: Enhancing Virtual Conferences with Gaze Awareness and Interactive 3D Photos." *UIST '21: ACM Symposium on User Interface Software and Technology*. https://www.youtube.com/watch?v=Fnl8hxxgGVQ.

Hess, Ursula, and Agneta Fischer. 2013. "Emotional Mimicry as Social Regulation." *Personality and Social Psychology Review* 17 (2): 142–57. https://doi.org/10.1177/1088868312472607.

Hinton, Geoffrey E., Li Deng, Dong Yu, George E. Dahl, Abdel-rahman Mohamed, Navdeep Jaitly, Andrew Senior, Vincent Vanhoucke, Patrick Nguyen, Tara N. Sai nath, and Brian Kingsbury. 2012. "Deep Neural Networks for Acoustic Modeling in Speech Recognition: The Shared Views of Four Research Groups." *IEEE Signal Processing Magazine* 29 (6): 82–97.

Hirsch, Joy, Xian Zhang, J. Adam Noah, and Yumie Ono. 2017. "Frontal Temporal and Parietal Systems Synchronize within and across Brains During Live Eye-to-Eye Contact." *NeuroImage* 157: 314–30. https://doi.org/10.1016/j.neuroimage.2017.06.018.

Insel, Thomas R. 2017. "Digital Phenotyping: Technology for a New Science of Behavior." *JAMA* 318 (13): 1215–16. https://doi.org/10.1001/jama.2017.11295.

Itier, Roxane J., and Magali Batty. 2009. "Neural Bases of Eye and Gaze Processing: The Core of Social Cognition." *Neuroscience & Biobehavioral Reviews* 33 (6): 843–63. https://doi.org/10.1016/j.neubiorev.2009.02.004.

Kiilavuori, Helena, Veikko Sariola, Mikko J. Peltola, and Jari K. Hietanen. 2021. "Making Eye Contact with a Robot: Psychophysiological Responses to Eye Contact with a Human and with a Humanoid Robot." *Biological Psychology* 158: 107989. https://doi.org/10.1016/j.biopsycho.2020.107989.

Kobayashi, Hiromi, and Shiro Kohshima. 2001. "Unique Morphology of the Human Eye and Its Adaptive Meaning: Comparative Studies on External Morphology of the Primate Eye." *Journal of Human Evolution* 40 (5): 419–35. https://doi.org/10.1006/jhev.2001.0468.

Kotliar, Dan M. 2020. "Data Orientalism: On the Algorithmic Construction of the Non-Western Other." *Theory and Society* 49 (5–6): 919–39. https://doi.org/10.1007/s11186-020-09404-2.

Kuang, Beibei, Xueting Li, Xintong Li, Mingxiao Lin, Shanrou Liu, and Ping Hu. 2021. "The Effect of Eye Gaze Direction on Emotional Mimicry: A Multimodal

Study with Electromyography and Electroencephalography." *NeuroImage* 226: 117604. https://doi.org/10.1016/j.neuroimage.2020.117604.

Lamm, Claus, Henryk Bukowski, and Giorgia Silani. 2016. "From Shared to Distinct Self–Other Representations in Empathy: Evidence from Neurotypical Function and Socio-cognitive Disorders." *Philosophical Transactions of the Royal Society B: Biological Sciences* 371 (1686): 20150083. https://doi.org/10.1098/rstb.2015.0083.

Liebling, Daniel J., and Sören Preibusch. 2014. "Privacy Considerations for a Pervasive Eye Tracking World." *Proceedings of the 2014 ACM International Joint Conference on Pervasive and Ubiquitous Computing: Adjunct Publication*, 1169–77. Seattle: ACM. https://doi.org/10.1145/2638728.2641688.

Madipakkam, Apoorva Rajiv, and Marcus Rothkirch. 2019. "The Unconscious Processing of Social Information." In *Transitions between Consciousness and Unconsciousness*, edited by Guido Hesselmann, 92–117. New York: Routledge. https://doi.org/10.4324/9780429469688-4.

Mainzer, Klaus. 2017. "From Augmented Reality to the Internet of Things: Paradigm Shifts in Digital Innovation Dynamics." In *Augmented Reality: Reflections on Its Contribution to Knowledge Formation*, edited by José María Ariso, 25–40. Berlin: De Gruyter. https://doi.org/10.1515/9783110497656-002.

Mirzoeff, Nicholas. 2014. "Visualizing the Anthropocene." *Public Culture*, 26 (2): 213–32.

Mirzoeff, Nicholas. 2011. *The Right to Look*. Durham, NC: Duke University Press.

Mirzoeff, Nicholas. 2002. "The Subject of Visual Culture." In *The Visual Culture Reader*, edited by Nicholas Mirzoeff, 3–23. New York: Routledge.

Müller, Philipp, Michael Dietz, Dominik Schiller, Dominike Thomas, Guanhua Zhang, Patrick Gebhard, Elisabeth André, and Andreas Bulling. 2021. "MultiMediate: Multi-modal Group Behaviour Analysis for Artificial Mediation." *Proceedings of the 29th ACM International Conference on Multimedia*, 4878–82. New York: Association for Computing Machinery. https://doi.org/10.1145/3474085.3479219.

Nag, Anish, Nick Haber, Catalin Voss, Serena Tamura, Jena Daniels, Jeffrey Ma, Br yan Chiang, Shasta Ramachandran, Jessey Schwartz, Terry Winograd, Carl Fei nstein, and Dennis P. Wall. 2020. "Toward Continuous Social Phenotyping: Analyzing Gaze Patterns in an Emotion Recognition Task for Children with Autism through Wearable Smart Glasses." *Journal of Medical Internet Research* 22 (4). https://doi.org/10.2196/13810.

Nake, Frieder. 2015. "Surface, Interface, Subface: Three Cases of Interaction and One Concept." In *Paradoxes of Interactivity: Perspectives for Media Theory, Human-Computer Interaction, and Artistic Investigations*, edited by Uwe Seifert, Jin Hyun Kim, and Anthony Moore, 92–109. Bielefeld: transcript.

Nancy, Jean-Luc. 2000. *Being Singular Plural*. Redwood City, CA: Stanford University Press.

Niedźwiecka, Alicja. 2020. "Look Me in the Eyes: Mechanisms Underlying the Eye Contact Effect." *Child Development Perspectives* 14 (2): 78–82. https://doi.org/10.1111/cdep.12361.

Noah, J Adam, Xian Zhang, Swethasri Dravida, Yumie Ono, Adam Naples, James C. McPortland, and Joy Hirsch. 2020. "Real-Time Eye-to-Eye Contact Is Associated with Cross-Brain Neural Coupling in Angular Gyrus." *Frontiers in Human Neuroscience* 14 (19). https://doi.org/10.3389/fnhum.2020.00019.

Nomura, Tatsuya, Takayuki Kanda, Tomohiro Suzuki, and Sachie Yamada. 2020. "Do People with Social Anxiety Feel Anxious about Interacting with a Robot?" *AI &SOCIETY* 35 (2): 381–90. https://doi.org/10.1007/s00146-019-00889-9.

Oppenheimer, J. Robert. 1954. Prompting Apparatus, Patent Publication US2883902A. https://patents.google.com/patent/US2883902A/ (accessed August 18, 2022).

Palsson, Sveinn, Eirikur Agustsson, Radu Timofte, and Luc Van Gool. 2018. "Generative Adversarial Style Transfer Networks for Face Aging." In *2018 IEEE/CVF Conference on Computer Vision and Pattern Recognition Workshops (CVPRW)*, 2165–68. Salt Lake City, UT: IEEE. https://doi.org/10.1109/CVPRW.2018.00282.

Paripally, Gopal, Craig Clapp, and Tony Pham. 2012. Reflective and Refractive Solutions to Providing Direct Eye Contact Videoconferencing. US Patent 8,970,655, filed December 14, 2012, and issued March 3, 2015. https://patents.google.com/patent/US8970655B2/en?q=teleprompter+mirror&oq=teleprompter+mirror+a (accessed August 18, 2022).

Parisi, Luciana. 2020. "Negative Optics in Vision Machines." *AI &Society* 36: 1281–93. https://doi.org/10.1007/s00146-020-01096-7.

Pasini, Marco. 2019. "Style Transfer with GANs on HD Images." *Medium*, July 23, 2019. https://towardsdatascience.com/style-transfer-with-gans-on-hd-images-88e8efcf3716.

Pathirana, Primesh, Shashimal Senaratha, Dulani Meedeniya, and Sampath Jayarathna. 2022. "Eye Gaze Estimation: A Survey on Deep Learning-Based Approaches." *Expert Systems with Applications* 199: 116894.

Peterson, Mike. 2020. "FaceTime Eye Contact Correction Feature to Launch with IOS 14." Apple Insider, June 22, 2020. https://appleinsider.com/articles/20/06/22/facetime-eye-contact-correction-feature-to-launch-with-ios-14.

Popp, Alla. 2020. "#alieneffect Facefilter Workshop für Anfänger*innen. Spy on Me #2 Online Program." YouTube video, March 20, 2020. https://www.youtube.com/watch?v=lkp6zkyYh8Y.

Sartre, Jean Paul. 1956. *Being and Nothingness: An Essay on Phenomenological Ontology*. Translated by Hazel E. Barnes. New York: Philosophical Library.

Schukin, Dave. (@schukin). 2019. "How iOS 13 FaceTime Attention Correction works: it simply uses ARKit to grab a depth map/position of your face, and adjusts the eyes accordingly. Notice the warping of the line across both the eyes and nose."

Twitter, July 3, 2019, 12:07 p.m. https://twitter.com/schukin/status/1146359923158089728.

Sedgewick, Robert. 1984. *Algorithms*. Reading, MA: Addison-Wesley.

Senju, Atsushi, and Mark H. Johnson. 2009. "The Eye Contact Effect: Mechanisms and Development." *Trends in Cognitive Sciences* 13 (3): 127–34. https://doi.org/10.1016/j.tics.2008.11.009.

Shimojo, Shinsuke, Claudiu Simion, Eiko Shimojo, and Christian Scheier. 2003. "Gaze Bias Both Reflects and Influences Preference." *Nature Neuroscience* 6 (12): 1317–22. https://doi.org/10.1038/nn1150.

Siegert, Ingo, and Oliver Niebuhr. 2021. "Case Report: Women, Be Aware That Your Vocal Charisma Can Dwindle in Remote Meetings." *Frontiers in Communication* 5. https://doi.org/10.3389/fcomm.2020.611555.

Silverman, Kaja. 1996. *The Threshold of the Visible World*. New York: Routledge.

Smith, Brian A., Qi Yin, Steven K. Feiner, and Shree K. Nayar. 2013. "Gaze Locking: Passive Eye Contact Detection for Human–Object Interaction." *Proceedings of the 26th Annual ACM Symposium on User Interface Software and Technology*, 271–280. UIST '13. https://doi.org/10.1145/2501988.2501994.

Srnicek, Nick. 2017. *Platform Capitalism*. New York: John Wiley & Sons.

Steyerl, Hito. 2014. "Proxy Politics. Signal and Noise." *e-flux journal* 60. https://www.e-flux.com/journal/60/61045/proxy-politics-signal-and-noise.

Strouse, Gabrielle A., and Jennifer E. Samson. 2021. "Learning from Video: A Meta-Analysis of the Video Deficit in Children Ages 0 to 6 Years." *Child Development* 92 (1): e20–38. https://doi.org/10.1111/cdev.13429.

Suwajanakorn, Supasorn, Steven M. Seitz, and Ira Kemelmacher-Shlizerman (2017). "Synthesizing Obama: Learning Lip Sync from Audio." *ACM Transactions on Graphics (ToG)* 36 (4): 1–13.

Thies, Justus, Michael Zollhöfer, Marc Stamminger, Christian Theobalt, and Matthias Nießner. 2016. "Face2Face: Real-Time Face Capture and Reenactment of RGB Videos." *2016 IEEE Conference on Computer Vision and Pattern Recognition (CVPR)*: 2387–95.

Tolentino, Jia. 2019. "The Age of Instagram Face." *The New Yorker*, December 12, 2019. https://www.newyorker.com/culture/decade-in-review/the-age-of-instagram-face.

Tomasello, Michael, Brian Hare, Hagen Lehmann, and Josep Call. 2007. "Reliance on Head versus Eyes in the Gaze Following of Great Apes and Human Infants: The Cooperative Eye Hypothesis." *Journal of Human Evolution* 52 (3): 314–20. https://doi.org/10.1016/j.jhevol.2006.10.001.

van Loyen, Ulrich. 2015. "Herzensergießungen. Süditaliens böser Blick." *Zeitschrift für Kulturwissenschaften. Begeisterung und Blasphemie* 9 (2): 89–95. https://doi.org/10.25969/mediarep/13940.

Vidal, Mélodie, Jayson Turner, Andreas Bulling, and Hans Gellersen. 2012. "Wearable Eye Tracking for Mental Health Monitoring." *Computer Communications* 35 (11): 1306–11. https://doi.org/10.1016/j.comcom.2011.11.002.

Wang, Ting-Chun, Arun Mallya, Ming-Yu Liu. 2021. "One-Shot Free-View Neural Talking-Head Synthesis for Video Conferencing." *CVPR*. https://nvlabs.github.io/face-vid2vid.

Wegenstein, Bernadette. 2010. "Body." In *Critical Terms for Media Studies*, edited by Mark B. N. Hansen and W. J. T. Mitchell, 19–34. Chicago: University of Chicago Press.

Wood, Erroll, Tadas Baltrusaitis, Louis-Philippe Morency, Peter Robinson, and Andreas Bulling. 2017. "GazeDirector: Fully Articulated Eye Gaze Redirection in Video." *arXiv:1704.08763 [cs.CV]*. http://arxiv.org/abs/1704.08763.

Zhang, Jichao, Meng Sun, Jingjing Chen, Hao Tang, Yan Yan, Xueying Qin, and Nicu Sebe. 2019. "GazeCorrection:Self-Guided Eye Manipulation in the Wild Using Self-Supervised Generative Adversarial Networks." *arXiv:1906.00805 [cs.CV]*. https://arxiv.org/abs/1906.00805.

Zhang, Jichao, Jingjing Chen, Hao Tang, Wei Wang, Yan Yan, Enver Sangineto, and Nicu Sebe. 2020. "Dual In-Painting Model for Unsupervised Gaze Correction and Animation in the Wild." *Proceedings of the 28th ACM International Conference on Multimedia*, 1588–96. Seattle: ACM. https://doi.org/10.1145/3394171.3413981.

Performing Video Conferencing and VR for a "Real Virtual Life"

A Warm Welcome to Distant Socializing!

Martina Leeker

For Alfred & Inge, in gratitude.

While face-to-face contact was impossible during the pandemic, performing arts practitioners were able to continue their work by using technological applications for telecommunication and telepresence. There are two constituent systems: video conferencing and virtual reality (VR), especially VR rooms from the world of gaming, which are part of desktop VR. Such tools enabled theater and performance practitioners to engage in a range of activities remotely, including streamed live performances, online rehearsals, and the realization of a genuine digital format called nettheater. The focus of this paper is the cultural effects of these formats, specifically their contribution to digital cultures. The thesis explored here is that they are generating virtuality as its own reality, a "real virtual" condition. Real virtual manifested strongly during the pandemic due to the expansion of digital cultures into a living in "distant socializing." A historical contextualization of performing telepresence makes visible a regime of the real virtual. It generates a soft adaptive and anticipative relationalization, supporting digital cultures for which techno-human cooperation is constitutive. The regime tilts between technocratic subjection and a fragile agency of happily "gaming around"[1] in digital entanglements, unfolding a subject of virtuality within an episteme of contingency.

1 The term "gaming around" describes the transformation of the cultural technique of playing, the relocation of its interplay of rules and agency, as well as of immersion and distance/reflexion, into the realm of the digital and its operativity, dealing with decision instead of choice and with connectivity instead of relating.

Introduction

In this paper, I explore performances and installations engaging with telecommunication and telepresence[2] enabled by, among other things, technologies of video conferencing. The focus lies on the specific constitution and relevance of telepresence, which is defined in a general way by Shen and Shirmohammadi: "Telepresence, also called virtual presence, is a technique to create a sense of physical presence at a remote location using necessary multimedia such as sound, vision, and touch" (2008a, 849).[3] It was the pandemic situation that made telepresence existential and, as a result, highly valuable. This means putting physical existence under the technological conditions of digital applications (virtualization), making it thereby its own viable reality: a "real virtual" as a "living in distant socializing." This understanding of virtuality, generated prominently by performing telepresence, expands its existing meanings and practices[4] and becomes a crucial, not yet fully researched, constitutive part of digital cultures. Against this background, a historical perspective might help to specify the media-cultural effects of telepresence, and performing with telepresence, that come from its special interpretation and configuration of virtuality.

For further and deeper exploration, the field of performing telepresence and its specificity must be described at the outset as it is initially quite counterintuitive and, additionally, expands—for good reasons—the book's focus on video conferencing. The specificity of telepresent performances, also called nettheater (see Heinrich-Böll-Stiftung and nachtkritik.de 2020), consists of the fact that it combines technologies of video conferencing with systems for virtual reality (VR).[5] "Video confer-

2 The notion "telepresence" was coined in 1980 by Marvin Minksy, who used it in the sense of acting at a distance via smart robotic devices (Shen and Shirmohammadi 2008a). It became a well-known concept in the 1990s, during a boom in the VR performance and telematics (see Decker and Weibel 1990). Even if it may appear antiquated, the notion is useful again due to the relevance of telepresence in the situation created by the pandemic. It is also relevant today because the term is replaced by the notion "digital liveness" (see section 2), thus obscuring important connotations and the normalization of "telepresence," including its technological conditions.

3 Shen and Shirmohammadi explain the difference between telepresence and virtual presence from a technical point of view: "telepresence is a networked paradigm by nature, whereas virtual presence does not have to be networked and can run completely locally" (Shen and Shirmohammadi 2008b, 267).

4 Virtuality has previously been seen as something beyond reality, either as a "hyperreality" (Baudrillard 2004) or as a not yet actualized reality (Esposito 1998; Klappert 2020). Further explication in this introduction.

5 Shen and Shirmohammadi note: "Virtual Reality is the technology that provides almost real and/or believable experiences in a synthetic or virtual way" (Shen and Shirmohammadi 2008b, 962).

encing" involves not only the various software applications such as Zoom or Jitsi, which were an essential and notorious element of the pandemic. The term has been in use since the 1970s as a generic label for technologies that enable distant operations via the live-transmission of video images of persons through networked infrastructures (e.g., satellites in the 1970s, now the Internet) (Shen and Shirmohammadi 2008a; Shen and Shirmohammadi 2008b). Virtual reality (VR) systems, the second strand and category forming telepresence in nettheater, come in as desktop VR (Shen and Shirmohammadi 2008b)[6] known as graphical video games, being formed in contrast to video conferencing systems within spaces generated by data only.

The main point and argument is that nettheater's amalgamation of these two strands of technology, and their prehistory[7] and epistemology, transforms each component, resulting in the formation of its own reality (real virtual) of living in distant socializing. First, the adaptation of video conferencing becomes itself virtual within the intermingling with VR, which is quite counterintuitive at first glance, as video conferencing is referred to as (tele)communication, whereas virtuality is, as mentioned, more commonly linked to data spaces, without reliable links to the physical. In other words, the understanding of virtuality is expanded. Second, virtual environments of desktop VR, such as VRChat or Mozilla Hubs (see Sauerländer 2020; Diesselhorst 2021), are transformed within performing telepresence into media of telecommunication. To better understand the specificity of this transformation, it should be noted that this strand shows two subgroups: immersive VR, and desktop VR; the latter is used in nettheater.[8] Whereas immersive VR is a very solipsistic and only locally available setting, offering visitors the experience of immersive encounters within highly speculative data driven virtual environments that appear in VR glasses, desktop VR is used for performing games online, which

6 This notion is proposed by Shen and Shirmohammadi, distinguishing it from immersive VR, used with, for example, head-mounted displays. The authors offer this definition: "Desktop VR uses a computer monitor as display to provide graphical interface for users. It is cost-effective when compared to the immersive VR as it does not require any expensive hardware and software and is also relatively easy to develop" (Shen and Shirmohammadi 2008b, 963).

7 In section 1, the history of video conferencing systems as technologies for telepresence since the 1970s is unpacked. These systems have undergone different modifications, changing from a "third-space-telepresence," as it is named in this paper, to a ubiquitous digital liveness (see section 2). For the prehistory of VR systems/desktop VR, see the following footnote.

8 Due to limited space, the history of VR systems/desktop VR, building the second constitutive category and strand of telepresence as real virtual, cannot be covered adequately. The technology belongs to the large domain of gaming cultures, starting in the 1970s with text-based MUDs, also in theater with MOOs, the ATHEMOO system (Burk 1999), for example, for digital theater role play, followed by graphical MUDs, so-called massively multiplayer online games (MMOs) (see Pepe 2020). This line of desktop VR was implemented 2014/2017 in the platform VRChat.

also encompasses social exchange. The crucial point is that once desktop VR is used in performances, its focus shifts from gaming together to, on the one hand, exploring conditions and possibilities of imagination for absent things, persons, and situations[9] and, on the other, the researching and testing of modes of telepresent liveness. This shift concerns the traditional formation of theater and performances which is predicated on copresence—being at the same time in the same space—as well as on staging absence, to be filled by audience imagination. Once theater and performance become virtual, their traditional constitution is reinvented under the technological conditions, transforming technology at the same time. Within nettheater, virtuality not only orientates toward telepresent corporality and sociality, but more importantly, it makes an essential contribution: the transformation of desktop VR into a real virtual, enabling an existence in distant socializing, as shown by theater during the pandemic.

To sum up, the encounter of video conferencing and desktop VR offers not only the ability to act and communicate with and at a distance but transforms virtuality into a real virtual of being in, and living in, distant socializing. This unfamiliar understanding of virtuality that we see emerging from performing telepresence goes beyond traditionally held concepts. Existing concepts of virtuality fall into two camps, both of which agree that the virtual is something special beyond reality. The first camp sees virtuality as another domain and sphere, threatening to collapse the difference between the real and the virtual. As such, it allegedly generates a "hyperreality" (Baudrillard 2004), which exists purely technologically without any reference to reality. The second camp sees virtuality as a potential (Esposito 1998; Klappert 2020) waiting for its actualization, forming a culture and epistemology of being-in-contingency (Klappert 2020).[10] As such, according to Klappert, virtuality is always real.

Within performing telepresence, however, virtuality is no longer interpreted as the menacing "other" of the "real" reality or as a more adequate model of reality rooted in the actualization of options but as its own reality (real virtual) of distant socializing. Thus, the performative intermingling of both strands (telepresence and VR) under pandemic pressure crystallizes the objective of telepresence and virtuality as influenced by telepresence. That objective, held since the 1970s, is to become its own fully implemented reality (real virtual) equal to non-telepresent realities. The

9 In theater, a chair could be, for example, a queen's throne. For a discussion on imagination versus representation in video gaming and theater, see also Pepe 2020.

10 Furthermore, Annina Klappert rejects the conceptualization of virtuality with the help of the opposites real/virtual, and unfolds it by pairing virtual/actual, which she distinguishes from the description of the virtual as "the possible." The possible, according to Klappert, is always preconfigured, whereas the actual is constituted in transformation and differences, coming up as an event, an innovation, being unforeseeable and therefore leading to an existence in an epistemology of contingency (Klappert 2020, 11–69).

contribution of nettheater to this "becoming real" is working on a specific body, a mentality, a subject, and an order of sociality for telepresent existences, providing a technologically installed and supported liveness. This complex process becomes understandable only if the intermingling and interplay of video conferencing and desktop VR is considered.

Furthermore, the pandemically accelerated digitalization shows what had been valid since the 1970s: analog events and moments of being are virtual in digital cultures because they are configured in the conditions of technical virtuality (see also Kasprowicz 2020). Therefore, it becomes clear that performing virtuality does not correspond to a history of losing reality (Baudrillard 2004) or to the existing in the permanent transformative actualization of only virtually present options (Esposito 1998; Klapperer 2020). On the contrary, this performing is a training for making the virtual a liveable reality and sociality.

This contextualization and insight help understand what kind of regime produces performing telepresence. It builds the basis for a symbiotic techno-human cooperativity as the constitution of today's digital cultures. This cooperativity is organized as permanent adaptation and anticipation, evoking a regime of soft relationalization,[11] which binds the human and technological together, as though they are genuinely an agency of action and decision. The performances with telecommunication and telepresence are operating physical, perceptive, and social training for this adaptation, tilting between voluntary subjection under technocratic needs, and happy "gaming around"[12] in and with technological entanglements.

To make this context and constitution concrete, the focus for the following analysis of telepresent performances since the 1970s lies on the techno-cultural conditions they generate, concerning: (1) mentality, (2) sociality, and (3) epistemology. The first example refers to telematic performances with the live transmission of video images via broadband technology (category/strand: video conferencing) since the 1970s, which also represents, as mentioned above, the prehistory of today's video conferencing in Zoom and similar applications. Artists Kit Galloway, Sherrie Rabinowitz, and, later, Paul Sermon brought geographically dispersed people together real-virtually, forming a doubled transgressive body (see also Rieger 2019), and auto-

11 This regime of relationalization is even seen as a "better" view on the human, as unpacked in today's media studies and cultural studies by, for example, Karen Barad, Rosi Braidotti, and Bruno Latour (for detailed elaboration on this status and the consequences of this hype on "entanglement" see Leeker 2021). It is said that humans can no longer be seen as entities that are at a remove from their environment, but that they are insolvably entangled in and with their environment, becoming something only through intermingled cooperation. This discourse is ennobled as a better understanding of the humanity that could help save the earth and generate more equal societies, because entanglement means a consciousness of togetherness and responsibility.

12 See footnote 1.

s(t)imulation (Leeker 1995) as (1) performatively enhanced mentality. This category of telematic performances is combined today with those of desktop VR, which also becomes, within this context, an environment for the real virtual in distant socializing. Whereas historic performances with video conferencing had the character of sporadic events, contemporary Zoom performances of the *punktlive* theater collective unleash an extension of telepresence's scope, also using social media as enablers of distant socializing in the real virtual. This integration makes distant socializing a part of everyday life, configuring an existence in a ubiquitous digital liveness in today's networked infrastructures. It is within this extension that a (2) telepresent sociality is formed, constituted by hyper-egocentric subjects of virtuality, compensating the dissolution of the individual into data doubles in today's platform economies and their data mining. Becoming social in the real virtual is envisioned as both a warmly welcomed normality and a highly connected recursive loneliness at the same time. Finally, concerning (3) the epistemology of performing telepresence for the real virtual, a performative intellectual "gaming around" in "let's play" becomes interesting. Taking place on twitch.tv, a platform for performing video gaming as live events, "let's play" finally integrates the two strands of video conferencing and desktop VR into an environment of performative and unforeseeable knowledge production, which is like a game of distant socializing. This configuration of the real virtual establishes an epistemology of overburdening within feedback loops of "gaming around" with linguistic fragments, making knowledge and understanding a happy performance of contingency.

The following analysis section is structured in two parts. Part 1 extensively examines the history and presence of the (1) mentality of the real virtual in performing telepresence in order to show their relevance. Due to limitations of scope, part 2 will focus more informally—with less depth and no historical reconstruction—on the (2) sociality and (3) epistemology of the real virtual, as generated in performances. Despite its brevity, the second section can be seen as an important contribution to the research in this paper, giving an initial but fragmentary and incomplete overview of the real virtual as a constitution of digital cultures in a deep desire for a telepresence existence.

1. Mentality of the Real Virtual: Doubled Bodies and Auto-S(t)imulation

Technologies of video conferencing and performing telepresence not only affect corporality, the body, and its sensuality but also produce their own mentality, intermingling the physical, the perceptive, the spiritual, the emotional, the mental, and the cognitive. This builds a set of people's perceptions, thoughts, emotions, attitudes, and positions, conditioning their being-in-the-world and self-understanding as the specific telepresent mentality.[13] This mentality of real virtual is based on the fact that human agents have two transgressive bodies (Rieger 2019)[14] at their disposal, which are intermingled.[15] There is, on the one hand, the physical body in analog spaces. On the other hand, that physical body is connected to a virtual body, an avatar, as seen in performances with VR chatrooms,[16] or a video image (Zoom). The most interesting aspect is the question of how these performances deal with the transgressive physicality and which effects this handling evokes concerning the telepresence mentality.

Kit Galloway and Sherrie Rabinowitz's *Image as a Place* (1970s)

Figure 1: Kit Galloway, Sherrie Rabinowitz, Hole in Space *(1980), real-time videoconference connection between public spaces in New York and Los Angeles*

Source: https://anthology.rhizome.org/mobile-image (accessed August 11, 2022).

A historical reconstruction helps to answer these questions, detecting and understanding what is going on concerning the mentality of the real virtual. To do so, the accepted historical narrative must be modified. That narrative says that *Hole in*

13 The notion "mentality" is used to mark this complex situation and constitution.

14 Stefan Rieger calls this constitution "*Grenzverschieblichkeit*" (Rieger 2019, 85ff.).

15 A convincing example of this constitution is the performer CodeMiko (see Kooboto 2020).

16 The platforms used in theater and performance include Mozilla Hubs, VRChat, and Gather.town (see Diesselhorst 2021).

Space (1980),[17] an installation by artists Kit Galloway and Sherrie Rabinowitz, is the starting point of telepresence experience in so-called real time. The artists launched the piece in 1980, connecting people from New York and Los Angeles by live video image transferred via satellite.

Reactions to the event were very touching, and still are today, because it dealt with a real virtual presence. The affective drive becomes visible, as persons from across the country flirted, made appointments, and introduced newborns to extended family.

But, more informative for the constitution of telepresence's reality and its cultural and political effects is the *Satellite Arts Project* (1975–1977),[18] an installation created and developed by the same artists in the 1970s. It features dancers separated by up to 5,000 kilometers moving together in so-called real time via transferred video images. The essential and important aspect of the installation is the struggle with time delay and latency in the transmission of video images (see also Paulsen 2013; Paulsen 2017), which ask for a specific bodily and cognitive-sensorial adaptation in order to synchronize the interaction with other people (see also Distelmeyer 2021). For example, Kit Galloway counted beats, beckoning with his hand in order to adapt to the delay and synchronize the time, to give the impression of waving reciprocally with another person (see also Paulsen 2013, 102). Another groundbreaking invention, which helped to realize "the 'simultaneous now' of satellite telecommunication" (Paulsen 2013, 108), was the substitution of a split screen with a screen that fed the two images into one by video-keying. Whereas the split screen separates the screen into two halves and makes invisible those parts of a body on the screen where a person goes beyond the range of visibility of a camera (Paulsen 2013, 104–105), the "mixed image" (Paulsen 2013, 103) enabled the dancers to share one (!) common space. They could dance together, instead of, for example, pantomiming touching each other by trying to overcome the frontier of the screen. Artist and writer Steven Durland describes the effect of performing telepresence and distant, virtual socializing as the "image becoming a place" (Durland 1987), requiring, as Kris Paulsen says, "that the dancers negotiate all of their embodied senses through their collective image on the screen" (Paulsen 2017, 98).

---

17 Information on the art project can be found on the following websites: http://www.ecafe.com/museum/history/ksoverview2.html (accessed May 19, 2022); https://anthology.rhizome.org/mobile-image (accessed May 19, 2022). Video excerpts from *Hole in Space* can be viewed on: https://www.youtube.com/watch?v=SylJJr6Ldg8 (accessed May 19, 2022); https://www.youtube.com/watch?v=QSMVtE1QjaU (accessed May 19, 2022).

18 For information on *Satellite Arts Project*, see https://anthology.rhizome.org/mobile-image (accessed May 19, 2022).

Figure 2: Kit Galloway, Sherrie Rabinowitz, Satellite Arts Project (1977), telecollaborative dance, relating via two-way satellite over long distance

Source: https://anthology.rhizome.org/mobile-image (accessed August 11, 2022).

The thesis in this paper is that the telecommunicative and telepresent performance propel the realization of a real virtual mentality, which is constituted by the fact that its virtuality, that is to say, its technological status, becomes psycho-physically real through embodiment. Or, to put it differently: the aim is a "real" liveness within the virtual. This is achieved by the activation and the training of, as I call it, auto-s(t)imulation (Leeker 1995), as seen in Kit Galloway's methods of adaptation to time delay and latency in the transmission of video images. It deals with simulating the missing parts in physicality (smell, touch, and real time, for example) during a techno-human encounter by stimulating, or more specifically, activating your own body and imagination. In this context, the importance of the invention of a third communal space, which overcomes the split screen, becomes understandable. It aims to enable training to deal with the gaps and delays of telecommunication, by flowing into a space of immersion, which smoothes differences and helps to compensate and adapt by minimizing the difficulties through the illusion of being-in. In this way, physicality is adapted to a techno-human cooperation, being able to fill gaps and overlook differences. Kris Paulsen gives a hint of the upcoming virtual mentality as real virtual when she says:

> Galloway and Rabinowitz hypothesize an ethics of engagement with others in mediated environments. They imagine what it might be like to be simultaneously real and virtual, self and other, subject and object, seer and seen, here and there, now and then. (Paulsen 2013, 99)

Paul Sermon's *Third Space* (1990s)

But, finally auto-s(t)imulation is about more than just a temporal state, and capability that is activated in cases of an event during a telepresent transmission. On the contrary, the psycho-physical adaptation to telepresence's physicality becomes a permanent configuration, and constitution of mentality, which functions beyond events in telepresence as well. That is, the training generates a physicality that is constituted in the real virtual and *habitualizes* it as new, ordinary mentality, thereby adapting human agents perfectly to the technological conditions of digital cultures.

This becomes visible in performative installations by Paul Sermon, who has worked exclusively with telecommunicative and telepresent performances since the 1990s with the help of chroma-keying.[19] He continued the setting invented by Kit Galloway and Sherrie Rabinowitz and named the mixed space beyond split-screen *third space*. The crucial point of Sermon's performative installations is the turning to left and right in video images (Leeker 2019).[20] If you want to touch a person that is on your right side in physical space, you have to move to the left to perform the action in the video image (Leeker 2002; Leeker 2019, 10). I personally experienced the effect of this training when this conversion was automatized over two weeks during an intense workshop with Paul Sermon and dancers at the *Choerografisches Zentrum* in Essen in the summer of 2001 (Leeker 2002, 244–305). My son visited me during this workshop and sat beside me on my right. When I decided to give him a hug, I turned to the left side, unconsciously using the trained mode, anticipating the conversion for the virtual presence. It was a vivid example of the power of adaptation within the regime of relationalization for techno-human cooperativity. It produces an all-over anticipation of a technological situation as a kind of joyful and voluntary obedience.

The background for this normalization of the virtual body-shaping is Paul Sermon's mapping the virtual third space (Leeker 2002), originally on a video screen, into physical space. Sermon's *Telematic Dreaming* (1992)[21] is a good example. The installation featured a bed, set up in each place, onto which were projected images of remotely located people who were then invited to touch each other.

The interesting point is that Paul Sermon made the physical surroundings themselves a screen, so that both the physical surroundings and the virtual became transgressive, hybrid zones in between the real and the virtual. The virtual is real, and the real virtual. In these environments, it makes sense that the virtually reconfigured body takes over.

---

19 For an overview over Paul Sermon's artworks, see his website http://www.paulsermon.org/ (accessed May 19, 2022).

20 This turn can be adapted easily in today's video conferencing via Zoom using the "Mirror My Video" function in the video settings.

21 http://www.paulsermon.org/dream/ (accessed May 19, 2022).

Figure 3: Telematic Dreaming, Paul Sermon, 1992

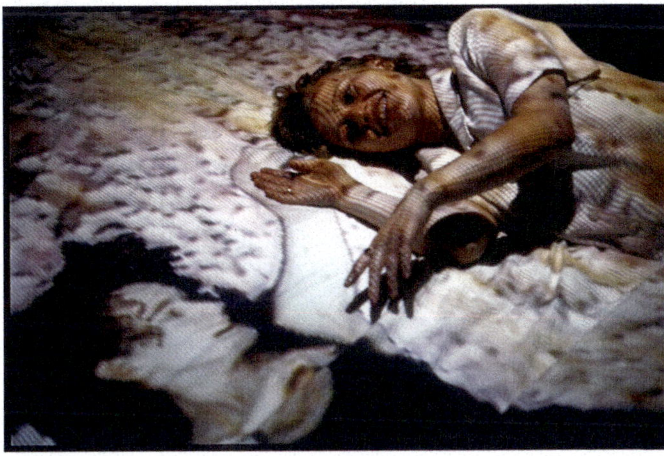

Source: http://www.paulsermon.org/dream/.

Performing VR Chatrooms (2020s): VR for Telepresence via Telecommunication

Today's telepresent performances follow from this prehistory and also integrate the second strand of telepresence: desktop VR using, for example, the platform VRChat with its VR chatrooms. These performances are given without any knowledge of the prehistory or thoughts about it, taking the training in auto-s(t)imulation for granted, which could be seen as a clue to its hyper-normalization. This may be the condition for the easy shift in performances with encounters in VRChat rooms via avatars during the pandemic, which are far more abstract and technically demanding to create and use than performing in and with video conferencing. Vivid examples are the VR environments created by Roman Senkl and Nils Corte[22] and virtual theater collective *CyberRäuber*,[23] which spectators either explore by navigating through the environments and encountering snippets of theater or by becoming part of a performance seen through VR glasses, which presents stages and actors that evolve through the movements of spectators wearing the glasses.

One task facing artists according to VR producer and programmer Nils Corte is the development of sophisticated concepts that might help users experiencing virtual spaces for the first time (Diesselhorst 2021). It seems that under the pressures of the pandemic, performances expanded and strengthened the becoming real of

22 For an overview, see http://nils-corte.de/ (accessed May 19, 2022).

23 http://wp11159761.server-he.de/vtheater/de/home/ (accessed May 19, 2022).

virtuality (real virtual) by enabling a progression to the next step. This step refers to enlarging the ability to deal with the two transgressive bodies (the physical and the virtual) when avatars have become the vehicle for encounters. Furthermore, within this setting, the impression and effects of a vivid telepresence are generated differently than in video images of the persons, as done by Galloway and Rabinowitz, or Paul Sermon. Whereas video conferencing is engaged by touch and proprioceptive sensations, the liveness (*Lebendigkeit* in German) in VRChat rooms is mainly generated via chatting—human agents can speak with each other via the avatars—as well as with the help of written chat functions. This chatting attests that the persons are here and there at the same time, sharing a common event. This sensation and impression of telepresence as ultimately a veritable digital liveness is enabled and supported by the fact that the visitors have to learn to navigate through VR via keyboard or VR glasses. In either case, they have to engage their physical bodies in order to deal with the technological challenges, adding physicality to the virtual, just as they were once trained via auto-s(t)imulation. So, the digital liveness of telepresent performances in desktop VR is one of a physicality of imagination.

It has to be mentioned that these performances in the virtual rooms of VRChat are surrounded by others, using them as journeys through virtual landscapes and encounters for speculating about other worlds, asking, for example, what a more diverse society might look like and how societies could be organized more equally. Instead of focusing on training for transformation into a transgressive body or auto-s(t)imulation, today's real virtual is also about changing the world. Perhaps it is all about magic moments, as if the real virtual could make speculation become reality.

Fronte Vacuo's Ethics of Digital Liveness (2021): Let's Become Hesitants

Figures 4–6: Humane Methods: 4.01 The Ether Sessions (2021)

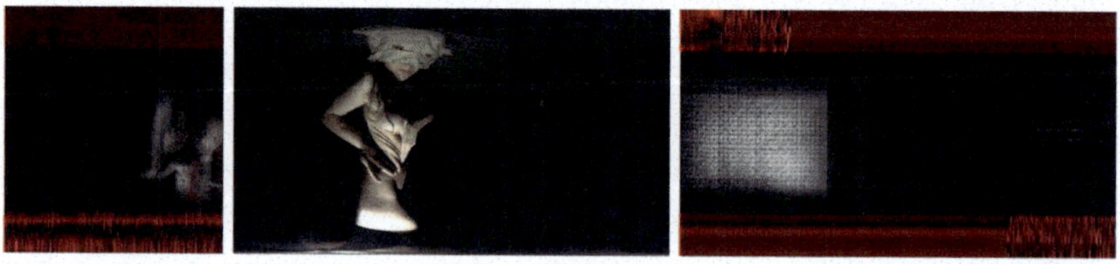

Source: Stills from performance video. A work in progress by Fronte Vacuo, directed by Andrea Familari, performed live and online at *TOOLS Festival*, Theater Rampe, Stuttgart.

Currently, more advanced and sophisticated formats for telepresence and real virtual are being developed in theater and performance—specifically, one very particular interactive offer. Spectators are asked, for example, during the streaming of live performances, to control a 360-degree camera from their own device to create individual views and edits. Spectators become directors of their reception of the performance, playing around with dis/appearances of all participants. With this format, performing telepresence also shifts to a further exploration of digital liveness. Whereas performing telepresence is always partially linked to its technological constitution, namely transmission, and its operative mode, namely remote communication and interaction, digital liveness is an attempt to create a digital version of the emphatic presence in live performances and its idealistic enhancement as epiphanic appearance (Fischer-Lichte 2008). The background of this shift may be on the one hand the normalization and becoming commonplace of telepresence, giving it more scope for creative shapes. On the other hand, the amalgamation of teleprescence with desktop VR, as just described, could charge it with metaphysical potentials. This charging was demonstrated in an experiment undertaken by the *Fronte Vacuo* collective (Marco Donnarumma, Andrea Familari, and Margherita Pevere) in 2021, using a 360-degree camera to live stream during the *Tools-Festival* at Theater Rampe in Stuttgart (Germany).[24]

Marco Donnarumma and Margherita Pevere performed a piece called *Humane Methods: 4.01 The Ether Sessions* (2021),[25] which was conceived by Andrea Familari, who also curated the video directing and the live AI algorithm. Spectators were invited to control a 3D camera while following the performance, which allowed them (the spectators) to scale the space between their viewpoint and the virtual presence of the performers, bringing them into view. That is, they were able to "go" nearer to performers or stay at a distance. At first glance, this intriguing version gave control and power of agency to the participative spectators. But at the same time, this luxury offer became itself the intrigue as the performance, even in its mediatic state of being too near, forced the spectators to decide on their position in relation to the almost naked performers, who were exposed in a highly vulnerable corporality and non-privacy. Participant-spectators oscillate between shame, shyness, and, inevitably, voyeurism. Viewers were challenged with the question of whether they would have liked or been able to assist a live performance of this experiment. This "gaming around"[26] with distances, being a (self-)reflection on telepresence, propelled it into the sphere of digital liveness and its specific qualities, such as constantly oscillating between distance and nearness. Could it be possible that digital liveness allows more nearness and asks for a humbler responsibility than analog liveness? Could control and power

---

24 https://theaterrampe.de/stuecke/humane-methods-4-01-the-ether-sessions/; https://www.spectyou.com/de/video/humane-methods-4-01-the-ether-sessions-tools-festival (accessed May 19, 2022).

25 https://theaterrampe.de/stuecke/humane-methods-4-01-the-ether-sessions/ (accessed May 19, 2022).

26 See footnote 1.

of agency in the virtual be extremely painful because a liveness, presence, and nearness are provided that you would not dare ask for, or which you could not endure in analog-physical performative encounters? Does gaining control mean also losing control? The real virtual turned into an intriguing spectacle of being- and becoming-too-near and made the regime of continuous adaptation and anticipation a question of individual responsibility. The technological became a question of ethics. It is as if theater and performance transformed into a training center for forms and modes of social reliability and credibility of digital liveness and the real virtual. They provide a lonely sociality that comes from a mode of hesitation, combining it with the sensing of metaphysics in telepresent encounters.

Training for Techno-Human Cooperation and Digital Liveness

This historical and systematic overview on telepresence shows that virtuality in video conferencing and VR environments was not primarily about the loss of the real in favor of the virtual (Baudrillard 2004), or about inventing other, allegedly "better" worlds by speculating in VR. On the contrary, it deals with the adaptation of human agents to technological conditions by auto-s(t)imulation, and the doubled transgressive body. Through training, both the body and mentality are transformed into appearances of a technological in-betweenness—that is, they are shaped by a techno-human constitution, existing as a space of relations between humans and technology. As such, body and mentality are prepared for permanent adaptation and anticipation, sensing and guessing others' behavior, whether they are human or nonhuman agents. This becoming real of the virtual is a basic and indispensable condition of digital cultures engaged in techno-human cooperation, as shown within the normality of distant socializing during the pandemic. It may even be that the pandemic organization of social life in distant socializing was possible only because humans had been trained in this mode of existence by performing telepresence beforehand.

A second point becomes obvious. It concerns the general constitution of liveness, which may be evident under the impression of digital existence but is nevertheless worthy and important to mention here explicitly, also referring to pre-digital conditions: neither in the prehistory of digital cultures nor in the contemporary situation was there ever an analog liveness in the sense of a physical copresence in immediacy, or a "real real." We have never been live or real but always live in a mediated way (Hammelburg 2020; Leeker 2019) and in a virtual condition (see also Kasprowicz 2020).[27] Telepresence and liveness, understood as feeling humans' socio-physical existence, is always mediated.

Meanwhile, in 2022, humans are here and there, now and then (Paulsen 2017; Hammelburg 2020) quite "naturally," as in today's digital cultures forms of so-called analog liveness are intermingled with the digital ones (see also Hammelburg 2020). This does not refer to a cultural-pessimistic view, mourning the loss of a paradisiac, allegedly unmediated liveness. It is, on the contrary, an insight into an insoluble technological condition of human existence and the acknowledgement of a diversification of multiple forms of presences, which are each in a mediated state.

2. Sociality and Epistemology of the Real Virtual: Hyper-Excited Egos and Knowledge on the Flight

The mentality of the real virtual is accompanied by an order of sociality and an order of epistemology in performing telepresence, which is explored in this section in a more essay-like manner without a deeper historical reconstruction, as noted in the introduction.[28] The telepresent performance *Möwe.live* (2021) by the *punktlive* collective is presented as an example of digital sociality, formed by asocial pseudo-subjects finding their self-understanding in social media. These subjects fall into deep depression and, at the same time, once they detect that this subjectivation is an illusion and betrayal, hide their existence as providers of data. "Let's play" offers a telepresent format that is informative for an understanding of the epistemology of performing telepresence and the resulting real virtual. Namely, within "gaming around" with digital applications on the internet, binding references to collectively shared frames and values of research and knowledge are dissolved. Finally, both performances support the techno-human cooperation in digital cultures congenially. On the one hand, the status of pseudo-subjects helps to reduce resistance against the binding in techno-human entanglement. The "let's play" performances smooth the effort and stress of permanent and unforeseeable techno-human adaptation by designing non-understanding and non-knowledge as fun and pure joy of contingency.

27 In his text, Dawid Kasprowicz unpacks the virtuality of labor in techno-human collaboration, which is constituted as such because it is generated in the simulation of machine-communication and interaction, being mapped onto bodies afterward.

28 See footnote 8.

Sociality of the Real Virtual: Ego-Excitement in *punktlive*'s Performing Social Media

The *Möwe.live*[29] performance by *punktlive* collective,[30] which premiered at Theater Stuttgart in 2021, is a telling example for today's nettheater, contributing its own form of telepresence to the constitution of digital cultures in our current, digitally enhanced living conditions. The performance combines video conferencing—using Zoom, a contemporary technology enabling telepresence by transferring video images of distant persons in so-called real time that became widespread during the pandemic—with social media. This coupling extends the reach of telepresence mentality in today's networked infrastructures, installing the narrative of existence as permanent and normalized distant socializing. This broadening of the real virtual transforms the concept of telepresence that held sway from the 1970s to the 1990s. The latter, which could be called now for better differentiation from the umbrella term "telepresence," "third-space-telepresence," linked more to occasional events of transmission, is modified to an everyday, highly networked telepresent existence. With this expansion, telepresence becomes a digital liveness, which is equal to analog copresence.

The constitution of sociality under this condition is explored in the following section by looking at the methods and relevance of the ubiquitous expansion, the upcoming type of subject, and its modes of being together with others.

The plot of *Möwe.live*, directed by Cosmea Spelleken, invents the circumstances of an existence in digital liveness, tending to an almost complete transformation of analog sociality into digital systems. The description says:

> A summer filled with carefree days at the lake: anything is still possible, dreaming of the future. A young Kostja, Nina, Masha and Kostja's mother, Arkadina, with her new lover, Trigorin, spent a summer together in a holiday home in France. Now just memories, the experiences of that summer were recorded only in Trigorin's video diary and numerous photos. The characters, connected via social media, follow what has become of the others. The pictures show shiny, happy lives, but they are deceptive. For all those involved must realize that their expectations of life are not necessarily compatible with reality.[31] (trans. ML)

---

29 For a first impression, see the trailer https://www.youtube.com/watch?v=OWzNgmYr-4E (accessed May 19, 2022).

30 https://punktlive.de/ (accessed May 19, 2022).

31 "Ein Sommer voller unbeschwerter Tage am See: alle Wege stehen offen und man träumt sich eine Zukunft. Der junge Kostja, Nina, Mascha und Kostjas Mutter Arkadina mit ihrem neuen Liebhaber Trigorin haben den Sommer gemeinsam im Ferienhaus in Frankreich verbracht. Die Erlebnisse von damals sind nunmehr Erinnerungen. Festgehalten nur in Aufnahmen aus Trigorins Video Tagebuch und zahlreichen Fotos. Verbunden über soziale Medien verfolgen die Figuren, was aus den anderen geworden ist. Glänzende, glückliche Lebenswege zeigen

The structure and dramaturgy of the performance configure the conditions for installing a ubiquitous digital liveness by, on one hand, expanding the scope of places and spaces for distant socializing, spreading it over interconnected platforms and devices and on the other hand, transforming those platforms and devices into media for personal use by individuals. Within this process, telepresence becomes ubiquitous and refers no longer only to communication but also to identification, as the telepresent images and figures are taken as a part of oneself, building an indispensable basis for a permanent distant socializing.

Telepresence's ubiquity is installed by the setup of the piece in a combination of Zoom calls in which characters meet; the showing of pre-produced films of past physical social being-together; and the activities of each character on their laptops, such as writing emails that are never sent, scrolling around in folders of photos, and checking social media such as Twitter, Instagram, and Facebook to meet other characters and make exchanges with them (see also: Fischer 2021).

Furthermore, in the performance, the laptops become spaces of persons, as they are staged as media of intimacy and personality in which a person's life is stored and activated (Fischer 2021). "Show me your desktop and I will tell you who you are," writes journalist Jan Fischer (2021, trans. ML). A specific contribution of the performance is to open these spaces to the public, showing a private desktop within distant socializing. This is realized by giving each character an account on Instagram,[32] feeding it with input from the performers' laptops during the performance and using these digital presences for communication with other characters of the piece. The audience can also communicate with the theatrical figures on Instagram in real time during each live performance. The crucial point is that these social media interactions are shown within the performance on the laptops of the characters. So, the performance not only "generates" a device as part of a person but also makes it an interface for distant socializing by distributing the "me-in-a-device" over social media. With this techno-dramaturgy, telepresence is occupying more and more of the internet. Thus, this performance works on the enlargement of the real virtual to the techno-sphere of a whole networked world and turns telepresence into a life in a normalized digital liveness: distant socializing is now, as the performance put it, everywhere, at any time.

---

die Bilder, doch sie trügen. Denn alle Beteiligten müssen feststellen, dass ihre Erwartungen ans Leben nicht unbedingt mit der Realität vereinbar sind." (https://www.staatstheater-nuernberg.de/spielplan-21-22/moewe-live/11-12-2021/1930) (accessed May 19, 2022).

32 For the Instagram accounts, see https://www.staatstheater-nuernberg.de/spielplan-21-22/moewe-live/11-12-2021/1930 (accessed May 19, 2022): @kostja.treplejow, @nina_sarajetschnaja, @mascha_schamrajew77, @boris.a.trigorin, @irina.arkadina_official, @flying.katja.

This shift confronts the specific theatrical live presence that had previously been dependent on the copresence of actors and spectators at the same time, in the same space (Fischer-Lichte 2008) as an indissoluble condition of human sociality. In the real virtual, on the contrary, as in all-over digital liveness, sociality can be seen as a colorful and easy remix of different kinds of technical *components*. In the same movement, the analog social becomes a mediated past that is perfectly integrated into this potpourri of ever newly remixed realities once it is saved as a video file on a laptop or in social media. The result is that digital liveness becomes socially accepted and habitualized as fundamental for human sociality, as it had been for performative copresence.

It is within this extension and becoming commonplace of the real virtual that the process of subjectivation is launched, figuring as the basis of the constitution of sociality by performing telepresence as digital liveness. The interesting point is that within the performance there is an attempt to revoke the de-subjectivation (Rouvroy 2013; Rouvroy 2020) of human agents in social media and economic platforms, using them as a condition for the inauguration of a subject. This installs a twofold movement that simultaneously takes over social media as a place of subjects and brings them under the regime of technical operations.

The subjectivation is managed by the modes of actor's play in *Möwe.live*. Suddenly a very traditional "speaking theater" invades the virtual space of social media, grounded in a Stanislawski-like actor's play, generating coherent figures with a complex inner psychic structure (Stanislawski 2007). This aesthetic is almost unbearable, as it is highly subjective in the sense that each character in the play is just turning around itself, egomaniacally. The question is, what is the performance working toward with this ego excitement, running into a hyper-subjectivity. Possible answers come from an artist talk after the premiere in December.[33] It was noted that the egomaniacal actor's play corresponds not only with the constitution of the characters in Chekhov's *The Seagull* but also to social media's culture of the self. On Instagram, according to the artists, people create and excessively manifest their egos through images and short films.

But as seen in the description of the play above, these subjects are isolated and menaced by destruction. The characters cannot live up to the image they present of themselves in social media. On the contrary, they fail to, for example, become successful actors. As the performance uses clichés, it turns the characters of the play into quotations, patterns, empty shells that are uninhabited and uninhabitable. The consequences unfold within the piece, specifically within Kostja's development. He commits suicide at the end because he has understood that even his success with a piece of digital theater doesn't make him happy or complete.

33 There is unfortunately no documentation of the artist talk, only notes made by the author while listening to the discussion.

This ambivalence signals that the technological constitution is breaking through, showing that the interpretation of humans as they appear in social media as persons or subjects is highly problematic and naïve, because the "selves" generated are just an operative personalization of human actions from and for data mining with no subject. As media scholar Antoinette Rouvroy argues,

> Algorithmic governmentality does not produce any kind of subject. It affects, without addressing them, people in all situations of possible criminality, fraud, deception, consumption, ... which are situations where they are not requested to "produce" anything, and certainly not subjectivation. Rather, algorithmic governmentality bypasses consciousness and reflexivity, and operates on the mode of alerts and reflexes. (Rouvroy 2013, 153)

Against this background, the only decision Kostja can make in this piece in his function as a digital place marker is to end his life. It becomes obvious that the performances are serious and dangerous games. Their performing the real virtual is creating the illusion of a self, hiding the technological condition, sparking, on the contrary, the amount of given data. The consequence, which corresponds to an alleged kind of way out, is finally a desire for death.

The thesis is that the *Möwe.live* performance gets its specific effect and relevance from foisting the status of subjectivity—as being related to a subject—and the power of subjectivation onto data mining on the internet, which has, as previously stated, no interest in persons, selves, or subjects. On the contrary, the subjects are generated during their selfish staging, whereas the technical systems are working with purely operative, algorithmic correlation and distribution. In this context, the performance can only generate the illusion of an individual use, and of a consistent subject, concealing the technological a-subjectivity. In this sense, *Möwe.live* is not only about a misinterpretation of the technological conditions of digital cultures but also about a betrayal of their status, hiding them behind an allegedly successful process of *individualization*. Beyond the pseudo-subjectivity, the persons are purely and operatively data-collecting and data-connecting points that desire to be continuously connected and networked.

So, the performers adapt, without recognizing it, what Antoinette Rouvroy (2020) called "algorithmic governmentality."[34] Capturing and redefining social media as environments for an existence in ubiquitous distant socializing also means making video conferencing part of the algorithmic operations, such as collecting and connecting data on the internet. So *Möwe.live* makes video conferencing a

34 She writes: "Algorithmic governmentality is the idea of a government of the social world that is based on the algorithmic processing of big data sets rather than on politics, law, and social norms. ... With big data, the idea is to generate hypotheses and classification criteria from the data" (Rouvroy 2013, np).

question of being performed by algorithms while performing them. Furthermore, by combining Zoom and social media, it became clear in the performance that in digital liveness the technologically generated self is a fragile and illusionary construction; a fact that was not always obvious in endless Zoom meetings, black-boxing the technological conditions of image processing. Thus, the opportunity that *Möwe.live* offers is to make clear that data, not persons, are performing in distant socializing.

The sociality of the ubiquitous real virtual and its digital liveness popping up under these conditions is of a specific kind: Our society, represented by the performers, is no longer oriented toward a democratic community of subjects but toward an assemblage of fragments and quotes. It is about lost persons who are identifying themselves with empty data, filling them with sense, needing each other only in order to be perceived and acknowledged. Sociality becomes a ritual of producing correlating quotes and patterns and connecting to technological conditions, which primarily helps the regime of relationalization, elaborating the adaptivity for techno-human cooperativity.

It is certain that subjectivation in today's digital cultures is linked to the exchange with data-mining and the becoming-social of its effects (e.g. profiles, strategies to make people give data, adaptation to connectivity within techno-human cooperations). This subjectivity is distributed over technological devices. A "gaming around" within this constellation is to be undertaken, though it should be done without the illusion of being a subject within social media but instead by recognizing that your participation is part of a performance of and with data flow.

Epistemology of the Real Virtual: Let's-Play-Knowledge

In March 2021, the German *Dramaturgische Gesellschaft* annual conference was on "Digitality and Performing Arts." It was set up as a "let's play"[35] on Twitch, replacing the traditional format of talks, inputs, and discussions as media of knowledge generation and exchange.[36] "let's play" is about playing computer games and making comments while doing so.[37] During the conference, artists, cultural studies scholars, and programmers formed teams. These teams were asked to reflect on gaming itself as a cultural technique and to discuss the aspects and issues of theater and digitality during their sessions.

---

35 For the history and aesthetics of the "let's play" format, see Ackermann 2017.

36 The concept for this event was elaborated by artists, and pedagogues Sarah Fartuun Heinze, Christiane Schwinge, and Friedrich Kirschner. It consisted of twelve "let's play" sessions, distributed over three days. For further details, see http://konferenz-2021.dramaturgische-gesellschaft.de/ (accessed May 19, 2022).

37 See, for example, Gronkh, a well-known figure in "let's play" (Gronkh-Wiki 2018).

With this performing telepresence of "let's play," the categories of video conferencing and of VR systems/desktop VR and their history, which were presented as two strands earlier in this paper, are finally converging.[38] The reason is that the twitch.tv platform usually transmits live performing video games (adventure games, or role plays, see Pepe 2020) as an application of desktop VR. With this amalgamation, the "let's play" of the events at *Dramaturgische Gesellschaft* converted gaming into a fundamental element of the epistemology of digital liveness. The events became a serious game for knowledge and knowledge production of the real virtual.

Figure 7: Let's play, Judith Ackermann and Christiane Schwinge

Source: Dramaturgische Gesellschaft Annual Conference: DiG IT ALL/LET'S PLAY, March 26–28, 2021, concept and moderation: Sarah Fartuun Heinze (theater pedagogue, culture educator), Christiane Schwinge (Initiative Creative Gaming, PLAY Festival Hamburg) and Friedrich Kirschner (Spiel && Objekt master's degree program), https://dramaturgische-gesellschaft.de/blog/lets-play-judith-ackermann/, http://konferenz-2021.dramaturgische-gesellschaft.de/2021/02/22/lets-play-2/?block=2, https://dramaturgische-gesellschaft.de/blog/dig-it-all-gekommen-um-zu-bleiben/ (accessed August 11, 2022).

A short description should give an impression of the performance. In this setting and framework of rules, an impressive explosion of sentences, words, exclamations, notions, and gestures was activated, which involved shifting from one glimpse of ideas and fragments of thought to another, with no connection or logical reference

38 See also footnote 8.

to each other. Even the subjects of understanding had been dissolved as the individuals of the gaming teams were busier creating avatars as actors in the games being played, also assisted by the wishes of the spectators, than following a coherent line of thought.[39] It seems as if within this process, the avatars, as a technological part of the doubled transgressive body in the real virtual, were getting ready to be an active part of thinking and understanding.

A deeper analysis and reflection of these sessions, concerning the games, or the performance of performing them, is beyond the scope of this paper.[40] Therefore, the focus will be on some initial ideas on the epistemology of this setting, coming out of, and at the same time realizing a real virtual. How could playing around fit in with the production of knowledge and cognition under the conditions of the telepresence and digital liveness within distant socializing?

An answer lies in the hyper-performativity of the "let's play" performance. The players had to create the whole performing of the gaming from scratch, without being prepared for what would come. Generating knowledge within digital liveness in virtual environments became an effect of a complex being-in-a-situation in which the agents had no distance or time to think, reflect, or prepare. Doing, reflecting, acting, deciding, performing, and affection-effects happened all at once. In that constellation, knowledge mutated into dealing with and confronting the joy of non-knowledge. Furthermore, within the flow of performing gaming, subjectivity became obsolete because it was not subjects that acted and decided but an assemblage of factors—for example, the gamers, the software of the games, the spectator-participants offering comments, votes, and questions, and the technology of this gaming. So, the "let's play" builds on what was found within the discussion of digital subjectivation: the subjects became connecting points for data flow and data analysis. While in *Möwe.live* this constitution resulted in death, in performing the gaming, it became pure fun and enjoyment.

With the gaming-a-game, contingency becomes the ground of epistemology, constituted in pure processuality and flow, causing uncertainty, unpredictability, *and uncontrollability*. With non-knowledge and contingency, the real virtual becomes a celebration of the regime of continuous adaptation and anticipation. Or, in other words, "let's play" trains adaptation and anticipation as a condition of knowing, understanding, and communication, instead of looking for facts, proofs, experimentation in spelled out assumptions, and lines of thought. It is, on the contrary, about a

---

39 For a good example, see the recordings of Judith Ackermann's and Christiane Schwinge's gaming of Sims Character Editor at the Dramaturgische Gesellschaft Conference, 2021: https://dramaturgische-gesellschaft.de/blog/lets-play-judith-ackermann/ (accessed May 19, 2022).

40 For deeper insights and analysis, see a blog (Vogelsang 2021a) and comments on Twitter (2021b) by Arne Vogelsang, who was a live commenter for the event.

consistently beginning, transforming, flowing, running, scraping, and being overwhelmed and overburdened. The focus is on a performance of knowledge, not on gaining knowledge. The virtual real, generated in telepresence and digital liveness, gets its own epistemology of "gaming around," producing a subject in between overwhelming and moments of concentration.

Resume: Adaptation and Anticipation of Techno-Human Cooperation

To sum up: telepresent performances helped understand the mentality, sociality, and epistemology of digital cultures in which distant socializing constitutes its own reality—being equal to analog live settings and overcoming the restrictions of liveness by, for example, removing the need to be copresent in the same place at the same time. This becoming real of the virtual was enabled by performing telepresence within different formats of video conferencing (both historical formats and today's Zoom) as well as in desktop VR, making them a real virtual. Furthermore, the examples show that distant socializing is gradually becoming more normalized and habitual, combining the corporal, the mental, the social, and knowledge production of telepresence and digital liveness.

Concerning the value of this performing telepresence for the regime of digital cultures, it could be assumed that training in the mentality, sociality, and epistemology of the real virtual builds a perfect interplay with the regime of adaptive and anticipative techno-human relationalization. For example, training in auto-s(t)imulation is established in techno-human adaptation and anticipation, preparing the generation of a real virtual body and mentality. This fundamental relationalization is supported by the invention of an illusionary subject, which is not only able to give sense to data mining but is also existentially bound into a pure digital existence. The epistemology of the real virtual is adaptation and anticipation. This becomes obvious in "let's play" knowledge, where reacting to what is coming up in the performativity of improvisation is primary—you never know what will come; you cannot plan. The impact of this regime of relationalization becomes clear in the question, How can you critique or change something that is the condition of your existence? Why be critical toward fun and contingency? Who should be the executor of critique if there is no-body? Moreover, the regime of relationalization sells itself, as if there is for the first time a good power play.

So, a new form of critique is necessary. The required critique, which is now emerging, must be able to deal with the hyper-entangled being-involved in techno-human cooperations, constituted by a permanent tipping movement between technocratic subjection and "gaming around," providing humans with moments of sporadic and instable agency (Leeker 2022). Performing telepresence as enabler of the real virtual is then readable as a cultural technique for a new type and epistemology of "skilled player" (*Versierter Spieler* in German, Leeker 2022), who is simultaneously part of the game, being absorbed by it, and configuring it by performing. This concept is not a version of cultural pessimism, mourning the loss of modern ideas such as human autonomy. On the contrary, it encapsulates the irreducible entanglement of humans, technical devices, data, algorithms, companies, politicians, mass media, and the market as the promising mode of today's digital existence: A warm welcome to distant socializing, and: Let's play!

Acknowledgements

I would like to thank Olga Moskatova for her constructive critical reading, and special thanks to Janet Leyton-Grant for copyediting support.

References

Ackermann, Judith, ed. 2017. *Phänomen Let's Play-Video: Entstehung, Ästhetik, Aneignung und Faszination aufgezeichneten Computerspielhandelns*. Wiesbaden: Springer VS.

Baudrillard, Jean. 2004. *Simulacra and Simulation*. Ann Arbor: University of Michigan Press.

Burk, Juli. 1999. "ATHEMOO and the Future Present." In *Theatre in Cyberspace*, edited by Stephen Schrum, 109–34. New York: Peter Lang.

Decker, Edith and Peter Weibel, eds. 1990. *Vom Verschwinden der Ferne: Telekommunikation und Kunst*. Köln: DuMont.

Diesselhorst, Sophie. 2021. "Mit einer URL in die digitale Zauberwelt: Worin liegt das Potenzial von Mozilla Hubs?—Ein Interview mit den VR-Theater-Spezialisten Nils Corte & Roman Senkl / minus.eins." *Nachtkritik* (12. Februar 2021). https://nachtkritik.de/index.php?option=com_content&view=article&id=19168:worin-liegt-das-potenzial-von-mozilla-hubs-ein-interview-mit-den-vr-theater-spezialisten-nils-corte-roman-senkl-minus-eins&catid=101&Itemid=84 (accessed May 20, 2022).

Distelmeyer, Jan. 2021. "Programmatische Verhältnisse Wer oder was lebt in Zoom? Fragen an die neue Normalität von Videokonferenzen." *Cargo* 49. https://www.cargo-film.de/heft/49/essay/programmatische-verhaltnisse/ (accessed May 20, 2022).

Durland, Steven. 1987. "Defining the Image as Place: A Conversation with Kit Galloway, Sherrie Rabinowitz & Gene Youngblood." *High Performance: A Quarterly Magazine for the New Art Audience* 37: 52–59. http://www.ecafe.com/museum/hp_gy_1987/hp_gy_1987.html (accessed May 20, 2022).

Esposito, Elena. 1998. "Fiktion und Virtualität." In *Medien, Computer, Realität. Wirklichkeitsvorstellungen und neue Medien*, edited by Sibylle Krämer, 269–295. Frankfurt am Main: Suhrkamp.

Fischer, Jan. 2021. "Möwe.live-punkt live. Als alles noch gut war." *Nachtkritik*. https://www.nachtkritik.de/index.php?option=com_content&view=article&id=20374:moewe-live-punktlive&catid=247:staatstheater-nuernberg&Itemid=40 (accessed May 20, 2022).

Fischer-Lichte, Erika. 2008. *The Transformative Power of Performance: A New Aesthetics*. London/New York: Routledge.

Gronkh-Wiki. 2018. "Let's Play Minecraft." https://gronkh-wiki.de/wiki/Let%27s_Play_Minecraft (accessed May 20, 2022).

Hammelburg, Esther. 2020. "Live Event-Spaces: Place and Space in the Mediatized Experience of Events." In *Locating Imagination in Popular Culture: Place, Tourism and Belonging*, edited by Nicky Van Es, Stijn Reijnders, Leonieke Bolderman, and Abby Waysdorf, 215–29. London: Routledge.

Heinrich-Böll-Stiftung, and nachkritik.de, ed. 2020. "Netztheater: Positionen, Praxis, Produktionen." https://www.boell.de/de/netztheater (accessed May 20, 2022).

Kasprowicz, Dawid. 2020. "Virtualisierte Arbeitsplätze: Medienhistorische Annäherung an die Interaktion und Kollaboration zwischen menschlichen und maschinellen Agenten." In *Handbuch Virtualität*, edited by Dawid Kasprowicz and Stefan Rieger, 129–47. Wiesbaden: Springer.

Klappert, Annina. 2020. *Sand als metaphorisches Modell für Virtualität*. Berlin: De Gruyter.

Kooboto. 2020. "Ok, Who is CodeMiko?" https://www.youtube.com/watch?v=CsQjxEd-gsw (accessed May 20, 2022).

Leeker, Martina. 1995. *Mime, Mimesis und Technologie*. Berlin: Alexander.

Leeker, Martina. 2002. "Dance on Telematic Stages: Paul Sermon." In *Dance and Technology: Moving towards Media Productions*, edited by Söke Dinkla and Martina Leeker, 244–305. Berlin: Alexander.

Leeker, Martina. 2019. "Performing Technology." In *Handbuch Virtualität*, edited by Dawid Kasprowicz and Stefan Rieger, 1–18. Wiesbaden: Springer.

Leeker, Martina, 2021. "Pain, Befallen-ness, Conflict, Mediocrity: Dark Entanglements for Artistic Research and Critique in Digital Cultures." In *Throwing Gestures: Protest, Economy and the Imperceptible*, edited by Florian Bettel, Irina Kaldrack, and Konrad Strutz, 293–314. Wien: Verlag für moderne Kunst.

Leeker, Martina. 2022. "Be part, play the game! Vorschlag für ein Modell zu Bildung in digitalen Kulturen." *Kulturelle Bildung online*. https://www.kubi-online.de/index.php/artikel/be-part-play-the-game-vorschlag-modell-bildung-digitalen-kulturen (accessed May 20, 2022).

Paulsen, Kris. 2013. "'Image as Place': The Phenomenal Screen in Kit Galloway and Sherrie Rabinowitz's Satellite Arts 1977." *Leonardo Electronic Almanac* 19 (2): 98–111.

Paulsen, Kris. 2017. *Here/There: Telepresence, Touch, and Art at the Interface*. Cambridge, MA: The MIT Press.

Pepe, Felipe. 2020. "ROBLOX is a MUD: The History of MUDs, Virtual Worlds & MMORPGs." https://felipepepe.medium.com/roblox-is-a-mud-the-history-of-virtual-worlds-muds-mmorpgs-12e41c4cb9b (accessed May 20, 2022).

Rieger, Stefan. 2019. *Die Enden des Körpers: Versuch einer negativen Prothetik*. Wiesbaden: Springer.

Rouvroy, Antoinette. 2013. "The End(s) of Critique: Data Behaviourism versus Due Process." In *Privacy, Due Process and the Computational Turn: The Philosophy of Law Meets the Philosophy of Technology*, edited by Mireille Hildebrandt, and Katja de Vries, 143–68. New York: Routledge. https://publicityreform.github.io/findbyimage/readings/rouvroy.pdf (accessed May 20, 2022).

Rouvroy, Antoinette. 2020. "Governmentality and the Death of Politics." *Green European Journal*. https://www.greeneuropeanjournal.eu/algorithmic-governmentality-and-the-death-of-politics/ (accessed May 20, 2022).

Sauerländer, Tina. 2020. "Virtuelle Räume selbst gestalten. Wie die SocialVR-Plattform Mozilla Hubs WebVR und VR verbindet." *Netzpiloten Magazin*. https://www.netzpiloten.de/virtuelle-raeume-selbst-gestalten-mit-mozilla-hubs/ (accessed May 20, 2022).

Shen, Xiaojun, and Shervin Shirmohammadi. 2008a. "Telepresence." In *Encyclopedia of Multimedia*, edited by Borko Furht, 849–52. Boston, MA: Springer.

Shen, Xiaojun, and Shervin Shirmohammadi. 2008b, "Virtual and Augmented Reality." In *Encyclopedia of Multimedia*, edited by Borko Furht, 962–70. Boston, MA: Springer.

Stanislawski, Konstantin Sergejewitsch. 2007. *Die Arbeit des Schauspielers an sich selbst und an der Rolle*, edited by Bernd Stegemann. Berlin: Henschel.

Vogelgesang, Arne. 2021a. "Konferenzbeobachtung." https://dramaturgische-gesellschaft.de/blog/arne-vogelgesang-konferenzbeobachtung/ (accessed May 20, 2022).

Vogelsang, Arne. 2021b. "Thread zum Let's Play." https://twitter.com/foyglgezang/status/1375829094978437122 (accessed May 20, 2022).

"In Eight and a Half Seconds the World Has Changed"
An Interview with Telecommunication Art Pioneer Bill Bartlett

Tilman Baumgärtel

Figure 1: Bill Bartlett

Source: Private.

Bill Bartlett was among a small group of international artists who used the telecommunication media that became available to them in the late 1970s and early 1980s to conduct artistic actions, performances, and collaborations. For this purpose, they experimented with early computer networks, satellite connections, and later fax and ARPA-net email. But the most important and most accessible media for these activities was the little-known Slow Scan Television (SSTV) that they used to send pictures in real time from and to art spaces and studios in North America and Europe (Bartlett 2019).

SSTV was an analogue picture transmission method, used mainly by amateur radio operators, to transmit and receive static pictures via radio in black and white and without sound. SSTV has also been called "narrowband television." It was employed during the space missions Apollo 7, Apollo 8, Apollo 9, and Apollo 11 to send pictures from space to NASA, but later also by ham radio enthusiasts who wanted to create their own low-band television stations and to set up early precursors of today's teleconferences.

The ways in which artists used this technology predate contemporary video conferences in terms of not only technology but also the methods they developed to interact and collaborate. A lot of the artistic work that was done via SSTV addresses the ways of "making contact" with remote partners and associates via telecommunication media. Since there was no inherent audio connection, communication via gestures, symbols, visual clues, props, and methods such as sign language became an important part of their artistic exploration and vocabulary.

Today, these experiments are not very well remembered, partly because of their time-based and ephemeral nature, partly because they did not fit into an art world that has increasingly focused on *commodifiable* paintings and objects that can be traded on the art market. For a long time, there were technical issues, too: the SSTV performances by Bartlett and other artists were recorded as modem-style sound signals on audio tapes with a recording device called Robot. These analogue files tapes were only recently decoded using historic equipment and modern computers and are now available in digital format.

These experiments belong to a little-known prehistory of the internet-based art of today. The best-known attempts to induct art into the global media at that time were by Nam June Paik, who made television as well as history with the global satellite projects *Good Morning, Mr. Orwell* (1984), *Bye Bye Kipling* (1986), and *Wrap around the World* (1988). Like Paik, other artists tried to get access to international television satellites and global media networks early on: in 1980, the artist duo Mobile Image (Kit Galloway and Sherrie Rabinowitz) in their work *Hole in Space: A Public Communication Sculpture* had people in New York and Los Angeles interact live with each other via satellite video connection; "satellite sculptures" almost became a genre in its own right in the following years, with artists as diverse as Ingo Günther, Peter Fend, Dennis Oppenheim, Sharon Grace, Liza Bear/Willoughby Sharp, Wolfgang Staehle, and General Idea participating (for an overview of this period of media art, see Baumgärtel 2000). From today's point of view, these artistic endeavours preceded the video conferences of today not only on a conceptual level in a technical environment that wasn't accessible to most people. They also initiated modes of interaction and forms of communication that have since become commonplace in our use of Skype, Zoom, FaceTime, and other platforms.

An important instance of the artistic attempt to "open up" the very expensive and inaccessible international media networks was the conference "Artists' Use of

Telecommunication" in which Bill Bartlett participated at the San Francisco Museum of Modern Art, where it was physically held, but in which artists in other cities and countries were connected via satellite using the time-sharing computer system of the insurance company I. P. Sharp. "Remote participants" in this symposium included Gene Youngblood (Los Angeles), Hank Bull (Vancouver), Douglas Davis (New York), Norman White (Toronto), and Robert Adrian X (Vienna). All the mentioned artists were pioneers of the use of telecommunication in the arts.

In the late sixties, Bill Bartlett was involved with the initiative *Experiments in Arts and Technology* (E.A.T.) after he had received his bachelor of fine arts at the Otis Art Institute in Los Angeles. E.A.T. was a nonprofit and tax-exempt organization that was established in 1967 to develop collaborations between artists and engineers (NTT InterCommunication Center 2003). Engineers Billy Klüver and Fred Waldhauer and artists Robert Rauschenberg and Robert Whitman had collaborated in 1966 on the groundbreaking performance art event *9 Evenings: Theatre and Engineering* in New York. They subsequently formed E.A.T to continue to combine the aesthetics interests of artists with the technical expertise of technicians, which enabled artists such as John Cage, Lucinda Childs, Öyvind Fahlström, Deborah Hay, Yvonne Rainer, and David Tudor to create work they would not have been able to do without technical support. The pinnacle of the activities of E.A.T. was the Pavilion for the Osaka World Fair in 1970, which was sponsored by Pepsi and was a psychedelic *Gesamtkunstwerk* with contributions by artists such as David Tudor, Fujiko Nakaya, Robert Breer, and Robert Whitman (Kluver et al. 1972). Bartlett was a volunteer at the Los Angeles office of the organization and worked on the construction of an early version of the pneumatic Pavilion in the US.

After moving to Victoria, Canada, he cofounded the Northwest Coast Institute of the Arts (today the Victoria College of Art) and joined the artists initiative Open Space, where he was instrumental in setting up an exhibition space that is still operational today. As artistic director, he got involved with artists who were using Slow Scan TV and computer networking. That led to the development of ARTEX, the first online system for artists, that was used for a number of groundbreaking online art projects. He organized twenty-two telecommunication events between 1978 and 1983, including *Sat-Tel-Comp*, *Pacific Rim*, and *Interplay Workshop*, where the artistic use of Slow Scan Television was part of the aesthetic tool set.

The Slow Scan technology that Bartlett used around 1980 was only able to send individual, sequential pictures over copper telephone lines. When ISDN became available, it became possible to actually "stream" moving images around the globe without access to expensive telecommunication satellites, for example during the art project *Piazza Virtuale* by the German media art group *Van Gogh TV* at documenta 9 in 1992 (Baumgärtel 2021), which simultaneously used a version of SSTV to connect with studios in countries where ISDN television was not available (Van Gogh TV 2021). Hence, the use of the low-resolution, static transmission of black-and-white images via the Robot encoder and copper telephone cables was a step toward a television that was run and transmitted by artists.

At the same time, it allowed their users an early form of the kind of tele-interaction that we take for granted today, especially after the Covid pandemic made video conferencing ubiquitous. Whereas these interactions were originally limited to facial expressions and gestures, as there were no audio connections in the first iterations of the technology, the creative ways to overcome those constraints are also a precursor of the ways that the world adapted Zoom and other platforms to their needs during the Corona pandemic.

Tilman Baumgärtel

Tilman Baumgärtel: In the late 1970s and early 1980s, you were involved in a number of art projects that used Slow Scan Television (SSTV). I would like to begin the interview with a technical question. Can you explain, first of all, what Slow Scan Television was and why it attracted you as an artist?

Bill Barlett: Slow Scan Television was developed by a fellow from Eastern Canada in 1958. His name was Copthorne Macdonald, and he developed a method to send pictures via ham radio. Depending on the connection you send a black-and-white still picture in eight and a half seconds, but if the connection was slow, it could take up to a minute to wire a picture from one location to another. These pictures consisted out of 120 lines, and each line had 120 pixels that would appear line by line on the monitor of the receiver, kind of like a fax. Depending on the speed, these sequences of still pictures were like the different phases of a filmed movement. It was a freeze-frame that was scanned down in an eight-and-a-half-second scan that would actually replace the previous image.

I was interested in this technology because it provided a way to communicate with others via a 3 kHz telephone channel. It wasn't exactly like having your own television station, but it was a way to transfer somewhat moving pictures from your place to another place. We were using a standard video camera to record ourselves. You would take the image, convert it into an audio signal with the Robot Research 530 Phone Line Transceiver, and all of this beep-beep-beep-beep-sound would be sent through a normal phone line, through ham radio, whatever way you can send a voice. You could send those pictures via shortwave, VHF, and UHF radio in your vicinity, and if you hooked it up to the telephone network, you could send these pictures to any other telephone connection in the world.

Tilman Baumgärtel: Did you have a background in media art before you started to work with SSTV?

Bill Barlett: When I was getting my bachelor of fine arts degree at the Otis Art Institute in Los Angeles in 1970, I was interested in both sculpture and figurative art and life drawing. While I was in Los Angeles, I got involved in performance art. And I was a volunteer for Experiments in Art and Technology (E.A.T.), an organization that the artist Robert Rauschenberg and the engineer Billy Klüver started in New York in the 1960s and that was about putting artists together with engineers. They opened an office in Los Angeles. Somehow, I got into volunteering to run their office in LA, and it gave me an opportunity to look through all the files. Art critic Gene Youngblood and a lot of people that were active in the area were quite taken by E.A.T.

I also worked on the construction of the large Spherical Mirror Dome, that E.A.T. presented at Expo '70 in Osaka, Japan. I was one of the artists working on the design model. They set up a big workstation in a dirigible hanger in a military base down south of Los Angeles with guards with machine guns guarding the place and designed the gores and put them together. It was a pneumatic structure, a ninety-five-foot mylar dome mock-up that was sponsored by Pepsi. We would cut gores from rolls of mylar and tape them together. The dome sort of exploded and had to be redesigned, but it was a fabulous experience to work on that.

Tilman Baumgärtel: In the 1970s, you were involved with the nonprofit arts organization Open Space in Victoria, Canada, and later founded your own group, the Direct Media Association, that developed global art events that used SSTV and pre-internet computer networks to connect artist around the world. At this time only a very small group of artists were involved with network and telecommunication art. How did you become part of this scene?

Bill Barlett: One of the programs that we had for the Open Space Arts Society in Victoria was—my wife actually coined the term—the Collaboratory. It was a cross between collaboration and laboratory, a kind of workshop. We came up with the Collabratory idea at just about the time when I went to a session at the LA Institute of Contemporary Art (LAICA), where one of the featured speakers was filmmaker and artist Liza Bear. She had done the *Send/Receive* project about six months before, where the artists set up a two-way satellite link between New York and San Francisco using a satellite that was co-owned by NASA and the Canadian government. The resulting program was broadcast on Manhattan Cable's public access channel.

Tilman Baumgärtel: *Send/Receive* (1977) was a pioneering piece of telecommunication art; the artists involved include Terry Fox, Sharon Grace, and Carl Loeffler in San Francisco, and Liza Bear, Willoughby Sharp, and Keith Sonnier in New York. Bear and Sharp were early champions of the artists' use of telecommunication media. It is probably impossible today to imagine how hard it was for people who did not work for a television station to broadcast material to other people. Today, with YouTube and Zoom and streaming it has become so common ...

Bill Barlett: Actually, they had a lot of technical problems with the satellite when they did *Send/Receive*. I was introduced to Slow Scan TV at the same time Liza Bear was, probably at LAICA. The next Collaboratory was called *Sat-Tel-Comp*, which stood for Satellite Telephone Computer, and we invited Liza Bear to come to Open Space from October 30 to December 9, 1978. Prior to that, in May and June of 1978, we started a series of small interactive link-ups in preparation. One was with Willoughby Sharp in New York, one was with Sharon Grace in San Francisco, one was done locally in Victoria, and one was multipoint with artists in Victoria, Vancouver, Toronto, and Memphis, Tennessee.

Figure 2: Sat-Tel-Com performance at Open Space (1978). Jim Starck, Bill Bartlett with mirror, Jim Lindsay, Chas Leckie, Susan Cormin, Daryl Lacey (Video Inn)

Source: Open Space Archives, Victoria, BC, Canada.

Tilman Baumgärtel: Who were these collaborators in these different cities?

Bill Barlett: When I was working at Open Space, I went to a lot of art spaces in North America. There was the organisation for artist-run centers (ANNPAC), and I did make a point of networking and getting out across Canada visiting these centers representing Open Space. They would eventually be the link for the telecommunication projects. That played a big part in the organization of these SSTV projects because it didn't just happen overnight with making the contacts. They came from a lot of legwork.

Tilman Baumgärtel: If you look at the pictures from these events, there is a lot of equipment ...

Bill Barlett: Yes, it looks like a lot of technology, but it's actually not very sophisticated. Once you had the equipment, the Robot for SSTV, monitors, camera, recording device, phone lines, it was not very difficult to operate.

Tilman Baumgärtel: You used this kind of technology to do performances at the Vancouver Art Gallery, the San Francisco Museum of Modern Art, MIT in Boston, and many artist centers around the world. You had to introduce all these places to the SSTV technology ...

Bill Barlett: ... and sometimes it was very hands-on. There's one photo of me where I basically take an old handheld telephone, unscrew the speaker part, put some boxer clamps or whatever onto the wires, and then feed the audio signal right through the Robot transceiver into the phone link.

Tilman Baumgärtel: Well, that's what the early hackers did when they did not have a real modem ...

Bill Barlett: That's right. It was just another hack. You could have more complicated setups, where you had one monitor that showed the SSTV scan and the other monitor showedwhat the camera was seeing. But in any case, it was really low tech. Inexpensive. And we could do it almost at any place as long as we could get matching equipment at the other end, which we usually borrowed from the equipment dealers. You could do it with only one person if you wanted. I invested into the basic equipment to make it happen, but overall it was not very expensive. And because of all the networking, slowly an infrastructure for this kind of activity started building up. So, one event literally built on another.

Tilman Baumgärtel: What were these performances like that you did via SSTV?

Bill Barlett: These activities were very informal. We would just sit down and talk about what we should do. You'd sort of make a plan on what you want to do, knowing the Robot took eight and a half seconds per image, and work in a progression of images. Look into the camera, respond to the last image, and finish the piece. I could do some visual sign or symbols or something. Then I could put a sign up that said, "Over to you." And then the other party would know that they could start sending or I could turn the camera to you. It was just a steady stream of images. You did it because you could watch, see what was on the monitor ...

Remember, at first, we had no audio connection, there was only this stream of freeze-frame pictures. So, anything we wanted to communicate, we had to do with gestures or signs or text. There was some very clever use of sign language. I found that the close-up of doing something facial was easy to relate to, to faces and hands. Most of what I did was based on that. There's one piece where I just taped over my whole face. There were dance pieces that created involvement. But I think there's so much movement in dance that takes place in between one frame and the next that you might miss an entire movement of how that body got from this place to that place. In eight and a half seconds the world has changed.

Apple eating worked well with this medium, because you could actually see something shrink down by the bites of eating. Most of the things I did were somewhat humorous. The more I worked with it, the more I thought it was a very personal kind of thing. I loved working with my glasses. I took a picture where I was wearing my glasses. Then I move it slightly, and it captures after the eight and a half seconds later how I've moved my glasses. The next one I might take my glasses off. And then finally pull it away.

I have a real love of photo booths, where you have no photographer. You're in there doing it yourself. I have a huge collection of photo booth images and tried to incorporate them into my visual art and things that I'm doing today. That is not so different from the stream of single images that you transmitted via SSTV. I keep changing, and I'm creating different pictures. And it's black and white. It's really stunning. People are talking about how it slows you down as you're watching these SSTV performances. It can seem boring at first, and yet the more you get into it, you get into the cycle. It really does suck you in. You just say, "Wow."

Tilman Baumgärtel: So, the limitations of this medium were also a challenge to your creativity?

Bill Barlett: Yes. Later we started to do these multipoint events, where we used a telephone conferencing service out of Denver, Colorado, that could provide audio link and the video link. So, each event was slightly more sophisticated. More events, more parties.

Tilman Baumgärtel: Did you see this as an art activity or rather an experiment with this new medium?

Bill Barlett: I think of it as art. My point of view was that of the initiator, as someone that brought people together. I had an interest in doing the *Sat-Tel-Comp* Collaboratory program. It really fit into that. I had an affinity for what we were doing. I saw myself as a facilitator of these activities, and it was an awful lot of work to put it all together, to get people involved and to make sure that there was something going on.

Tilman Baumgärtel: I assume these activities were so ephemeral that they were not recorded?

Bill Barlett: No, they were recorded. Because SSTV was an audio link, we could store the information, these beeping noises, onto an audio cassette connected to the Robot transceiver. It also played the recordings back. So, what we're faced with today is all of these audio cassettes, with all these beeps on them, but you need the Robot to decode these records.

Tilman Baumgärtel: Were these activities announced to the public? Was there an audience?

Bill Barlett: At first, there was virtually no audience. It was the group of artists that were there. Or if there were people in the gallery, where the event was taking place, we'd say, "Come on in. This is what we're doing." It was very spontaneous. The audience became part of the performances very quickly.

Figure 3: Bill Bartlett with the Robot Research 530 Phone Line Transceiver on the right side

Source: Open Space Archives, Victoria, BC, Canada.

Tilman Baumgärtel: What was the typical reaction of people when they first encountered SSTV?

Bill Barlett: It is very intuitive. You step in front of that camera and suddenly by having the two monitors you can see exactly what's taking place. And it doesn't take much of an instructional manual to picture yourself. It's just so simple.
As things progressed, we got into the use of satellites again. NASA provided the ATS-1 satellite for the PEACESAT educational program around the Pacific Rim, and there was a terminal at Simon Fraser University in Vancouver that we could link into. Slow Scan TV was the perfect way of using them, because it was dirt cheap. It was just like talking on the telephone. We did quite a number of events with them. Because of that, there are hundreds of people that I've worked with that I never met personally but only worked with at a distance.
In 1979, we did a series of events that were called *Pacific Rim Slow Scan*, where we connected the Vancouver Art Gallery with locations in New Zealand, Santa Cruz, and the Cook Islands via the PEACESAT Satellite. I think we did something like twenty link-ups to the South Pacific. I provided one loaned Robot 530 that was air-expressed from island to island. The installation was from April 20 to May 21, and we had transmissions each Friday as the Robot traveled from site to site the rest of the week. One day I come to the Art Gallery and our link-up was with the Cook Islands, and one of the guys there said: "Bill, we have a present for you this morning." And the camera

pans down on these two guys in the water. And out of the water they brought this humongous sea turtle. I believe that their unit was in a grass shack on the beach. So, you could move our equipment anywhere as long as you could plug it into a telephone jack.

So, that got us really out in the public. There were some nice articles written about it, and that got me invited to a big event on art and telecommunications at the San Francisco Museum of Modern Art. It was called "Artists' Use of Telecommunications Conference." That brought together many artists working with pre-internet online media.

Fig 4: Jim Starck performing via SSTV

Source: Screenshot of performance documentation.

Tilman Baumgärtel: You were also involved in some of the artistic uses of online communication that were also subject of this conference in San Francisco. Tell me about that ...

Bill Barlett: In September 1978, I went to a conference in Toronto called "The Fifth Network." It was for independent video producers, and that's where I met the artists Robert Adrian X and Norman White. Norman was friends with Ian Sharp, who was the founder of the computer company I. P. Sharp that offered time-sharing services, which was one of the first online uses of computers. They ran the computer network of the Toronto Stock Exchange. What appealed to Norman was that they had a mailbox system for their techs out in the field so they could send email-type messages to their headquarters. So, Norman said, "Wouldn't it be great if we try to set up an online network for artists?"

Tilman Baumgärtel: But at that time not everybody had a computer.

Bill Barlett: Actually, that's very true. I went out and bought a Texas Instruments data terminal. They made a model with a suction modem built in, but mine wasn't that fancy. It had a separate modem, but I could literally go to a pay phone and take the receiver off the pay phone and plug it in and do my data connections. The message would go to their data center in Victoria, and then it would move from there. I. P. Sharp set up a mailbox for us artists that was called Artex.

Tilman Baumgärtel: That sounds like an early social medium for artists. Who else was on this network?

Bill Barlett: There was a small group of artists, including Robert Adrian X, Roy Ascott, Max Neuhaus, Norman White, Don Foresta, Eric Gidney, Western Front ... Artex allowed to send online messages to every member of the group, and it included a program that they called Confer, where you could do an online thing together—a chat if you will. As time went on, the group got bigger and more people wanted to share more. We paid by the character at that time, and we paid for everything going out and everything coming in. Three messages at that time would cost about $9. So, the more people got involved, the more expensive it became. If you CCed out a project proposal that was 18 pages long, everyone receiving this document would have to pay for the word count, whether or not you wanted to receive it.

Tilman Baumgärtel: It must have been tremendously expensive to participate in this network.

Bill Barlett: It *was* tremendously expensive. That was one reasons why I eventually got uninvolved. In one letter, Robert talked about how he just cleared up a $20,000 debt. So, it did add up a lot. We eventually worked with the programmers to bring up something that was a little more cost effective. But it reached a point where the financial burden became too much.

Tilman Baumgärtel: You were involved with the online art project *The World in 24 Hours* by Robert Adrian X, which he did at the Ars Electronica in Austria in 1982, which brought together all these different means of pre-internet telecommunication media. Following the midday sun around the planet, artists from all over the world sent work via SSTV, fax, telephone, computer networks to Robert Adrian X in Linz in Austria, where it was presented. That was a truly global performance, probably the largest of these early network and telecommunication art projects of that time ...

Bill Barlett: Yes, I participated from Western Front, an art space in Vancouver. But just before that I got a job with Canada Post. I was tired. I did a hell of a lot to build things up. And there was a lot of stress from working with a lot of artists, with organising things. It was really important to me to do whatever was necessary to make sure all the loose ends fit in coordinating a project. So, I became a postmaster on the small island that I lived on. And it really became a lifetime, fully pensioned job. I eventually became a trainer for Canada Post and spent six years involved with the Stamp Advisory Committee. It was a great job, and it provided a good, steady income. Looking at today's internet and interactive communications technology, what we worked on in the late 1970s seems primitive, but that's how things progress.

Acknowledgements

The author would like to thank Doug Jarvis, guest curator at Open Space, for facilitating this interview at Ars Electronica in 2019.

References

Bartlett, Bill. 2019. *Interplay: Exploring Telecommunictions, Art and Collaboration.* Unpublished Manuscript.

Baumgärtel, Tilman. 2000. "Net Art: On the History of Artistic Work with Telecommunications Media." In *net_condition: Art and Global Media*, edited by Peter Weibel and Timothy Druckrey, 152–62. Cambridge, MA: MIT Press.

Baumgärtel, Tilman. 2021. *Van Gogh TV's "Piazza Virtuale": The Invention of Social Media at documenta IX in 1992*. Bielefeld: transcript (open access: https://www.transcript-verlag.de/media/pdf/e0/fc/36/oa9783839460665Gcc2tWTcGf9ig.pdf).

Van Gogh TV. 2021. "Use of Slow Scan TV and ISDN Videophones." https://vangoghtv.hs-mainz.de/?portfolio=use-of-slow-scan-tv-and-isdn-videophones&lang=en (accessed May 7, 2022).

Klüver, Billy, Julie Martin, and Barbara Rose, eds. 1972. *Pavilion by Experiments in Art and Technology*. New York: E. P. Dutton & Co.

NTT InterCommunication Center (ICC). 2003. *E.A.T.: The Story of Experiments in Art and Technology*. Tokyo: Intercommunication Center

Working | Cooperating

Things in the Background

Video Conferencing and the Labor of Being Seen

Alexandra Anikina

When your interlocutor excuses herself and exits the frame of a video conference, leaving the camera on, where does your gaze go? Is it drawn toward the bookshelf in the background? Toward the photograph or the poster on the wall? Toward an accidental cat walking past the camera? Or do you switch tabs and check your emails?

The background (fig. 1) is always involved as a silent or not-so-silent participant in the visual culture of video conferencing. A home office has been a longstanding feature in the rise of freelance and enterprise economies. Still, the Covid-19 pandemic forcefully launched the so-called work-from-home experiment on an unprecedented scale, making many homes visible to the public through video conferencing. The act of seeing directly into other people's homes brought a new, technological dimension of vulnerability to the idea of a home office, already a site of precarious domestic and immaterial labor. The background brings up the questions of choice, or an impossibility to choose a communicational setting; of information divulged or hidden; of symbolic representation; of labor of watching and labor of being seen; and of what these questions mean for seeing the video conferencing as a practice of social production and reproduction.

The background is also a witness to the complicated networked architecture of the gazes in video conferencing. As Anne Friedberg points out, the screen produces voyeuristic "virtual windows" (2006) between the viewer and the looked-at. The frames—the screen and the application window—introduce "the rectangle of perspectival rendering" (2006: 38) that alters the very conception of space the communication takes place in. It might seem that if each interlocutor is equally involved in the live-streamed process of *seeing* and *being seen*, the situation is equal, unlike in live-streaming, blogging, or surveillance where the gaze is one-sided and does not necessarily take place in real-time. But the seemingly stable and equalizing point where the two gazes meet—on the screen—inevitably produces insight *into* the other person's life, seen on the background behind them. Video conferencing does not equalize the two gazes but rather introduces a mutable and live architecture that involves each participant in the acts of seeing and being seen.

Figure 1: The backgrounds of a video call. 2021

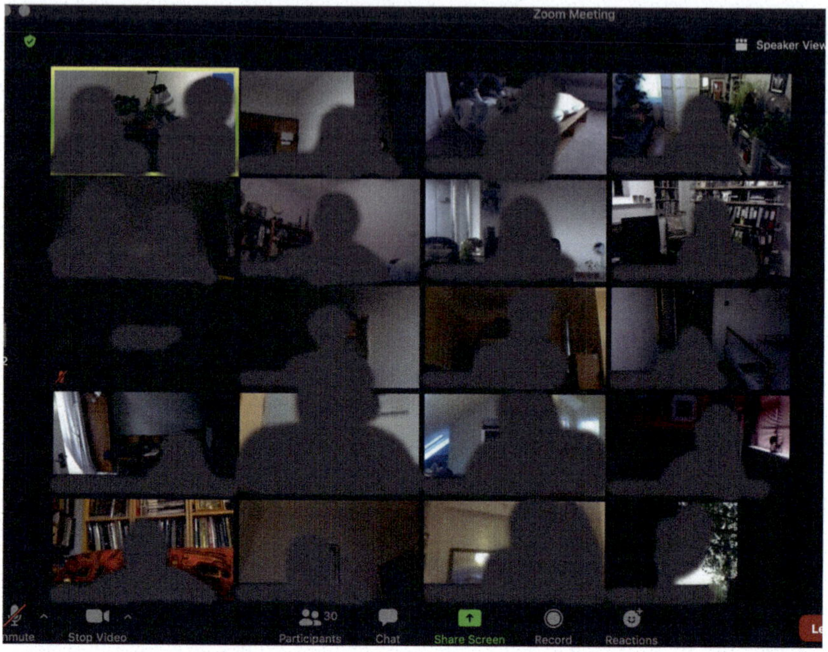

Source: Author.

In this paper, I see the background as a symptom of power relations appearing between the interlocutors as they open the video conferencing software. The main locus is the background of a domestic space that accommodates video conferencing labor. What it makes visible are the questions of precarity and immaterial labor revealed through a range of aesthetic procedures and accidental markers. Hardt and Negri (2004) define immaterial labor as networked, decentralized, mobile, and rooted in sociality and affect. Producing "communication, social relations and cooperation" is its key characteristic (Hardt and Negri 2004, 113). Video conferencing is a type of immaterial labor that involves, at the same time, the labor of watching and the labor of being watched (Andrejevic 2002), as well as surveillance, as it is most often mediated by proprietary platforms such as Zoom or Microsoft Teams. As I argue throughout the chapter, the background also introduces a dimension of labor of being seen. Video conferencing produces asymmetries that are embodied and technologically situated, and the idea of the gaze as an act of looking is also, by default, networked as it is dependent on the processes of encoding and transferring information, bandwidth, and imperceptible delays, as well as inequalities in access to internet connection and hardware.

The professionalization of the background—the ways in which the users adapt their living and working spaces to the newly formed architectures of video conferencing labor—is visible evidence of a pressure of representation. The focus on the tactics chosen by the users is also a call to consider the professionalization of background as a larger shift in visual culture. The users make considered aesthetic decisions for their representation: they might conceal, hide, or perform the background differently. These daily practices, often disregarded, constitute the central focus of this chapter precisely because they reveal points of vulnerability and agency. And while there is a clear difference between the "two-sided" and "one-sided" architectures of the gaze, there are many instances in which tactics developed by live streamers and bloggers become adopted into the more generalized practice of video conferencing.

What does it mean to include or acknowledge this architecture? What does the "professionalization" of the background involve in terms of our changing relationship to privacy, pressures of self-representation, and labor conditions? The architecture of the gazes needs to be examined in the context revealed by lived experiences and instances of performative immaterial labor.

"Credibility Bookshelf," or Symbolic Capital in the Era of Visibility

The history of the background in Western visual culture is strongly linked to representations of power. In early photographic portrait, it served to underline and enhance the social status of its aristocratic and bourgeois subjects by putting them in an appropriate environment. The photographic ateliers of the nineteenth century widely used painted backdrops, which featured natural landscapes, architecture, ruins, pastoral scenes, sharing a common field of reference with theatrical backdrops and tableaux vivants[1] that recreated scenes from plays and classical literary narratives (fig. 2). In doing so, they produced an idea of what an elevated aristocratic portrait could and should be for the sitters: a symbol of status, but also a cultural imaginary that they could occupy by birthright. As Lucy Lippard notes,

> The backdrop portrait creates a spatial dislocation into a magical elsewhere not provided by ordinary portraiture. The subject, having (usually) chosen the setting, extends her of his identity to meet this invented context. (1997, 8)

---

1 A popular aristocratic pastime between theatrical performance and a parlor game, in which the participants wore costumes, positioned themselves among the props and posed for the viewers, to appear as if in a painting.

In the case of aristocratic portraits, the tasteful furnishings, staircases, and idyllic landscapes were the prevalent choice.

Figure 2: Princess Alice, Grand Duchess of Hesse by Camille Silvy, albumen carte-de-visite, July 4, 1861
Figure 3: A late-nineteenth-century painted backdrop. Interior of Stafhell & Kleingrothe photographic studio. Medan, Sumatra, Indonesia, 1898

Source Figure 2: ©National Portrait Gallery, Photographs Collection.
Source Figure 3: Collection KITLV.

A portrait of Princess Alice, third child of Queen Victoria, provides a great illustration to one such tasteful dislocation (fig. 3). It was taken in 1861 by Camille Silvy, a photographer who was very popular among the aristocratic circles of London in the 1860s. He kept a record of everything the studio produced in the daybooks; his range of backdrops and props is easily recognizable behind the clients' figures. Princess Alice's backdrop is a painting of a park. The prop, a large stand almost concealed by the ivy leaves, allows for a natural posture and blends into the background. The painted architectural detail on the left is adorned with a letter "A" which potentially indicates that the backdrop was made specifically for the portrait and would not be used by other clients: the members of the Royal family were not expected to share even the illusionary space with others. The photograph presents its elegant subject in a setting completely appropriate to her character and essence, and fitting the format of carte de visite portraits that were exchanged socially and collected in albums.

However, as Julie Codell notes, "while photographic images negotiated older portrait conventions of body posture, gestures, props, and dress borrowed from painting, they destabilized earlier notions of identity" (2012, 493) as later Victorian photography started to borrow the aristocratic props and backdrops for the working class (Codell 2012, 494). Arjun Appadurai writes on colonial photographic backdrops being not simply passive props, but both instruments of creating imperialist cultural imaginaries and sites of experimentation with "visual modernity" (1997, 6–7). Already in that early era of photography, the increasing participation of self-portraits in the social processes underlines how the background reflects both the pressures and the defiances of self-representation.

Video conferencing inherits some of the symbolic procedures of this type of representation. In higher education settings, one can often see the bookshelf making an appearance in the academics' backgrounds. While the "virtual window" might simply be opening toward the working space of someone in research—a library, or its home-adapted version—within video conferencing, it also becomes a part of a portrait, a reminder of their credibility, a confirmation of books read (or at least bought). Such images, seen in various televised expert appearances, also contribute to a universalizing trope that the symbolic library is where academics belong, as if video conferencing from their kitchens would somehow undermine their expertise.

The pandemic has produced well-deserved sarcastic outlooks and popular analyses of such images, treating the background as the main site of reflection. The Twitter account Bookcase Credibility (@BCredibility), with the tagline "What you say is not as important as the bookcase behind you," tracks the appearances of experts, politicians, and other public figures on the backdrop of bookshelves. For example, to the news broadcast image of a British politician speaking in defense of a Downing Street party during the lockdown, the Twitter account enigmatically notes: "Michael Fabricant is going to need a lot more books than those if he wants to successfully defend the indefensible" (@BCredibility 2022) (fig. 4). Another Twitter account, Room Rater (@ratemyskyperoom), presents an equally sarcastic take, selecting as its objects of critique not only the bookshelves but also lighting, interior design, memorabilia, and decorative objects. The *New York Times* critic Amanda Hess takes note of the credibility tools in the age of working from home:

> It is remarkable how quickly the bookcase has become obligatory, how easily it has been integrated into the brittle aesthetic rules of authority. The appearance of the credibility bookcase suggests that the levers of expertise and professionalism are operating normally, even though they are very much not. (Hess 2020)

In some ways, the video conferencing background aligns with the art historical canon: visual studies scholar Mieke Bal observes that it is via the "cult of portraiture" in the Western European and North American contexts that ideological value systems are continually reified, and "the dominant classes set themselves and their heroes up as examples to recognize and to follow" (Bal 2003, 22). Where the figure in the foreground appears as an expert, as an authority figure, as the one who speaks for others, the viewer can read the "aura" of expertise from the iconographic clues in the background.

Figure 4: "Michael Fabricant Is Going to Need a Lot More Books than Those If He Wants to Successfully Defend the Indefensible," January 11, 2022

Source @BCredibility.

Becoming Vulnerable

If the painted backdrop of portrait photography in the nineteenth-century ateliers served, for the sitters, as a way to commemorate the best version of themselves, the backgrounds in video conferencing follow the increase in working from home and the communicators' desire to appear appropriate: not as homely subjects in their pajamas, but as experts in their place of professional occupation. However, at home, the video conferencing backgrounds also reveal signs of living and moving, inhabiting space and sharing it with others—humans, animals, plants, and machines. As the home office becomes open to the gaze, these signs often reveal home as a gendered space and a site of reproductive labor. During the pandemic, the increasing visibility of the background revealed a parallel increase in domestic violence and the unequal burden of domestic labor (Graham-Harrison et al. 2020). The backgrounds in the video conferencing become a witness to the tension between the public and the private, unfolding in the inhabited space altered by telecommunication.

As a form of virtual architecture comprising of screens, gazes, and backgrounds, video conferencing appears as an accidental and temporary apparatus that establishes itself only when connected. Anne Friedberg calls the screen a voyeuristic virtual window: "the screen is a component piece of architecture, rendering a wall permeable to ventilation in new ways: a 'virtual window' that changes the materiality of built space, adding new apertures that dramatically alter our conception of space and (even more radically) of time" (Friedberg 2006, 1). Likewise, there is accidental voyeurism in looking at someone's home, scanning for clues: like in the game genre "find a hidden object," the signs of life—shelves, possessions, photographs, pets, plants, the state of cleanliness or disarray—become socioeconomic clues to the person's life, hobbies, and interests.

The vulnerability of not being able or not wanting to reveal the contents of your home to the stranger can be directly seen in the ordinary gestures of telecommunication: Where does one position their camera preparing for a video conference? Unless a home office is already set up, it faces an inconspicuous corner of the room or a blank wall; rarely do we get to see the unwashed dishes, clothes on the sofa, and boxes left from moving. From my own remote teaching experience during the lockdown, students' reluctance to turn on their cameras is often connected to their living conditions. The home as a site of reproductive labor stands in stark contrast to the expectation to appear professional in a video conference in a work- or study-related setting. In shared flats, the lack of a quiet and neutral space makes the act of communication violent and intrusive.

Like reproductive labor, the labor of being seen is not always recognized as such, and therefore engaging in it often involves more of a grudging acquiescence than open consent. Video conferencing, in this sense, remains an architecture of intrusion, the presence of which can be felt every time we sit down in front of the computer screen and adjust its position slightly so that some of the elements of our houses are seen and not the others; set up the lighting so that our faces appear well-presented; double-check if the camera is off and if the microphone is on "mute." These subtle adjustments show how our bodies, our sources of light, and our significant objects become implicated in the lines of the gaze extending through the virtual window.

Furthermore, the architecture of gazes introduces an important consideration for the kind of vulnerable space it produces. While a two-sided act of communication, video conferencing connects two individual spaces. The two gazes meet, but they do not create a third, common space; they create two simultaneously existing situations of professionalization and vulnerability. In the following sections I also draw on the situations that are one-sided: sex cam work, lifelogging, and streaming; however, the experiences of vulnerability and resistance to it are equally applicable to the video conferencing context.

Becoming Professional

The background becomes vulnerable when it becomes a foreground. This is very visible in the situations when the person leaves their position in front of the camera. A media artwork by Addie Wagenknecht and Pablo Garcia provides a great example of such reversal: the website brbxoxo.com "searches online sexcam sites and only broadcasts feeds when the performers are absent" (Wagenknecht and Garcia 2015). In an artist talk, Pablo Garcia says that during the work on another project with sexcams, they noticed that sometimes the camera would keep running when the performers left. They started to collect "little video clips of just ... nothing ... but it's not 'nothing,' these are people's actual homes around the world" (Garcia 2015).

The context of sexcam work only underlines home as a site of labor; it raises the stakes for the aspects that are normally disregarded as insignificant when one thinks of working from home. It makes the architecture of the networked gaze more visible: even the word "room" is used on many sexcam platforms to signify not only the real but also the virtual room in which the encounter between the performer and the viewer takes place. One aspect is privacy: the background becomes not just a vulnerability but a real security risk, as distinctive visual markers, such as a view from the window, mean that the home address of the performer can be identified and made public (Cunnningham et al. 2018, 54). Another aspect is the direct correlation between the sex worker's presence on the screen and the monetary value—the background on its own is worthless.

Finally, when the virtual "room" is located in an actual home and not in a studio, the background contributes to the illusion of authenticity. Writing about sexcam workers, Angela Jones notes that "their rooms are often their bedrooms, and from the perspective of a consumer, everything about the experience appears real" (Jones 2020, 6). One of the respondents in Jones's study streamed not only her erotic performance but also her morning routine and her preparation for the live stream. Even in studios, the rooms are often styled as bedrooms, performing authenticity for the encounter.

The cases of credibility bookshelf, sex work, and education already point toward the significant *professionalization* of the background, meaning that these contexts involve a commodification of personal space and individuality, a construction of either neutral or performative space. Already professionalized background practices, such as those of vloggers, private tutors, language instructors, therapists, and many others, give us a good idea of what such neutrality means. The construction of background, then, constitutes an aesthetic procedure: a conscious attempt to restage the architecture of the gaze and to diminish the vulnerability.

Furthermore, it is also necessary to point out the literal professionalization of the background in its technological frameworks and infrastructures. For example, the applications used to enact workplace surveillance employ a range of methods—from keystroke logging, messages, regular screenshots, user prompts, and remote control to facial recognition (Roe 2021, Abril and Harwell 2021). This draws a fine line between management and explicit surveillance: Zoom's "attention tracking" feature that identified if the attending person's gaze was fixed on the screen or strayed elsewhere only remained for a couple of months before being removed amidst privacy concerns (Yuan 2020). Further concerns of privacy and vulnerability arise if one considers how power relations inherent to the acts of visibility and invisibility can be exacerbated by algorithmic processing, opening the way to data-harvesting infrastructures. In video conferencing software, the identifiable background (for example, via blurring filters) appears *for* someone, for a person. The military techniques of differentiating between the figure and the ground to identify the target continue their existence in the contemporary practices of video conferencing.

The Labor of Being Seen

The professionalization of the background reveals that there are different modalities to being seen and to whether something becomes an object of conscious attention. The labor of being seen is not entirely equivalent to a continuous performance for the gaze. The construction of the background has to do more with the decisions made toward the person's own privacy (in revealing, concealing, or performing a particular type of space), as well as with the decisions that radically alter or extend the ideas of public and private. Through the aesthetic decisions made by users, the background reveals how these ideas are navigated.

For this reason, video conferencing should be discussed not as an equivalent of a face-to-face conversation but rather as a set of aesthetic and cultural practices that the users of video conferencing platforms adopt and share. Unlike practices such as lifelogging, where a conscious decision to document and share one's personal life in its entirety, the decision to be or not to be visible is a social pressure that is not always resolved straightforwardly.

Lifelogging, however, presents an interesting delineation of the border between the labor of being seen and acts of performance for someone's gaze. Lifelogging is an intentional sharing of a private life through recordings, live video, or photographs, often using a wearable camera that shoots continuously or takes photographs at certain time intervals. Lifelogging rarely contains extended narrativization, and its defining categories are the date and time where the recording took place, making it closer to a public diary, a visual imprint of the person's life. The private lives on social media tend to give in to a pressure of representation, as the platform itself frames "performing the self" in specific ways (e.g., a "professional self" on LinkedIn, a lifestyle "self" on Instagram) (van Dijck 2013). Lifelogging, however, moves toward a documentary rather than a performative practice—even if we consider some degree of performing the self inevitable.

An important elaboration of the labor of being seen as opposed to performing for the gaze can be seen in the case of one of the early adopters of lifelogging—Jennifer Ringley, an American college student and web designer. Between 1996 and 2003, she broadcasted her life online uninterruptedly via her website, JenniCam. The webcam technology was based on updating the URL page with a new photograph every three minutes. The initial audience of JenniCam was only Ringley's friends, but quite quickly, the website's address went viral and started to receive millions of hits per day. In 1997 she added a paid subscription to cover the dramatically increased bandwidth costs. However, Ridley was adamant that the project, while self-supporting, should not be about revenue: "the site isn't up because I can make money and fame from it ... The site is up because YOU continue to enjoy it" (Ringley 1997). At some point, JenniCam received four million hits per day (Banet-Weiser 2012, 51).

What is particularly interesting about Ringley's lifecasting project is her acute understanding of the processes unfolding in the architecture of the networked gaze. On the website, she described the project as:

> 1: a real-time look into the real life of a young woman
> 2: an undramatized photographic diary for public viewing via internet (Ringley 1999b).

Even this brief definition already describes the webcam architecture as constituted from two sides: a look *into* and a space *for* viewing. Furthermore, Ringley was firm in positioning it as an experiment rather than a commercial enterprise or an artwork. For example, when appearing on David Letterman's show, she pointed out that JenniCam, although similar to TV and other broadcasting, is "something that is made for the medium" of Internet ("Jennicam's Jenni on Letterman's Late Show" 1998).

JenniCam did not conceal nudity or sex scenes, and for this reason, the media of the time framed JenniCam as exhibitionist and attention-seeking (see, for example, Weeks 1997, Weisman 1998). Antonia Hernández in her doctoral dissertation points out the misogynist nature of these characterizations (Hernández 2020, 54). However, in her public statements and on the website, Ringley was very clear in defining JenniCam as a social experiment primarily aimed at documenting and not performing. For her, the question of privacy was not the central point of the project: "Just because people can see me doesn't mean it affects me—I'm still alone in my room, no matter what. And as long as what goes on inside my head is still private, I have all the space I need" (Ringley 1999a).

There are, therefore, several salient points that we can glimpse from Ringley's descriptions of JenniCam that bear on the discussion of the backgrounds in video conferencing. First, she underlines the difference between the labor of being seen and performing for the gaze. Ringley stated: "I keep JenniCam *not* because I want to be watched, but because I simply don't mind being watched. It is more than a bit fascinating to me as an experiment" (Ringley 2000). Secondly, the authenticity of this encounter, while mediated and framed, remains, for her, a major reason to continue with the project:

> The concept of the cam is to show whatever is going on naturally. Essentially, the cam has been there long enough that now I ignore it. So whatever you're seeing isn't staged or faked, and while I don't claim to be the most interesting person in the world, *there's something compelling about real life* that staging it wouldn't bring to the medium. (Jennicam 1999a, my emphasis)

As Ringley herself points out, the duration of and commitment to her project change the conditions of engagement with her presence on the screen—both for her and for the spectators. She notes that for many viewers JenniCam acts as a shared space which they "put ... in the corner of their monitor and it's like having someone in the next room" (Allen 1999). Here, the consent to being *seen* for a long time, at any given moment, seems a bit similar to the kind of consent we implicitly give when we open a video conferencing application. Apart from the pressure of representation, a space opens for other aspects: senses of connection and curiosity.

After six years of continuous broadcasting, Ringley shut down the site in 2003, as PayPal, which she used to accept donations, severed ties with her for the reason of images containing nudity. The archive.org reveals the last images accessible from JenniCam before its closure in 2003 (fig. 5). One is taken on Friday, November 29, 2002, at 7:30 p.m., and is captioned "Gone to San Francisco for a day to enjoy sun in the park" (Ringley 2002). It shows her bedroom, lit only by a small lamp; the empty bed occupies most of the image. Another image is from Thursday, July 31, 2003, 1:45 p.m. It reveals her working space: a chair; a shelf with boxes, folders, and books; and a table corner (Ringley 2003). Ringley is absent from both images. These backgrounds, probably incredibly familiar to the followers of her website, seem all the more relevant now that her absence from them is made permanent.

Figure 5: Left: JenniCam, November 29, 2002. Right: JenniCam, July 31, 2003

Source: ©Jennifer Ringley.

The empty background also forces the viewer to confront their own alienation. Like in slow cinema, where, because of the slow pacing and long takes, "seeing becomes a form of labour" (Schoonover 2012, 66), the background constitutes the part of the image that we start to notice only after a while. The slowness invites more effort, more ethical and political engagement and reflection on the part of the spectator than in an act of quick consumption of messages and narratives.

The labor of being seen, therefore, situates the background as a specific mode of being present and even copresent. The *New York Magazine* writer Michael Wolff points out,

> I know people who keep Jenni in the corner of their screens as they work—she's a background presence, like radio. The fact that she keeps going, keeps showing up, keeps doing whatever inconsequential things she does, is not only reassuring but instructive—this is how people deal with time, this is how people fill their days. (Wolff 1999)

This reassuring presence is increasingly easy to find in contemporary audiovisual networks. For example, one could think of the YouTube phenomenon LoFi Girl—a YouTube live stream with a looped animation of a young woman studying that broadcasts relaxing music to over 10 million subscribers ("LoFi Girl" 2020). The labor of being seen, as a *copresence* that helps to relax and focus constitutes a part of daily practices of video conferencing—for example, in online study groups and cowriting sessions. Not only *in* the background, but *as* the background, these practices operate around the explicit acknowledgment of the labor of being seen.

Background Aesthetics: Hiding in Plain Sight

The result of the lived-in spaces confronting the pressure of self-representation and self-commodification are the aesthetic procedures—tactics of revealing, hiding, or confronting the gaze without any modification. This can be seen both in the meticulous arrangements of the background, such as in streaming setups and in the complete refusal to let it be seen—either by turning the camera off or in using a virtual background. In this sense, the virtual background in video conferencing, while seemingly offering an infinity of options, from a tropical beach to a fictional location (and, therefore, interpretable as a site of symbolic representation), seems to be more of a curtain for the real background.

The background, therefore, becomes a contested visual space. The process of professionalization disrupts the home space as one of lived reality and flattens it into the space of representation. With this, individual tactics of resisting, concealing, hiding, distracting, and showing enter the aesthetics of video conferencing. The background is a symptomatic landscape of immaterial labor that records the changes in contemporary working conditions. As modes of digital labor expand to include digital nomads, workers who "give up on 'settled' living and embark on nomadic world travel, and perform work from different locations around the world, taking advantage of digital infrastructures and coworking spaces" (Schlagwein and Jarrahi 2020, 3), the background reflects the multitude of liminal spaces where such work takes place. A digital nomad can be calling from the airport, car, or train, cafes and coworking spaces, and those cyber-natural landscapes of parks, beaches, and forests that boast a good 5G connection.

On a different scale, the analysis of the background of our homes also invites attention toward the homogeneity produced by the supra-individual forces: the industries of construction and interior design, economic conditions, and even remnants of centralized planning. In a 2014 essay, writer and software consultant Paul Ford discusses the fact that many YouTube videos seem to share a common background—one that he calls "the American Room":

> The curtains are drawn. Some light comes through, casting a small glow on the top left of the air conditioner. It's daytime. The wall is an undecorated slab of beige. That is the American room. (Ford 2014, n.p.)

He notes that the American room appears in enough videos to make him think about the standardized nature of American suburban living, mass-production, and standardization in construction and housing. As Ford points out via Lawrence Busch's 2011 book *Standards*,

> the standard height of ceilings could vary considerably in a world in which walls were constructed of plaster and individually cut laths. But once standard building materials, such as (in the United States) 2 × 4s, and 4 foot × 8 foot sheets of plasterboard were made available, the variation in the height of ceilings was sharply reduced even as the speed at which wooden homes could be built increased. (Busch 2011, 34)

As the conditions of life, production, alienation, and labor open up in the inhabited spaces revealed in the background of our communication, the professionalized "selves" that are presented stand against it even more visibly.

An explicit example of a background that accepts the necessity of televised labor and capitalizes on the affective clues of the body is a game streamer's room. As a site of active performance of their professional selves, the game streamer's room is their workspace—sometimes dedicated solely to their labor. The introduction of new financial instruments that allowed vloggers to monetize their activities more consistently (e.g., advertisements on YouTube) and the increasing popularity of streaming platforms (such as Twitch) contributed to the professionalization of this occupation. A game streamer is in the center of the image, sitting in an ergonomic chair, facing several screens, often surrounded by branded items and symbolic objects, posters, figurines and featuring soft, well-distributed lighting in neon colors. She is, herself, more a character in our collective viewing fantasy rather than an interlocutor in a video conferencing setting. However, thinking about the increasing acceptance of work-from-home settings and the game streamers as early adopters of such practices, it is worth considering this particular background as a situation of speculative design with a potential for influencing other architectures of the gaze and the screen.

The videos that share the "setups" of the streaming rooms narrate, in a lot of detail, the hardware specifications, comfort, air quality, sound isolation, and other functional aspects. However, they are equally attentive to the aesthetics of the resulting techno-space, the symbolic objects, and the general ambience. The YouTube channel TechSource is dedicated almost exclusively to discussing various gaming setups hosted by Ed Oganesyan. In nearly three hundred episodes of the series called "Setup Wars," he showcases different environments submitted by the followers. The setup videos reveal that the American Room has not disappeared but was instead disguised with neon ambient light. Behind IKEA furniture and state-of-the-art hardware, the light-colored walls, sofa, blinds—all point toward this ruse. Another video deliberately makes the American Room its starting point, revealing the host standing in front of the beige walls and carpet first and then cutting the video sharply to the same angle, only lit up by a multitude of LED lights in blue, green, and purple (TechTesseract 2020).

Conclusion

As I have explored through a variety of possible architectures, the arrangement of gazes is always mutable and temporal; in each of various virtual "rooms" that reveal the precarious conditions of contemporary communication, it is structured differently. As a symptom, an environment, and a site of developing tactics and habits against its vulnerability, the background has opened up multiple points of entry into the consideration of video conferencing acts. The professionalization of the background bears on not only the notions of what constitutes the workspace and self-representation but also larger shifts in visual culture. The labor of being seen needs to be distinguished and acknowledged as a different modality from performative labor; interspersed with architectures and structures both human and technical, intra- and supra-individual, it reveals the specific background aesthetics, with a background as a witness to, and a trace of, a variety of tactics and attitudes that reflect the individual users' capacity to situate themselves within the architecture of the gaze.

Acknowledgements

The background became a central figure of this paper through the collective work with Mijke van der Drift and Neda Genova on the panel "Homing In, Zooming Out: Space Making as Pandemic Practice" for the 2020 annual conference of the German Society of Media Studies (Gesellschaft für Medienwissenschaft, GfM). My thanks go to them and to the editors Axel Volmar, Olga Moskatova, and Jan Distelmeyer for the comments and help.

References

@BCredibility. 2022. "Michael Fabricant Is Going to Need a Lot More Books Than Those If He Wants to Successfully Defend the Indefensible." January 11, 2022. https://twitter.com/BCredibility/status/1480871039584423938.

Abril, Danielle, and Drew Harwell. 2021. "Keystroke Tracking, Screenshots, and Facial Recognition: The Boss May Be Watching Long after the Pandemic Ends." *Washington Post*, September 24, 2021. https://www.washingtonpost.com/technology/2021/09/24/remote-work-from-home-surveillance.

Allen, Jamie. 1999. "'Ed' of the Internet: JenniCAM Going Strong after Three Years." *CNN*, March 26, 1999. http://edition.cnn.com/SHOWBIZ/Movies/9903/26/jennicam.

Andrejevic, Mark. 2002. "The Work of Being Watched: Interactive Media and the Exploitation of Self-Disclosure." *Critical Studies in Media Communication* 19 (2): 230–48.

Appadurai, Arjun. 1997. "The Colonial Backdrop." *Afterimage* 24 (5): 4–7.

Bal, Mieke. 2003. "Visual Essentialism and the Object of Visual Culture." Journal of Visual Culture 2 (5): 5–32.

Banet-Weiser, Sarah. 2012. *AuthenticTM The Politics of Ambivalence in a Brand Culture*. New York: New York University Press.

Busch, Lawrence. 2011. *Standards: Recipes for Reality*. Cambridge, Massachusetts: MIT Press.

Codell, Julie F. 2012. "Victorian Portraits: Re-Tailoring Identities." *Nineteenth-Century Contexts* 34 (5): 493–516.

Cunningham, Stewart, Teela Sanders, Jane Scoular, Rosie Campbell, Jane Pitcher, Kathleen Hill, Matt Valentine-Chase, Camille Melissa, Yigit Aydin, and Rebecca Hamer. 2018. "Behind the Screen: Commercial Sex, Digital Spaces and Working Online." *Technology in Society* 53 (May): 47–54.

Ford, Paul. 2014. "The American Room." *Medium*. July 30, 2014. https://medium.com/message/the-american-room-3fce9b2b98c5.

Friedberg, Anne. 2006. *The Virtual Window: From Alberti to Microsoft*. Cambridge, Massachusetts: MIT Press.

Garcia, Pablo. 2014. "Adventures in Virtuality." Presented at the Artist Talk, Aksioma Project Space, Ljubljana. *YouTube*, April 2, 2014. https://www.youtube.com/watch?v=Zy2QxCx4X7Y&ab_channel=Aksioma.

Graham-Harrison, Emma, Angela Giuffrida, Helena Smith, and Liz Ford. 2020. "Lockdowns around the World Bring Rise in Domestic Violence." *The Guardian*, March 28, 2020. http://www.theguardian.com/society/2020/mar/28/lockdowns-world-rise-domestic-violence.

Hardt, Michael, and Antonio Negri. 2004. *Multitude: War and Democracy in the Age of Empire*. New York: Penguin Press.

Hernández, Antonia. 2020. *Sexcams in a Dollhouse: Social Reproduction and the Platform Economy*. PhD Thesis, Concordia University. https://spectrum.library.concordia.ca/id/eprint/987297.

Hess, Amanda. 2020. "The 'Credibility Bookcase' Is the Quarantine's Hottest Accessory." *The New York Times*, May 1, 2020. https://www.nytimes.com/2020/05/01/arts/quarantine-bookcase-coronavirus.html.

Jones, Angela. 2020. *Camming: Money, Power, and Pleasure in the Sex Work Industry*. New York: NYU Press.

"Jennicam's Jenni on Letterman's Late Show." 1998. *Letterman's Late Show*. CBS. https://www.youtube.com/watch?v=0AmIntaD5VE&ab_channel=shesnailie.

Lippard, Lucy R. 1997. "Frames of Mind." *Afterimage* 24 (5): 8–12.

"LoFi Girl: Lofi Hip Hop Radio—Beats to Study/relax to." 2020. *YouTube*, February 20, 2020. https://www.youtube.com/watch?v=5qap5aO4i9A&ab_channel=LofiGirl.

Ringley, Jennifer. 1997. "JenniCAM Guests." *JenniCam*. December 10, 1997. https://web.archive.org/web/19971210110930/http://www.boudoir.org/guests/index2.html.

Ringley, Jennifer. 1999a. "JenniCam—Frequently Asked Questions." *JenniCam*, January 28, 1999. https://web.archive.org/web/19990128161936/http://www.jennicam.org:80/faq.html.

Ringley, Jennifer. 1999b. "Jen-Ni-Cam." *JenniCam*, October 12, 1999. https://web.archive.org/web/20000621011841/http://jennicam.org.

Ringley, Jennifer. 2000. "What Is JenniCam?" *JenniCam*. October 17, 2000. https://web.archive.org/web/20030208075705/http://jennicam.org/j2kr/concept.html.

Ringley, Jennifer. 2002. "November 29, 2002." *JenniCam*, November 29, 2002. https://web.archive.org/web/20030216165906/http://www.jennicam.org.

Ringley, Jennifer. 2003. "July 31, 2003." *JenniCam*, July 31, 2003. https://web.archive.org/web/20031208210314/http://jennicam.org.

Roe, David. 2021. "Why Employee Surveillance Software Could Do More Damage than It's Worth." *Reworked.co*, September 22, 2021. https://www.reworked.co/employee-experience/why-employee-surveillance-software-could-do-more-damage-than-its-worth.

Schlagwein, Daniel, and Mohammad Hossein Jarrahi. 2020. "The Mobilities of Digital Work: The Case of Digital Nomadism." In *ECIS*. https://www.researchgate.net/profile/Daniel-Schlagwein-2/publication/340755802_The_Mobilities_of_Digital_Work_The_Case_of_Digital_Nomadism/links/5e9bf78ca6fdcca789245c52/The-Mobilities-of-Digital-Work-The-Case-of-Digital-Nomadism.pdf.

Schoonover, Karl. 2012. "Wastrels of Time: Slow Cinema's Laboring Body, the Political Spectator, and the Queer". *Framework: The Journal of Cinema and Media* 53 (1): 65–78.

TechTesseract. 2020. "TRANSFORM Your GAMING SETUP with RGB Lighting! How I Light My Gaming Room!" *YouTube*, October 2, 2020. https://www.youtube.com/watch?v=dZgABJ72y20&ab_channel=TechTesseract.

van Dijck, José. 2013. "'You Have One Identity': Performing the Self on Facebook and LinkedIn." *Media, Culture & Society* 35 (2): 199–215.

Wagenknecht, Addie, and Pablo Garcia. 2015. "Brbxoxo." International Film Festival Rotterdam. 2015. https://iffr.com/en/iffr/2015/films/brbxoxo.

Weeks, Linton. 1997. "Web Site for Voyeur Eyes." *Washington Post*, September 20, 1997. https://www.washingtonpost.com/archive/lifestyle/1997/09/20/web-site-for-voyeur-eyes/941cb4a0-10e7-4cbd-9e2f-2eebce80fa9e.

Weisman, Jonathan. 1998. "Her Life Is Just an Open Web Site Exhibitionism: With a Camera in Her Bedroom, a 21-Year-Old Woman Has Turned Her Day-to-Day Existence into a Hobby for Hundreds of Thousands of Web Crawlers." *The Balti-*

more Sun, May 13, 1998. https://www.baltimoresun.com/news/bs-xpm-1998-05-13-1998133113-story.html.

Wolff, Michael. 1999. "Must-See PC." *New York Magazine*, February 22, 1999. https://nymag.com/nymetro/news/media/columns/medialife/157/.

Yuan, Eric S. 2020. "A Message to Our Users." Zoom, April 1, 2020. https://blog.zoom.us/a-message-to-our-users.

People Who Stare at Screens

Winfried Gerling

Figure 1: Apple Ad: "Behind the Mac," 2018

Source: Screenshot from https://www.youtube.com/watch?v=quppef3bH-s.

In the summer of 2018, Apple released a short promotional video, "Behind the Mac."[1] The view of the people shown is partially obscured by a laptop. The video is in typical Apple color[2] and shows people interacting with the Mac in a relaxed atmosphere, even in public. Their view is always directed at the computer, and the surroundings are no more than a picturesque backdrop for the actors (fig. 1). Cultural diversity is especially emphasized through ethnic diversity. The video ends with the slogan "Make something wonderful behind the Mac."[3]

1 The video is part of a campaign in which the company wanted to show how users use the Mac to work in a creative and innovative way. The campaign featured twelve individual stories of how artists*, developers*, and many others are using the Mac in their respective fields.

2 An aesthetic informed by the iPhone's algorithmically normalized shots.

3 The accompanying music is an astonishing choice. "The Story of an Artist" by the late singer-songwriter Daniel Johnston is a song by a mentally unstable, reclusive outsider to the music scene about the artist who was rejected and misunderstood by friends and family.

Almost two years later, the ad campaign continues under the same title, though the mood in the video "Behind the Mac—Greatness"[4] is different.[5] Black-and-white photographs are used almost exclusively, showing internationally known creatives alone in mostly domestic settings.

Figure 2: Apple Ad: "Behind the Mac," 2021

Source: Screenshot from https://www.youtube.com/watch?v=8kF5x2D3rqo.

The impact of the pandemic is an obvious part of the atmosphere: alone but safe from the virus, you should continue to work productively with Apple and communicate with others within the walls of your home. With the knowledge of the production of her last album, which she is said to have sung entirely at home, this condition is particularly noticeable in the image of Billie Eilish (fig. 2). The slogan "Never stop making behind the Mac," which is inserted at the end, seems more like a perseverance slogan.

With the onset of what we now describe as a pandemic situation, many people's relationships with their screens and their environments changed permanently. This change is made vivid in the promotional videos.

Pupils and students were sent home from school and university, and workers whose presence in the company was dispensable were sent to the home office. Those who could and were allowed to escaped a potentially infectious world into the safety

4 People shown are Tom Hanks, Kendrick Lamar, Gloria Steinem, Billie Eilish, RuPaul, Tarana Burke, Serena Williams, Spike Lee, Stephen Colbert, Lisa Simpson, Pharell Williams, Donald Glover, Takashi Murakami, and Saul Perlmutter. https://www.youtube.com/watch?v=b3VcGKv9Cfw.

5 Again, the campaign begins by observing individual creative people like James Blake and Tyler Mitchell.

of their home environment (fig. 3 a).[6] The screen became more than ever the window to the world as social contacts in lockdown were deficiently reorganized via the software-based interfaces of home computers. Andre Gunthert (2020) aptly describes this situation:

> We rediscover it every day in our digital exchanges: the image is not synonymous with presence. Countless pragmatic signs separate the experience of audiovisual mediation from the experience of face-to-face, which are not or poorly reproduced by connected digital tools. I cannot touch or hug my virtual interlocutor. And the mosaic of screens of a videoconference only offers a disembodied and distant imitation of the physical encounter with its different levels of communication. But the image is no less irreplaceable when the circumstances prevent a direct contact.

Figure 3 a–b: Left: "Child homeschooling," 2020, right: Francis Miller: "Children at Classroom TV during a school strike in Minneapolis," 1951

Source: Photo by author; The LIFE Picture Collection/Shutterstock.

I will use photographic images of people in front of screens to historically develop how the screen, and with it the mediated face, has invaded domestic environments. In video conferencing, these faces are brought into a reciprocal relationship and become the standard of privileged, contagion-free communication under pan-

6 With projects like Classroom TV shown here (Fig. 3 b), similar concepts existed much earlier.

demic conditions. A reduction of the body to the face[7] is central to communication in video conferencing.

Focusing on photographic testimonies,[8] I will address the screen as the subject of near human surroundings, rather than more generally addressing the *screen*[9] as canvas or projection. I begin this story with the television because it is about the promise of live broadcasting as a form of telepresence and tele-actuality.

In this respect, the text is about images of humans in front of the television and the computer.

The fact that these two technologies are inseparable today is due to the alliance they have formed. A computer (laptop, smartphone, desktop, etc.) is as much a TV as the TV is a computer. Both are now capable of running programs, sensory acquisition, and data processing.

The images in question here are stills of a relationship between people and their screens, which have themselves become mobile. They draw attention to a communicative attention to the screen that has changed considerably in the past ninety years or so. It is a long way from analog transceivers to the touchable interfaces of networked universal computing machines. The arrangements of the apparatus, which regulate the relationship of the human being to the screen and thus also shape the communication relationships among each other today, are to become visible in this way. It is worked out how the screen is linked to the spheres of the private and the working world and connects them in a new way in the (post)pandemic situation.

The photographs of Lee Friedlander, in particular, were the occasion to reflect on this relationship. Friedlander was very prescient in first drawing on the medial intrusion of the human face into private spaces, and later very attentively observing the instructions and observations of the monitor to the workers sitting in front of it.

An approach via photographs of people in front of screens is informative for the development of video conferencing since the images can be used to show how the screen establishes itself differently as a counterpart to be communicated with in the home and in the office and how the office and the home enter into an instructive

---

7 This reduction is recognized at an early stage: "The face is a surface ... The face is produced only when the head ceases to be a part of the body, when it ceases to be coded by the body" (Deleuze and Guattari 1987, 170).

8 However, these can also be screenshots or screencasts, which for me belong to photographic practices (Gerling 2018).

9 It is not possible here to go into the long history of terms often used synonymously with *screen*, such as the display as something unfolding, and the monitor as something monitoring (Gerling 2022). The noun *scren* already exists in Middle English and from the end of the fifteenth century. In the early twentieth century, *to screen* is also used as a verb to indicate the "process of filtering and excluding unwanted effects" (Frohne 2013, 257). On the complex history of the computer display, see Thielmann 2018.

unity with the introduction of the PC. These images are, in the best sense, testimonies of a subjugation into the physical bond with the screen, even if it becomes mobile.

Little Screens

Figure 4: Cover page: Funkschau, August 1935

Source: https://archive.org/details/funkschau-1935-heft-34.

In the early images of people in front of screens, the screen stands as a bulky object in space and is defined as a counterpart without a return channel.[10] One sits down in front of the device, which glows like a campfire (McLuhan 1964, 359).

In 1935, at the beginning of German television history, there are very few televisions in the living rooms of the first television nation,[11] and so recordings like the one

10 The historical approach I develop in the following is based on a text that unfolds a wide-ranging photographic history of images of people in front of the screen (Gerling 2023).

11 In 1935, driven by the National Socialists, regular public television began in Germany. Two hours per evening; three evenings per week. After radio, Germany also wanted to demonstrate its leading role in television.

shown above (fig. 4) are probably better understood as attempts to convey these apparatuses as a new medium, rather than as documents of everyday use at the time. With 1800 Reichsmark acquisition costs per device, it could hardly advance to the *Volksempfänger*.[12]

Figure 5: Germany's first Fernsehstelle (TV viewing station) set up on April 10, 1935, at the Reichspostmuseum

Source: Museum für Kommunikation https://twitter.com/mfk_berlin/status/850999324381204480.

The image from the first television station in the Reichspostmuseum in Berlin (fig. 5) testifies to the attractive character of the new technology and at the same time to a spatial situation that is uncertain in terms of the arrangement of the viewers. The small screen with 180 lines and low contrast doesn't allow for a larger viewing distance, but still a cinema situation is emulated.

For the time being, most people are reached in the public television parlors. Unlike the cinema, they also allowed live broadcasts. This is the beginning of a culture we now call "public viewing," and it reached an early peak with the 1936 Olympics as a propaganda tool of the National Socialists (Kubitz 1997, 22).

12 It is also during this period that the first experiments are made with the television as a means of video telephony (see Tollmann 2020).

Although designed for the home,[13] it was not until the years after World War II that the television became widely available in Western industrialized nations, especially in the United States. The television program aligns with the daily routine and needs of a growing white middle class. As Lynn Spigel (1992) has shown for the US, this development is closely linked to the sprawl of large cities that have become too crowded and to suburbanization.[14] This is associated with corresponding cultures, which in turn are reflected in television programming (Spigel 2013).

Pictures from that time show white families gathering around the TV and devoting themselves to new program together (fig. 6). They are stereotypical images of establishing a medium for family community. It is focused on staging an accessible culture for the suburban community.

Figure 6: Harold M. Lambert: A happy family cheerfully sits in their living room and watches a televised clown and puppet show, 1957

Source: https://www.cheatsheet.com/entertainment/anna-duggar-shocks-counting-on-fans-by-revealing-her-kids-watch-tv.html.

As early as the mid-1950s, every second household in the US had a television, which also became a central reference medium for current events such as sports and politics. TVs could be found in bars, pubs, and drinking establishments and quickly became standard equipment in hotels and motels too.

13 The production of the apparatus was discontinued in Germany in 1939 with the start of the war. For the history of television in Germany, see in detail Kubitz (1997).

14 Between 1947 and 1953, the number of people living in suburban areas of the United States increased by 43 percent (Rubin and Scott 2013, 454). In Europe, this trend sets in somewhat later.

Here, the television becomes determinant in the perception of politics and other national events (fig. 7).

Figure 7: Paley Matters: A typical American family gathered around the TV, which displays John F. Kennedy's face, to watch the debate between Kennedy and Richard Nixon during presidential election, 1960

Source: https://medium.com/retro-report/the-presidential-debates-will-be-weirdly-educational-this-year-e6a038135e8e.

John F. Kennedy (1959) writes one year before his election as president: "The searching eye of the television camera scrutinizes the candidates—and the way they are picked. Party leaders are less willing to run roughshod over the voters' wishes and hand-pick an unknown, unappealing, or unpopular candidate in the traditional 'smoke-filled room' when millions of voters are watching, comparing and remembering." The first televised debate on September 26, 1960, between Kennedy and Richard Nixon then drew some 70 million viewers in the US to their screens.

Figure 8: Jacques Lowe: "John F. Kennedy, with his brother Robert and Robert's wife, Ethel, behind him, watching election coverage at Hyannis Port, Mass. on the morning of Nov. 9, 1960," 1960

Source: *The Kennedy Years*, Viking Press, 1964.

With reference to McLuhan, Nicholas Mirzoff (2015, 148) describes the short phase of the global village shaped by TV that begins here:

> The period of the global village was, in retrospect, quite short. It extended from the death of Kennedy to the 9/11 attacks. In this period global television audiences watched dramatic events like the first moon landing (1969), the wedding of Charles and Diana (1981), the fall of the Berlin Wall (1989) and the 9/11 attacks (2001). So in the course of just fifty years watching a world-changing event became a routine consequence of technology, available to hundreds of millions of people who might have little understanding of how technology works. People who were alive at the time can all recall TV broadcasts when President Kennedy was killed, or the 9/11 attacks occurred. Today, news breaks as much through Facebook, Reddit, Twitter and other such applications as it does through television bulletins. Media no longer prize form as much as content.

Then, starting in the 1950s, artistic photographers turned to these screens. They belonged to a media environment that has become commonplace, which McLuhan

(1967, 26) describes in one of his most prominent turns of phrase: "any understanding of social and cultural change is impossible without a knowledge of how media work as environments." And further, he adds, how these environments become active: "Environments are not passive wrappings, but are, rather, active processes which are invisible. The groundrules, pervasive structure, and over-all patterns of environments elude easy perception" (McLuhan 1967, 69).

One of the first photographers to have an eye for the environmental activity of the screen is Robert Frank. Perhaps the most famous photographs can be found in his book *The Americans*. One photograph shows a television with the first televangelist Oral Roberts speaking into a deserted café (fig. 9 a). Another shows a television studio: the presenter disappears to the edge, behind a dark silhouette, while her duplication, limited to the face, appears in the control monitor (fig. 9 b).

Figure 9 a–b: Robert Frank: The Americans, 1958

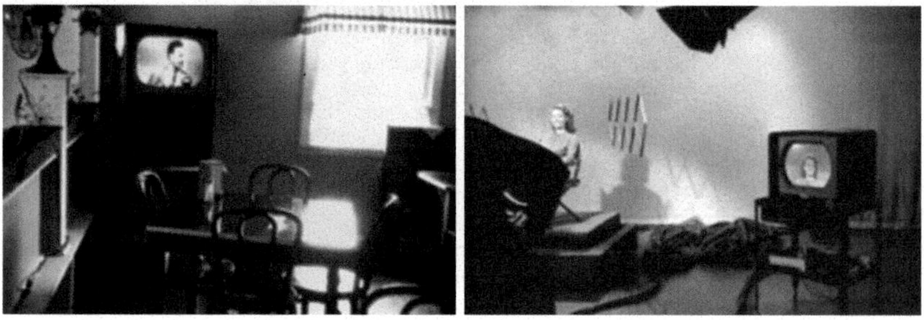

Source: Robert Frank *The Americans*. New York: Grove Press, 1959 (originally published as *Les Américains*. Paris: Robert Delpire, 1958).

Common to both images is that faces appear on the screens, invading and visually occupying spaces.[15]

In 1961, Lee Friedlander began photographing a series of images that focused entirely on this invasiveness of the medium: *The Little Screens*. They are shots of American living rooms and bedrooms that, like Frank's TV images, mostly present faces on the screens, which invade deserted, pragmatically furnished domestic environments and develop an even stronger presence in the space (fig. 10 a–b) than in Frank's images. With the picture from Washington (1962), which shows only one eye on the television screen, the function of observing is anticipated as no longer unidirectional (fig. 10 c).

15 This kind of invasiveness is newly discussed and perceived fifty years later with the webcam built into the laptop as a control view. It's not for nothing that these cameras are often taped shut by their users today.

Figure 10 a–c: Lee Friedlander from The Little Screens: Left: "Florida," 1963, middle: "Philadelphia," 1961, right: "Washington," 1962

Source: Saul Anton: *Lee Friedlander: The Little Screens*, Afterall Books, 2015.

An important exception to this pictorial program is an idiosyncratic self-portrait: In a spectacular turn, Friedlander directs the camera to the floor and shows only his legs and feet, which find their eerie reflection in the TV (fig. 11).

Figure 11: Lee Friedlander from The Little Screens: "Pennsylvania," 1969

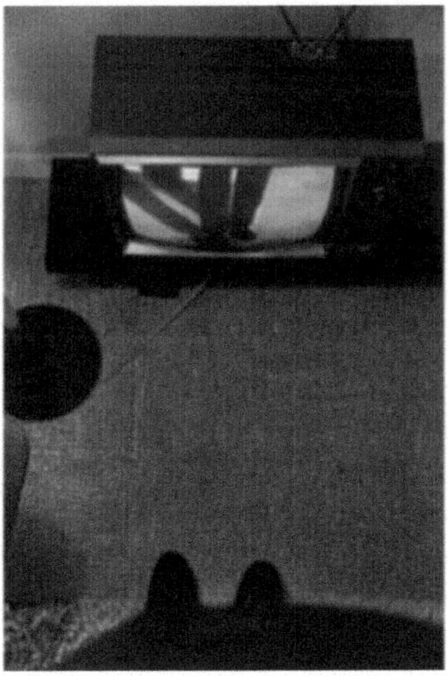

Source: Saul Anton: *Lee Friedlander: The Little Screens*, Afterall Books, 2015.

One could interpret this photograph as a culmination of Foucault's (1984, 4) well-known formulation:

> The mirror is, after all, a utopia, since it is a placeless place. In the mirror, I see myself there where I am not, in an unreal, virtual space that opens up behind the surface; I am over there, there where I am not, a sort of shadow that gives my own visibility to myself, that enables me to see myself there where I am absent: such is the utopia of the mirror.

In this image, the television embodies the utopia of the electronic mirror, which will only be realized in the computer. This electronic mirror has been normalized with the front camera in the smartphone and the camera in the laptop display[16] and has preconfigured the visual conditions for video conferencing. The self-image in these mirrors is, with selfie, video telephony, and video conferencing, an image that is always already intended for others and that is un-mirrored for them. In the "unreal, virtual space" of video conferencing these self-images are results of, as Christian Andersen and Søren Pold point out in this book, entanglements of faces and interfaces.

At Work

The early images of people working with screens are shots of the inventors posing in front of the screens and presenting the screen as the result of their research. They are documents of a scientific achievement that establish the screen as a special object and stage it as something desirable.

Most of them are posed images, as exemplified by the picture of Manfred von Ardenne, the inventor of electronic image transmission, from his laboratory in 1932 (fig. 12). These photographs show men working on technical equipment, next to rather than in front of the screens, because they are to be exhibited as new technology. The screen is often staged like another protagonist.

In 1954, David Sarnoff stands proudly in front of the first flat screen, an invention of his company (RCA), on which a picture of Jane Russel can be seen (fig. 13).[17] The technical object thus becomes doubly charged and paints a picture of a technology conceived by white men and whose "male gaze" (Mulvey 1975, 11 ff) frames as a double desire as a matter of course. "The male protagonist is free to command the stage, a stage of spatial illusion in which he articulates the look and creates the action" (Mulvey, 13).

---

16 The basis of this possibility is the webcam, first integrated by Apple since 2005 as the so-called iSight camera in their laptops and desktop computers such as the iMac. The front camera on the cell phone was introduced as early as 2003 with the Sony Ericsson Z1010 for business video telephony.

17 Russel had contracts with RKO Pictures of which Sarnoff was chairman for a time.

Figure 12: Manfred von Ardenne, 1932

Source: Ullstein Bild.

Figure 13: David Sarnoff, 1954

Source: Everett Collection Inc.

Figure 14: Lisa team at Apple: Paul Baker, Bruce Daniels, Chris Franklin, Rich Page, John Couch, and Larry Tesler, ca. 1982

Source: https://www.mac-history.net/apple-history-2/2019-02-09/parc-scientist-larry-tesler-recalls-jobs-famous-xerox-visits.

This does not change significantly with the early pictures of the PC's development, but the PC will decisively change the relationship to the screen.

With the establishment of the PC, the work behind screens quickly changes into a serving work with or at screens, as evidenced by the pictures of people taking a seat in front of a screen and working with it (fig. 15). The term *workstation* is symptomatic of this.

Figure 15: Larry Tesler at his Xerox Alto workstation, 1973

Source: Xerox Parc, https://www.latimes.com/business/story/2020-02-21/larry-tesler-dead-steve-jobs-personal-computer.

In a history of images of work on or with the screen, images of women would have to be given their own place. They are more often seen as an object on the screen than in front of the screens (Comstock 2014).

As in many histories, images showing women in positions of responsibility in development tend to be underrepresented. For example, very few images show the conditions of manufacture[18] of the early entertainment industry or women working with calculating machines.

One of the few exceptions is a *Cosmopolitan* article by Lois Mandel (fig. 16). However, computer pioneer Dr. Grace Murray Hopper tries to appeal to the magazine's readership by equating programming with housekeeping: "It's just like planning a dinner. You have to plan ahead and schedule everything so it's ready when you need

18 In the United States in the mid-1950s, nearly all workers in the electronics industry were female. Starting in the 1960s, manufacturing shifted to Latin American and Asian countries because labor was much cheaper there. This does not change with the production of the new technology of computers: for example, Fairchild Industries, a manufacturer of computer chips, employed female Navajos from the mid-1960s onward to produce integrated circuits (Donovan 2016).

it. Programming requires patience and the ability to handle detail. Women are 'naturals' at computer programming."

Figure 16: "The Computer Girls"

The Computer Girls

Source: *Cosmopolitan*, April 1967.

As recently as 1978, in an advertisement for the Apple II, a clear role is attributed to women.[19] The master of the house works relaxed in the modern equipped kitchen

19 Ignoring the reality that until the mid-1980s, nearly 40 percent of people working in the computer sciences were women. It was not until the introduction of the PC that the dominant notion of the white male nerd prevailed (cf. Thompson 2019). Another blind spot in this story is the part African American Women play in the context of Computer Sciences. They played a not insignificant role at NASA beginning in the 1950s with Katherine G. Johnson, Dorothy Vaughan, Melba Roy Mouton, Mary Jackson, and others (cf. http://blackwomenincomputing

at the screen—which is still a TV—while his wife is fittingly cutting apples in the background (fig. 17).

Figure 17: Apple Ad

Source: *Byte Magazine*, Jan. 1978.

One thing this promotional image does show, however: while the TV usually found its permanent place in the living room, the PC had not yet found its place. As Sophie Ehrmanntraut (2019, 152) writes in her study of the discourse history of the personal computer:

> The PC has been staged as a friend of the family, helping children with learning, parents with household chores, and bringing the family together to play. PCs should not dictate, should not set limits, but should empower their users. ... the companies [had to] lower their expectations of the market, or the users their expectations of the magical capabilities of the computer. Many computer laymen first had to learn that computers didn't do anything on their own.

The notion of the computer as a machine that brings people together and should be operable without special knowledge was coined by Mark Weiser (1999, 693–694):

> The program was at first envisioned only as a radical answer to what was wrong with the personal computer: too complex and hard to use; too demanding of at-

.org/who-we-are). Their percentage share was and still is very low (see: https://en.wikipedia.org/wiki/African-American_women_in_computer_science).

> tention; too isolating from other people and activities; and too dominating as it colonized our desktops and our lives. We wanted to put computing back in its place, to reposition it into the *environmental background*, to *concentrate on human-to-human interfaces* and less on human-to-computer ones. [emphasis W.G.]

The computer should fit into an everyday environment as an intelligent machine (Weiser 1991).

However, it is already established as a workplace before the PC successfully enters the private sphere. This is shown in the image of an early industrialization of screen work by Allan Sekula (fig. 18), who in his critical investigation into the normative arrangement of schools *School Is Factory* (1978–80), photographs new forms of training to low keypunch work[20] on the computer.

Figure 18: Allan Sekula: School Is Factory, 1978–80

Source: *Allan Sekula—Photography against the Grain: Essays and Photo Works 1973–1983*, MACK Books, 2016.

20 "The junior college delivers a lot of students, mostly women, to surrounding corporations with a need for clerical and low-level computer workers. Keypunch is the lowest level of computer work, rivaling the assembly line in its brain-numbing routine" (Sekula 2016, 203). It should be noted that this work did not take place on electronic displays but on specially constructed machines that produced analog paper output in the form of punched cards (Da Cruz 2001).

Lee Friedlander follows up here. Some of the images from the Silicon Valley of the eastern US are shot frontally from the perspective of the screen and stage the screen as a counterpart that has its operators firmly in its sight (fig. 19 a–b).[21]

Figure 19 a–b: Lee Friedlander: Left: "At Work"; right: "Boston," 1985

Source: *Lee Friedlander—At Work*, Steidl, 2002.

At the same time, the serial of the open-plan office is shown as the production site of a cognitive capitalism that puts its workers into interchangeable modular environments.

A strange uncertainty arises when looking at these pictures: Who is actually looking at whom?

The operators in Friedlander's images look as if they are being monitored by the monitor. The meaning of the Latin word *monitor*—"reminder, admonisher, overseer"—is fulfilled in a special way.[22] Friedlander points to an aspect that will only be realized in the automated controls of the screen workers or the mutual control of the video conference. In video conferencing, with the same perspective however, it is the programmatic control that maintains command. Just as the participants see

21 This perspective has appeared regularly in some recent photographic work. As, for example, in the work *Immersion* (2008–2014) by Robbie Cooper, which shows people sitting in front of the computer being greatly focused on the game, the film, the football match or the porn website. https://robbiecooper.com/portfolio/immersion.
In some cases, the person facing the screen is shown somewhat less differentiated at the moment of greatest blunting:
Donna Stevens: *Idiot Box*, 2013, http://donnastevens.com.au/idiot-box/donna-stevens
Wolfram Hahn: *A disenchanted playroom*, 2006

22 Contemporary images of working at a computer screen suggest a different situation, one that does not separate work from leisure. These are images of a new economy that propagate the interpenetration of work, creativity, and leisure. See the Apple campaign mentioned at the beginning: "Behind the Mac."

themselves mediated by the software, they see the others as mediated by the apparatus.

Lars Tunbjörk's pictures, on the other hand, show the chaotic, materially exuberant nature of these workplaces (fig. 20 a–b). In doing so, he focuses on the one hand on the environment that people create for themselves in the office and on the other hand on the stress of the never-ending stream of data that doesn't even allow you to dispose of the old equipment before turning to the next monitor.

Figure 20 a–b: Lars Tunbjörk: Office, 2001

Source: *Lars Tunbjörk—Office / Kontor*, Journal, 2002.

This shows the flip side of the individualized user-related programs of today's platforms of the World Wide Web: with the individual diversifying their individuality for the lowest wages on multiple screens to click on website elements that artificially elevate the status of a customer or product.

Such precarious click jobs of an invisible digital economy can now be found in hundreds of thousands around the world.[23] So-called content moderators, who sort out everything we should not encounter on social media platforms, scan operators at Google Books (Bergermann 2016), or the Mechanical Turks at Amazon. Jeff Bezos euphemistically calls this work "artificial artificial intelligence" and describes the concept behind Amazon's profitable business model:

> Normally, a human makes a request of a computer, and the computer does the computation of the task, but artificial artificial intelligences like Mechanical Turk

23 Typically, the work is performed by women from lower social classes in Southeast Asia, especially from India, China, and the Philippines, but also people of color in the US at Amazon and Google. See the documentary *The Cleaners* by Hans Block and Moritz Riesewieck (2019).

> invert all that. The computer has a task that is easy for a human but extraordinarily hard for the computer. So instead of calling a computer service to perform the function, it calls a human. (Pontin 2007)

Figure 21: Image of a Chinese click worker used in many social media posts exploring these kinds of working conditions

Source: Origin unknown. https://www.clickguardian.co.uk/click-farms/.

A work that invisibly labors at the functioning of an energy-consuming digital reality in order to maintain its myth of purity and immateriality. These images testify that this work is not decoupled from life as "artificial intelligence" and an exuberant materiality.[24]

This materiality is one that is often forgotten when we talk about video conferencing. The workability of the infrastructures is taken for granted and the pandemic

24 Part of this functioning, especially in a pandemic world, is that goods reach us without resistance. A growth of this economy is closely related to it. Little was heard in April 2020 about Amazon employees going on strike over poor hygiene and spacing rules at Amazon's goods distribution centers (Blest 2020).

ensured a further spread (Pressmann 2021) of the necessary technical means: laptop, camera, (ring-)light (Mull 2020), and microphone.

Home Work!

Man's relationship with their screens have become different through the pandemic. The screen is meant to protect against infection (Moskatova 2020), but at the same time it reconstitutes, as Simon Strick (2012, 234) has pointed out,

> the interaction between body and machine in the paradigm of a smooth surface and touchless intimacy, which makes any possibility of illegitimate use and intrusion impossible. The inside of the technology—the code—is sealed and immunized, its use becomes simple, personal, productive and non-invasive thanks to metaphorical touchable images (icons).

Thus, our daily counterpart is protection and an insurmountable surface, the suggested closeness disappears behind the glass layer of the display. And the "visually mediated present presence" (Villi 2015) remains a presence that is no more than a social copresence on the screen governed by software conditions. Gazes that cannot meet one another (see Rapoport and Tollman in this book).

"Computers embody a certain logic of governing or steering through the increasingly complex world around us. By individuating us and also integrating us into a totality, their interfaces offer us a form of mapping, of storing files central to our seemingly sovereign—empowered—subjectivity. By interacting with these interfaces, we are also mapped" (Chun 2011, 9).

These conditions are regulated by Skype, Zoom, MS Teams, BigBlueButton, Jitsi,[25] etc. This creates images of people turning to the screen with communicative intent, which have also been used in various ways in the mainstream media since at least the beginning of the pandemic. The provisional aesthetics of the image is always part of the communication. Photographs, screenshots, or screencasts of people in front of or on screens reveal themselves as coming from the screen.[26] Of course, this is especially true for video conferencing.

The laptop is usually positioned slightly below the face on the table, looks up unfavorably to the protagonist, and the algorithmically blurred background allows the software identify through which the image is conveyed. The image on the monitor

---

25 Like Amazon, the winners of this form of disaster capitalism as an exploitation of the Corona crisis for private profit (Klein 2007).

26 The screenshots in the following section are made by the author. Similar images were also used in print media to depict everyday media life under pandemic conditions.

shows an unveiled face, which is/was often no longer possible under pandemic conditions in public. It thus displays an openness in the protection of the screen and, at the same time, the privilege of a work that does not have to expose itself to any risk of infection.

Figure 22: Alexandra Popp, a player on the VfL Wolfsburg football team, talking to Hermann Valkyser about the re-start of the women's Bundesliga in Germany (30 May 2020)

Source: Screenshot from live TV.

Through these images, one no longer simply sees through to the protagonist, but one registers a mediation of mediation. They are second-order images (Mitchell 1994, ch. 2, and Schneider 2022), media-reflexive images that consciously exhibit their status. The supposed immediacy of the transmission is destroyed in this gesture.

They are images of a media-"media witnessing," as I will call it by extension of the term introduced by Paul Frosh and Amit Pinchevski (2009). The content of the medium becomes the medium itself and not another medium, as McLuhan (2013, 22–23) and many following him have presented as the principle of medial mediation. Of course, this only works under the conditions of the computer as a meta-medium (Manovich 2013, 101 ff.), which now combines all forms of (tele)presence. These images become testimony to a relationship to the world in which personal relationships—to the world and its subjects—also reveal themselves as mediated by the screen.

Interview pictures are making their way into the news of relevant broadcasters in inferior quality and, similar to the shaky smartphone pictures of a few years ago, testify to a particular urgency, topicality, or authenticity (Grittmann 2018).

"Almost all of the interviewees tune in live via Skype. The picture is a bit jerky, most of them are seen in a frog perspective from the chin up, but it works. Before that, we did everything we could to avoid poor quality Skype switching. Now we've already gotten used to the aesthetics" (Fiedler 2020).

Figure 23: ZDF reporter Claudia Neumann speaks to Martina Voss-Tecklenburg, Germany's national women's football coach, about the Corona crisis and its impact in a Skype interview, March 27, 2020

Source: Screenshot from live TV.

Figure 24: "Hart aber fair" with Frank Plasberg, May 2021

Source: Screenshot from live TV.

The continuing to function is central to the transmissions at the beginning of the pandemic and generates an acceptance of this aesthetic. Since then, these images have been used naturally in talk shows to replace the presence of people in the studio. This is often used in stagings that place the monitor instead of the interlocutors in order to be streamed into the households in a doubled telepresence.

While the self-representation at the beginning of the pandemic still seems very provisional and unstaged, this changes in the course of 2020. Looking into the mirror image[27] of the computer awakens attention to the surroundings, to one's own face, and to the camera position (Mende 2020). The self-image and thus the background in video conferences professionalizes[28] with increasing use.

In the Mirror

In networked photography, the connection between photographer and recipient is discussed as a production of presence through mediality and technology (Distelmeyer 2021) and as an expression of a culture of *being there* (Vannini and Steward 2017, 152). In the real-time[29] mode of video conferencing, it can be described as a culture of *being with (Dabeisein)*,[30] which is a two-way relationship. *Being with* means that I know I am there on the screen and am in an active negotiating relationship with my counterpart. This includes the possibility of acting at a distance and entering a relationship of interaction with the physical world regulated by technology.[31]

Although I excluded the selfie as an image type for this study at the outset, selfie research is informative for the nature of self-(re)presentation or presentification and its modes of communication. The staging of the face/body against a picturesquely chosen background is programmatic in the selfie. What differentiates selfies in video conferencing, however, is that they are usually recorded with the laptop's camera and, accordingly, are not as mobile as those on the smartphone.

---

27 Kracauer (1927, 34) already wrote of the "photographers' face" that people had acquired in view of the cameras that were everywhere.

28 This can be seen in the selectable background images, the filters that, for example, smooth the skin and automatically brighten the image, etc.

29 In computer science, real-time aims at establishing an input-output connection that is as unmediated as possible. The distance is supposed to become meaningless for human perception (Otto and Haupts 2012, 6). However, this is a recurring problem in video conferencing, as different bandwidths and compressions cause users to experience the lag in different and sometimes isolating ways.

30 This becomes particularly clear when the smartphone is carried through rooms to give the other person an impression of the place where one is currently located.

31 A feature of the software of the networked computer that, in the context of the pandemic, photographers increasingly had to make use of for economic reasons in order to take professional photographs from a safe distance (Li 2020 and Hein 2020).

While in PowerPoint or Keynote presentations in physical space only the desk (background) of the respective computer is ever revealed, in VC the physical background of the desk is in the picture and wants to be made real or digitally enhanced.

The faces, unlike in Friedlander's *Little Screens*, are actually intruders into the private sphere. But they are also the alien mirror images that Friedlander anticipated. I look at myself like another person:

"Since photography is first of all dependent on its apparatus and then, only in a secondary manner, on the body of the photographer, it allows for views of the self to be severed from the body and framed from an external point of view, one that others may occupy just as well" (Keenan 2018, 72).

The aspect that I and others share the same point of view is central to self-image in video conferencing. Unlike the production of a selfie, I see myself in comparison with others, lined up in a two-dimensional tile grid[32] of the respective software. What Tom Holert (2006, 12–13) describes in the context of the visual doubling of a person and their visual representation in a background projection now applies to everyone in video conferencing: "The visual double becomes a kind of control image, that viewers can—in the medium of photography—compare between two stages or aggregate states of the image, measure or test."

The relationship to the *screen-based device* during recording is described by Sabine Wirth (2018, 132):

> Furthermore, the selfie act includes certain knowledge of the operativity of user interfaces as well as a habitualization of media gestures. Taking a selfie encompasses practices like positioning one's body in relation to a screen-based device, fitting oneself into the framework of the smartphone screen, posing, smiling/not-smiling, checking for different angles and backgrounds ... using filters, playing around with formats and app functions—in short: operating a user interface. Thus, the act of taking a picture is ... in the case of the selfie between photographer and screen interface.

In video conferencing, it is precisely the background that takes on special significance, as shown by Alexandra Anikina in this volume.

The self-image of video conferencing becomes an interface and is an operational mirror image that reacts differently than a catoptric mirror image (Hagen 2021, 163ff). I can digitally influence and enhance it almost in real time, but then look at me not only like another person but also like a stranger. That is why it is possible to unreflect the self-representation in video conferencing or to hide the image

32 On the problem of two-dimensionality in video conferencing, cf: Palmer 2019 and referencing Covid 19: Palmer 2020.

altogether because of the intolerability of constantly seeing one's own face in comparison to others.[33] I no longer see myself where I am not, but only the processed mirror images of all the other participants in the video conference. Perhaps this is the only option to escape the domination of the camera: "The subject can only play an active role vis-à-vis the camera or the gaze regime if it resists the appropriation on the part of the images through which it willingly or involuntarily allows itself to be 'photographed.' Only in this way can it deal with them transformatively" (Silverman 1997, 50).

Techno-Imagination

I have attempted here to show how the story of human interaction and communication with electronic screen media can be told through photographic evidence. This relationship today goes further than Huhtamo (2004, 31) who very generally but aptly describes the relationship to the screen: "An increasing part of our daily lives is spent staring at screens." The inactivity associated with staring at the screen is confirmed on the one hand by the spatial bondage to the screen (Friedberg 1996, 28); on the other hand, the action that is now directly associated with the screen has long since ceased to be reducible to staring.

The images I discussed in the section "Little Screens" address the spatio-temporal perforation that began with television and a conditioning of people to these screens in the domestic environment. Related is the intrusion of alien faces into these private domestic spaces. "At Work" reconstructs how the computer screen monitors the conditions of a materially exuberant postindustrial labor. In the section "Home Work!" I show the pandemic driven temporary climax of the mutual real-time transmission of human faces with communicative intent. Here, the distinction between work and private spheres is becoming increasingly problematic due to video conferencing. "In the Mirror" is an attempt to describe the gaze relationships under the conditions of video conferencing as operative mirror images.

---

33 Platforms like Zoom have significant potential for gesture and emotion recognition. So far, the platforms hardly seem to be taking advantage of this. However, for the application of the "video filters" and the "studio effects," AI controlled face recognition and tracking is in use. Zoom is already evaluating gestures and translating them into digital symbols in a simple way, like the "thumbs up" symbol, which is an easily identifiable gesture and should be more visible as a symbol in such meetings (https://support.zoom.us/hc/en-us/articles/4407537406093-Using-gesture-recognition-). However, it is hard to imagine that the "face value" (McCosker and Wilken 2021, 34 ff) produced in this environment will not be used algorithmically in the future.

Historically, the community-building nature of television was followed by a phase of isolation in front of the screen that continues to this day. Thus the image in video conferencing testifies to a history of the postindustrial computer workspace in the domestic context and thereby to a new form of automatization of the gaze that is operationally machine-conditioned, even if cloudy backgrounds want to obscure this as a new immersive experience. "The question, then, is not what we can use the software to do, but what the software does to whatever it is being used to do" (Bucher 2012, 204).

Figure 25: Zoom: Immersive View

Source: https://blog.zoom.us/zoom-immersive-view-bringing-people-together-better/.

The community is to be formed in a collective techno-imagination behind the screens. This makes our sociality stronger than ever "coded by technology ... renders people's activities formal, manageable, and manipulable, enabling platforms to engineer the sociality in people's everyday routines" (Van Dijck 2013, 12).

References

Bergermann, Ulrike. 2016. "Web Extra. Digitus." *Zeitschrift für Medienwissenschaft*, online, September 26, 2016. https://zfmedienwissenschaft.de/node/943.

Blest, Paul. 2020. "Amazon Workers in New York Are Striking for the Second Week in a Row as Coronavirus Cases Pile Up. *Vice News*, April 6, 2020. https://www.vice.com/en_us/article/4age5j/amazon-workers-in-new-york-are-striking-for-the-second-week-in-a-row-as-coronavirus-cases-pile-up.

Bucher, Taina. 2012. *Programmed Sociality: A Software Studies Perspective on Social Networking Sites*. PhD dissertation, Faculty of Humanities, UiO.

Chun, Wendy Hui Kyong. 2011. *Programmed Visions: Software and Memory*. Cambridge, MA: MIT Press.

Comstock, Gordon. 2014. "Jennifer in paradise: the story of the first Photoshopped image." June 13, 2014. https://www.theguardian.com/artanddesign/photography-blog/2014/jun/13/photoshop-first-image-jennifer-in-paradise-photography-artefact-knoll-dullaart.

Da Cruz, Frank. 2001. "The IBM 026 Key Punch." *Columbia University Computing History*, http://www.columbia.edu/cu/computinghistory/026.html.

Deleuze, Gilles, and Félix Guattari. 1987. *A Thousand Plateaus. Capitalism and Schizophrenia*. Minneapolis: University of Minnesota Press.

Distelmeyer, Jan. 2021. "IT Sees: Speculations on the Technologization of the View and Its Distribution?" In *Versatile Images—Looking at the GoPro Movement*, edited by Winfried Gerling and Florian Krautkrämer, 63–78. Berlin: Kadmos.

Donovan, Bill. 2016. "50 Years Ago: The Highs and Lows of the Fairchild Operation at Shiprock." *Navajo Times*, April 21, 2016. https://navajotimes.com/50years/50-years-ago-highs-lows-fairchild-operation-shiprock/.

Ehrmanntraut, Sophie. 2019. *Wie Computer heimisch wurden. Zur Diskursgeschichte des Personal Computers*. Bielefeld: transcript.

Evans, Walker, and Lee Friedlander. 1963. *The Little Screens*. Harper's Bazaar.

Fiedler, Katharina. 2020. "Das Fernsehen in der Coronakrise. 'Wir wollen senden—egal was passiert.'" *Tagesspiegel*, March 30, 2020. https://www.tagesspiegel.de/themen/reportage/das-fernsehen-in-der-coronakrise-wir-wollen-senden-egal-was-passiert/25690622.html.

Foucault, Michel. 1984. *Of Other Spaces: Utopias and Heterotopias*. Architecture / Mouvement/ Continuité. Translated from the French by Jay Miskowiec.

Friedberg, Anne. 1994. *Window Shopping. Cinema and the Postmodern*. Berkeley: University of California Press.

Frohne, Ursula. 2013. "Screen." In *Kunst-Begriffe der Gegenwart von Allegorie bis Zip*, edited by Jörn Schafaff, Nina Schallenberg, and Tobias Vogt, 255–64. Cologne: King.

Frosh, Paul, and Amit Pinchevski. 2009. "Why Media Witnessing? Why Now!" In *Media Witnessing: Testimony in the Age of Mass Communication*, edited by Paul Frosh and Amit Pinchevski, 1–18. London: Palgrave Macmillan.

Gerling, Winfried. 2022. "Screenshotten." In *Historisches Wörterbuch des Mediengebrauchs*, edited by Heiko Christians, Matthias Bickenbach, and Nikolaus Wegmann, 3:430–453. Cologne: Böhlau.

Gerling, Winfried. 2023. "In-front-of-the-screen images—A photo essay." In *Screen Images: In-Game Photography, Screenshot, Screencast*, edited by Winfried Gerling, Sebastian Möring, and Marco De Mutiis, 93–134. Berlin, Kadmos.

Gerling, Winfried. 2018. "Photography in the Digital. Screenshot and In-Game Photography." *Photographies* 11(2/3): 149–67.

Grittmann, Elke. 2018. "Fotojournalismus und journalistische Bildkommunikation in der digitalen Ära." In *Handbuch Visuelle Kommunikationsforschung*, edited by Katharina Lobinger. Wiesbaden: Springer VS. https://doi.org/10.1007/978-3-658-0 6738-0_7-1.

Gunthert, André. 2020. *Le triomphe des images*, April 3, 2020. http://imagesociale.fr/ 8445.

Hein, Theresa. 2020. "Um sechs Uhr morgens saß ich da und hab den Lautsprecher angebrüllt." *sueddetsche.de*, June 17, 2020. https://www.sueddeutsche.de/kultur/ corona-zoom-fotografie-1.4934958.

Holert, Tom. 2006. "Und hinter ihr/ihm, in wesenlosem Scheine: Rückendeckungen im öffentlichen Bild." In *flypaper #2*, edited by Lutz Becker and Uwe Koch, Köln.

Huhtamo, Erkki. 2004. "Elements of Screenology: Toward an Archaeology of the Screen," in: *ICONICS: International Studies of the Modern Image*, 7: 31–82.

Kenaan, Hagi. 2018: "The Selfie and the Face. In Exploring the Selfie," edited by Julia Eckel, Jens Ruchatz, and Sabine Wirth, 67–77. London: Palgrave Macmillan.

Kennedy, John F. 1959. "A Force That Has Changed the Political Scene." *TV Guide Magazine*. November 14, 1959.

Klein, Naomi. 2007. *The Shock Doctrine: The Rise of Disaster Capitalism*. Toronto: Alfred A. Knopf.

Kracauer, Siegfried. 1977 [1927]. *Die Photographie.* In: *Das Ornament der Masse*. Frankfurt a. M.: Suhrkamp, 21–39.

Kubitz, Peter Paul. 1997. *Der Traum vom Sehen—Zeitalter der Televisionen*. Dresden: Verlag der Kunst.

Li, Grace Z. 2020. "A Photographer's Guide to Creating Portraits from a Distance. Visual Artists are Working from Home—With Help from FaceTime and Zoom." April 28, 2020. https://www.atlasobscura.com/articles/remote-photography-o ver-zoom.

Mandel, Lois. 1967. "The Computer Girls." *Cosmopolitan* (April 1967): 52–56.

Manovich, Lev. 2013. *Software Takes Command*. New York: Bloomsbury Academic.

McCosker, Anthony, and Rowan Wilken. 2021. *Automating Vision—The Social Impact of the New Camera Consciousness*. New York: Routledge.

McLuhan, Marshall, and Quentin Fiore. 1967. *The Medium Is the Message*. New York: Random House.

McLuhan, Marshall. 1995 [1964]: *Die magischen Kanäle.* Dresden: Verlag der Kunst.

McLuhan, Marshall. 1964: *Understanding Media: The Extensions of Man.* New York: McGraw-Hill.

Mende, Doreen. 2020. "The Live-stream's Split-screen, or, Urgent Domestic Politics." *Pink Mercedes/02*, April 5, 2020. https://www.harun-farocki-institut.org/e n/2020/04/05/the-live-streams-split-screen-or-urgent-domestic-politics-2/.

Mirzoeff, Nicholas. 2015. *How to See the World*. London: Penguin Books.
Mitchell, W. J. T. 1994. "Chapter 2." In *Picture Theory: Essays on Verbal and Visual Representation*. Chicago: University of Chicago Press.
Moskatova, Olga. 2020. "Networked Screens: Topologies of Distance and Media Regime of Immunization." *img journal*, 2 (2): 282–305.
Mull, Amanda. 2020. "Americans Got Tired of Looking Bad on Zoom." *The Atlantic*, November 19, 2020. https://www.theatlantic.com/technology/archive/2020/11/ring-lights-for-all/617143/.
Mulvey, Laura. 1975: "Visual Pleasure and Narrative Cinema." *Screen* 16 (3): 6–18.
Otto, Isabell, and Tobias Haupts. 2012. "Was ist ‹in Echtzeit›? Eine Einleitung." *AugenBlick. Marburger Hefte zur Medienwissenschaft*. 51. Bilder in Echtzeit. Medialität und Ästhetik des digitalen Bewegtbildes.
Palmer, John. 2019. *Spatial Interfaces*. August 30, 2019. https://darkblueheaven.com/spatialinterfaces/.
Palmer, John. 2020. *Spatial Software*. April 9, 2020. https://darkblueheaven.com/spatialsoftware/.
Pontin, Jason. 2007. "Artificial Intelligence, With Help from the Humans." *New York Times*, March 25, 2007. https://www.nytimes.com/2007/03/25/business/yourmoney/25Stream.html?smid=tw-share.
Pressman, Aaron: "PC sales have surged for at-home workers and learners during the pandemic." *Fortune*, January 11, 2021. https://fortune.com/2021/01/11/covid-computer-sales-lenovo-hp-dell-apple/.
Rubin, Joan S., and Casper, Scott E., eds. 2013. *The Oxford Encyclopedia of American Cultural and Intellectual History*. Oxford: Oxford University Press.
Schneider, Birgit. 2023. "'Shoot(ing) the Image'—A Look at Screen Images from a Meta-pictorial and Media-archaeological Perspective." In *Screen Images: In-Game Photography, Screenshot, Screencast*, edited by Winfried Gerling, Sebastian Möring, and Marco de Mutiis, 53–76. Berlin, Kadmos.
Sekula, Allan. 1984 "School Is Factory." In *Photography against the Grain: Essays and Photo-Works 1973–1983*, 199–234. Halifax: The Press of the Nova Scotia College of Art and Design.
Silverman, Kaja. 1997. "Dem Blickregime begegnen." In *Privileg Blick: Kritik der visuellen Kultur*, edited by Christian Kravagna, 41–64. Berlin: Edition ID Archive.
Spigel, Lynn. 1992. *Make Room for TV: Television and the Family Ideal in Postwar America*. Chicago: University of Chicago Press.
Spigel, Lynn. 2013. "Medienhaushalte. Damals und heute." *Zeitschrift für Medienwissenschaft* 9 (2): 79–94. https://doi.org/10.25969/mediarep/894.
Strick, Simon. 2012. "The Straight Screen: Begradigungsarbeiten am iPhone." *Feministische Studien* 30 (2): 228–44.
Thielmann, Tristan. 2018. "Der einleuchtende Grund digitaler Bilder. Die Mediengeschichte und Medienpraxistheorie des Displays." In *Display / Dispositiv: Ästhe-*

tische Ordnungen, edited by Ursula Frohne, Lilian Haberer, and Annette Urban, 525–75. Paderborn: Wilhelm Fink.

Thompson, Clive. 2019. "The Secret History of Women in Coding—Computer Programming Once had Much Better Gender Balance than It Does Today. What Went Wrong?" *New York Times*, February 13, 2019. https://www.nytimes.com/2019/02/13/magazine/women-coding-computer-programming.html.

Tollmann, Vera. 2020. "Poor Connections: A Long History of Videotelephony." *Cabinet Kiosk*, September 1, 2020. https://www.cabinetmagazine.org/kiosk/tollmann_vera_1_september_2020.php.

Van Dijck, José. 2013. *The Culture of Connectivity. A Critical History of Social Media*. New York: Oxford University Press.

Vannini, Philip, and Lindsay M. Steward. 2017. "The GoPro Gaze." *Cultural Geographies*, 24 (1): 149–55.

Villi, Mikko. 2015. "'Hey, I'm Here Right Now': Camera Phone Photographs and Mediated Presence." *Photographies* 8 (1): 3–22.

Weiser, Mark. 1991. "The Computer for the 21st Century." *Scientific American* 265: 94–104.

Weiser, Mark, Rich Gold, and John Seely Brown. 1999. "The origins of ubiquitous computing research at PARC in the late 1980s." *IBM Systems Journal* 38: 693–96.

Wirth, Sabine. 2018. "Interfacing the Self—Smartphone Snaps and the Temporality of the Selfie." In *Exploring the Selfie*, edited by Julia Eckel, Jens Ruchatz, and Sabine Wirth, 125–38. London: Palgrave Macmillan.

Video Conferencing and Performance Magic

Will Houstoun and Katharina Rein

More than a century after it was first imagined by electrical engineers and authors of science fiction, videotelephony became a broadly accessible communication tool with the internet and was widely adopted due to the Covid-19 pandemic. As self-isolation necessitated a shift away from in-person interaction, this not only replaced business meetings and birthday parties, but also the performing arts that could no longer take place in front of audiences in their usual spaces. Artists had to come up with ways to perform online, using the software that was available to them and to their audiences. In this way, video conferencing applications designed for business use became a virtual space for artistic performances. Among the artists who shifted their performances to online video conferencing platforms at the beginning of the Covid-19 pandemic were magicians.

Combining practical and theoretical knowledge, this chapter explores the connection between video conferencing and performance magic. It is equally informed by current performative practices of magic and its virtual adaptation, as well as by an academic approach from the perspective of media history. We have previously explored areas such as the shift from traditional in-person activities to video conferencing platforms in domains including performance, medical practice, and education (Houstoun and Thompson 2021; Kneebone, Houstoun and Houghton 2021; Houghton et al. 2021) as well as the use of media technologies in magic around 1900 (Rein 2015; Rein 2019; Rein 2023). This chapter combines these areas of expertise to explore how magic changes as it migrates to new transmission media.

In this chapter, we focus on a particular kind of magic—the kind that Simon During defined as *secular magic*, as opposed to *real* or *supernatural magic*. Secular magic, he writes, "is different from the magic of rituals, myths, and fetishes, as well as that of spirits, universal sympathies and antipathies, or of superstition or credulity. It is a self-consciously illusory magic, carrying a long history, organized around still-beleaguered lightness or triviality, which it also massively exceeds" (2002, 27). Because this kind of magic is performed for entertainment purposes, we call it performance magic. Moreover, it is received by audiences who are aware that they are witnessing illusions accomplished by techniques (of the body) or technology, while not understanding exactly how. For this reason, we exclude related but distinct performers like spiritualist mediums, who typically appear in a context in which the illusions are marked as the result of supernatural occurrences (on the distinction between spiritualism and magic as a matter of framing, see Lamont 2006).

The relationship between performance magic and video conferencing (as well as other technical media) takes place on at least three different levels. All three are interconnected and therefore touched upon in this chapter, while the focus is on the third one. The first is the "magification" of technical media. This level is related to cultural imagination and works both ways: On the one hand, technologies become the subject of "magical" fantasies; on the other, a sense of magic latches on to existing technologies. One expression of this is the suspicion of technical media's uncanny potential to function as spiritualist mediums—that is, to facilitate communication not only between the living but also with other worlds. This aspect has been well researched, for instance by Jeffrey Sconce (2000) and Anthony Enns (2015). Technology's spiritual and mystical connotations, too, have been discussed, for instance by Erik Davis (2004, 11), who writes that "the spiritual imagination seizes information technology for its own purposes. In this sense, technologies of communication are always, at least potentially, technologies of the sacred, simply because the ideas and experiences of the sacred have always informed human communication." This includes not only existing technologies but also fictional ones, with the boundaries between them often being blurred. Throughout history and fiction, various media have been imbued with supernatural qualities, becoming the object of fantasies of a utopian or dystopian, highly technical future. Videotelephony is one of those "magical" technologies that was the subject of fantasy long before it became reality. Not only has videotelephony been one of the central futuristic devices of science fiction literature and film from Jules Verne to *Metropolis* to *Blade Runner* and countless others, it has also sparked numerous scientific fantasies of "seeing by electricity" and its implications since Victorian times.

The second kind of relationship between magic and technology concerns their direct interplay in performative practices: Performers present both existing and imagined devices in their shows as "magical"—that is, apparently fulfilling magical functions such as restoring broken objects, reading minds, and teleporting people. Often, this performance practice directly taps into the cultural imaginary generated by technological progress. Technologies are also often employed covertly in order to create an illusion without audiences being aware of their existence, for instance when concealed transmitters convey messages that are allegedly received telepathically (see Rein 2015).

The third level of performance magic's relation to technology pertains to its own mediatization. Performance magic has been disseminated through textual, auditory, and visual media for a long time, displaying a remarkable ability to adapt to various production and distribution practices. While magicians traditionally appeared at fairground shows, in streets and marketplaces, they invaded theater and opera stages with great success in the second half of the nineteenth century, a move that resulted in the "golden age" of conjuring around 1900.[1] In the following century, magic was transmitted via letters, telegraphy, radio, television, and now the internet.

This chapter examines how illusionistic performance practices change as magic shows migrate across media, particularly in the context of video conferencing technology. While this primarily concerns the mediatization of magic that would traditionally be performed in person, it also involves the way conditions of a particular medium can be used for illusions that would not be possible in other settings. That, in turn, charges the given medium with magical potential, which then feeds into the cultural imaginary.

In order to understand video conferencing and its relationship to performance magic today, it is necessary to understand the technology's historical association with "magical" practices. The history of video conferencing technology takes us back to the late nineteenth century and the golden age of magic. In the late 1870s, electrical engineers first envisaged sending images, along with the sound that had recently become transmittable via telephone. While these early concepts partially fed into television technology, which switched to a one-way transmission model, experimentation with videotelephony continued throughout the twentieth century. We examine how magic ties into this history of technology by first migrating to television, despite being deemed unfit for this medium. We argue that a similar, equally striking media change occurred in 2020 as Covid-19 instigated magic's successful adoption of online video conferencing technology. For this reason, and due to the historical proximity of television and videotelephony, these two media and their relation to performance magic are both considered in this chapter. Both simulate telepresence that, in the realm of performance magic, has further been reflected in illusions that stage teleportation, different versions of which serve as case studies in this chapter.

The Magic of Transmission

In 1923 the crème de la crème of the North American magic business met at the McAlpin Hotel in New York for the nineteenth annual banquet of the Society of

1 There are different views on when exactly this golden age began, and how long it lasted. It is always placed around 1900. While, for instance, Mike Caveney (2009) proposed a period of 50 years, his is a US-centric perspective, beginning in the 1880s, when the popularity of magic made its way across the Atlantic. We, however, hold with a longer time span such as the one proposed by Jim Steinmeyer (2005), which extends from 1845 to 1936. Taking European magic history and its significant impact on US magic into account, it represents historical developments more accurately.

American Magicians. A certain "Doc Wilson," former dean of the Kansas City University of Medicine and Surgery and editor of *The Sphinx*, one of the pertinent magicians' trade journals, was invited as a guest of honor. He was not able to attend in person, star magician Harry Houdini announced at the banquet, but his address would be transmitted live[2] from Kansas City. For this purpose, Houdini had an authentic-looking radio cabinet installed by the Radio Corporation of America, complete with technicians in boiler suits pretending to be busy on the transmission. Suddenly, Wilson's voice emerged from loudspeakers installed around the dining room. Occasionally interrupted by static noise, he spoke of a vision of the future in which not only sound could be wirelessly transmitted across distances, but in which images, and even people, would be sent by radio waves. At the end of this address, Wilson surprisingly stated that he would test the latter claim right away, and subsequently "[t]o loudly crackling static, Wilson toppled out of the radio cabinet, eyes blinking, coat mussed, hair disheveled .... Houdini, almost losing his balance, reached over the table to shake Wilson's hand, having produced what in mediumistic terms would be a full teleportation" (Silverman 1996, 305). While in part anticipating future technologies, the conjuring of Doc Wilson also manifestly illustrates the connection between magic and media: starting from the wireless, a medium that was still relatively new yet real, the demonstration went on to imagine and stage the fantastic possibility of teleportation, drawing on the idea of technical media's supernatural abilities.

This presentation is an example of the second kind of relationship between magic and technology described above, in which both existing and imagined technological devices are presented in a show as fulfilling a "magical" function, while it also taps into the first level—that is, technology's perception as potentially magical in the cultural imaginary. This illusion references several themes of its time: the radio as the medium which went mainstream in the early twenties; spirit manifestations in the contexts of entertainment culture and Spiritualism (see Natale 2016), which reached a never-again attained peak in the aftermath of the First World War; and fantasies about radio waves transmitting something other than music and voices of the living. Such ideas were, at the same time, pursued by radio pioneers like Guglielmo Marconi, Manfred von Ardenne, and Oliver Lodge (see Hagen 2002, 232–35), while science fiction literature amalgamated electrical engineering and fantasy. One of the topics of this collective imagination was videotelephony (on late-nineteenth-century fantasies of "distant vision" see Burns 1998, 78–100). "Victorian engineers," Ivy Roberts (2019, 20) writes, "pictured mirror-like screens

2 In this article we refer to performances as being "live" when they have not been prerecorded but are transmitted in real time while they are taking place. We use the term "in person" to designate cases in which the audience and the performer share the same physical space—that is, are in the same room together, such as a theater hall.

and devices modeled after electrical telegraphs." In line with this project of "seeing by electricity," Jules Verne first described videotelephony in his short story "In the Year 2889" in 1889. Here, the "telephote" allows for the instantaneous "transmission of images by means of sensitive mirrors connected by wires" (665) along with the sound conveyed by a telephone.[3] Videotelephony subsequently became a staple of science fiction literature and film from Hugo Gernsback's *Radio for All* (1922), which envisions a "television and an automatic radiophone" (image caption "The Future of Radio," n. pag.) to the regulatory terrorism in the workplace depicted in Charlie Chaplin's *Modern Times* (1936) to *Back to the Future II* (1989), which imagines videotelephony as an everyday communication technology in the year 2015. Videotelephony recurs throughout history and fiction, for example in Hanna-Barbera's Space Age animated sitcom *The Jetsons* (1962–63), in which characters put on "morning masks" to hide their untidy appearance when called at an inconvenient time (Hanna and Barbera 1962).[4] These diverse examples show that, contrary to the narrative that sees videotelephony as a failed technology—at least until the advent of the smartphone made it mobile and more accessible (see, e.g., Held 2020), it was in fact one that was highly inspirational on various levels. Since its early days, videotelephony stimulated technological innovation, and it brought up issues of privacy and surveillance, which are prevalent in today's media culture. It also served as an ongoing source of inspiration for fiction and popular culture for over a century. The fact that, for over a century, numerous attempts to establish videotelephony as an everyday communication tool failed was, in fact, not followed by the technology's abandonment by engineers. Instead, energy continued to be invested into implementing its wide use in fiction as well as in reality.

Among the early Victorian inventors who were attempting to construct devices for videotelephony were the most renowned electrical engineers of the time: Edison announced a "Telephonoscope" in 1878, shortly after Bell presented his telephone to the public, and a "Far-Sight Machine" in 1889. Nikola Tesla, too, stated that he was working on an invention by the name of "visual telegraphy" which transmits images as well as sounds during phone calls (M'Govern 1899, 295–96). Edison's failure to deliver videotelephony got lost in the whirlpool of public excitement over two other inventions that he unveiled in 1891: the Kinetograph and Kinetoscope (see Roberts 2019, 22). This evidences that early endeavors to create videotelephony ultimately fed into the creation of cinema.

---

3 Published under the name of Jules Verne, this work is now believed to have been authored by his son Michel, probably based on the ideas of his father, who also mentions the "telephote" a few years later in *The Carpathian Castle* (1892, 177n).

4 This is in fact not an avatar that appears, as Tobias Held (2020, 59) writes, but a physical mask worn by the characters.

Another invention that arose from these attempts is television. While early mechanical-optical approaches did not prevail, the first successful demonstrations of fully electronic television devices took place in 1926. Along with other pioneers that shaped this new medium, the Bell Telephone Company was working on a device which was to transmit images along with voices, with the telling name of "Ikonoscope" (see Roberts 2019, 192–205). This device, Ivy Roberts shows in *Visions of Electric Media*, contributed to the paradigm shift in the conceptualization of television by turning away from a two-way communication device to a one-way transmitter. The first presentation of Bell's early television system took place in 1927, the year in which Fritz Lang's cinematic future fantasy *Metropolis* was released, which envisages videotelephony as an everyday communication medium of the rich and the privileged. Bell Labs' demonstration included the broadcasting of an address by Herbert Hoover, then Secretary of Commerce, from Washington, DC, to Bell Labs in New York, followed by a vaudeville act, which was broadcast from a studio in New Jersey (*New York Times* 1927; see also Burns 1998, 227–32; Roberts 2019, 196–99).

Hoover's voice emanating from loudspeakers and his image appearing on a screen may not have seemed any less magical than Houdini's "teleportation" did just a few years earlier, even though the speaker did not fully materialize in the end. In 1923, at the Society of American Magicians' annual banquet, the technological transmission of images, and even people, over vast distances, was presented as magic. A few years later, it had indeed become possible to send moving images in real time. The technology initially conceptualized as a two-way transmission, eventually split in two: television, in its practical implementation shifted to a one-way model, while the two-way approach resulted in videotelephony.

Performing Magic on Television

In "The Aesthetic of Astonishment" Tom Gunning (1995, 116) writes that it is "[t]he seeming transcendence of the laws of the material universe by the magical theatre" that "defines the dialectical nature of its illusions." Elsewhere, he ascertains that the effects achieved in a magic show heavily depend on in-person presence: when writing about the "optical uncanny," an effect that he attributes to magic tricks, Gunning (2008, 73) states that "[s]eeing an event—such as an elephant apparently vanishing before our very eyes, or a woman floating in midair—that contradicts rational expectations generates a different, perhaps more powerful, hesitation than reading an account of such an event." Magic, it seems, draws its strength, at least partially, from taking place directly within the reach of the spectators' sensory experience. In a similar vein, magician Thomas Fraps (2021, 57) writes that the kind of illusion that we encounter in performance magic differs from, for instance, optical illusions in that these illusions take place "in reality"—that is, in the space and time immediately shared by the audience. This "illusion of impossibility," as Fraps (2021, 60) calls it, "is unique to the art of magic, since it apparently happens in the here and now, in real time and space. There is no canvas, no screen, no paper on which the fiction is mediated by a painter, director, or writer. The canvas of a magician is the mind of the spectator and reality itself." Even the performance of an illusion, he says, cannot be perfected by the magician alone. Rather, it evolves over time, through interactions with audiences that take place during the performances (Fraps 2021, 54).

As a consequence of performance magic's strong link to in-person shows and its emphasis on audience interaction, it has often been argued that it is unfit for medial transmission. A glance at the history of performance magic, however, reveals that it has, in fact, not been exclusively successful in person. Famously, cinema, in its early days, was instantly invaded by magicians, both from artistic and commercial angles (see Barnouw 1981). Magicians played a crucial role in the history of early cinema: they were among the first to screen films in theaters (as part of their shows) as well as to bring film projections to the provinces and to other countries (on their tours); they were also among the first filmmakers, constructors of cinematic apparatus, and distributors of films (see Rein 2017 for a detailed account). The most famous example is the pioneer of cinematic special effects Georges Méliès, who was a well-respected magician and director of one of Europe's most renowned magic theaters before, during, and after his creative focus shifted to cinema. Another example is David Devant, widely regarded as the greatest magician in English history, who worked as Méliès's agent in the UK and was key in bringing early cinema to English venues away from metropolitan centers.

While this shift from stage practice to cinema as another illusionistic medium seemed like a logical extension of Victorian magic, the migration to television went less smoothly. Magicians began to appear on television in the 1950s, when they performed their acts as part of variety programs such as *The Ed Sullivan Show*. In 1960, Mark Wilson created the first magic show to be broadcast nationwide in the US, *The Magic Land of Allakazam* (1960–64). This program set the parameters for magic performances on television, including the ones illustrated by the example below. While, in early cinema, magicians quickly shifted from filming their stage effects in what Matthew Solomon (2010) calls *films of tricks* to inventing cinematic effects with which they were able to produce new illusions in *trick films*, when it came to television half a century later, camera tricks and special effects were something that made magic less convincing. Therefore, magicians put a lot of effort into persuading audiences that they were "merely" watching the transmissions of illusions, which could be performed on a theater stage in exactly the same manner. Television had to be presented as a medium merely distributing rather than changing its "content." However, as we know, media are not neutral transmitters. "[I]t is the medium," Marshall McLuhan writes in *Understanding Media* (1994 [1964], 9), "that shapes and controls the scale and form of human association and action. The content or uses of such media are as diverse as they are ineffectual in shaping the form of human association. Indeed, it is only too typical that the 'content' of any medium blinds us to the character of the medium." The main challenge performing on television posed for magicians was to make the defining role of the medium in the process of communication disappear and to create the illusion that television was a neutral transmitter.

This challenge is rooted in the view that performance magic was, and still is, often deemed unsuitable for mediatization. As described previously, this is because its effect is frequently attributed to its display of something impossible happening in the physical space shared by the spectators and that therefore follows the natural laws that they are familiar with. Performance magic, this narrative tells us, loses its power as soon as it is mediatized. However, the medium through which magic primarily reached its audiences in the twentieth century was television. Its great success on television since the 1960s remains in opposition to the aforementioned widespread claim. In the following section, we examine the ways in which this migration to a new medium has changed performance practices through the example of David Copperfield's "Portal" illusion. This will help us to understand, in the following section, how magic transforms when it changes media once again, invading the virtual space of video conferencing technology where it confronts the same claim concerning its mediatization.

Example: David Copperfield's "Portal" (2001)

On television, the lack of in-person interaction is compensated for by a constant confirmation of authenticity. Because performance magic on television is always already suspected of being accomplished by camera tricks or editing, "[c]redibility," magician and magic historian Jamy Ian Swiss (2007) writes, "is a supremely pressing issue" (see also Swiss 2022). Indeed, televised magic shows go to great lengths in order to establish authenticity, when, for instance, volunteers from the audience inspect props, floors and walls, serving as representatives of the larger audience. While this mechanism is also used at in-person shows, its role on television is even more important because the viewers at home have no chance to inspect the set-up on site. The volunteers serve as identification figures who experience the three-dimensional space in lieu of the viewers who only receive a two-dimensional picture of it. Some magicians even go so far as to make volunteers the central subjects of their illusions, thereby suggesting that the given feats could be accomplished with any and every person from the audience—including, by proxy, the television viewers.

In his "Portal" illusion, which was introduced in 2001 and featured in the television special *Copperfield: Tornado of Fire*, David Copperfield "teleports" himself and a spectator named Michael to the latter's father's home, Hawaii. The place of destination is allegedly transmitted in real-time—attested to by the word "LIVE" in the picture—and screened at the back of the stage throughout the performance.[5] Here, upon Copperfield's request, another person checks for a possible false bottom that might be hidden under the sand. The magician and Michael then ascend a platform, which is raised above the audience—suggesting that they cannot leave it without being noticed by the surrounding spectators. A curtain closes around the pair. When it is pulled away fifteen seconds later, they are gone. Another forty seconds pass, and Copperfield and Michael appear on the screen, apparently live in Hawaii.

While the story of this illusion is described quickly, the largest part of the performance comprises a complicated array of actions designed to establish credibility. In his introduction, Copperfield tells of letters he received, among them a particularly touching one from a father longing to meet his estranged son. While the father lives in Hawaii, Copperfield explains, his son is in the audience tonight. The volunteer, subsequently introduced as Michael, is thus provided with a backstory that is verified by childhood photographs appearing on a screen. The choice not to use a random audience member was a careful one, designed to seem fair yet believable (see Young, Britten, and Copperfield 2020). Copperfield then selects further spectators by throwing a ball into the audience and asking those who catch it to point out more volunteers. Thus, four more people join Copperfield and Michael on stage, where they leave personal marks—initials, hobbies, etc.—on a large sheet of paper. The result, along with the group, is documented by a polaroid photograph, which Copperfield and Michael take with them to Hawaii. Along with one of the spectator's initials that were written on Copperfield's arm, this proves that the footage has not been edited before the broadcast. Before vanishing, Copperfield further stresses his show's spontaneity and coincidence by announcing that he performs this illusion nightly, with different spectators.

After the two reappear on the "Hawaii" screen, Copperfield invites Michael to run into the ocean until his legs get wet. He then pulls up his sleeve to reveal the initials on his arm, while Michael removes the polaroid photograph from his pocket. The complex, mise en abyme–like structure of this authenticating procedure manifests itself in the form of this drawing, documented by a photograph, shown to a camera that transmits it to a screen on a stage that is filmed for a television show. To top up this spectacular media circus, Copperfield now hands Michael a camera and asks him to "take it to that guy over there," pointing to Michael's father who is about to walk into the picture frame and with whom Michael remains on the beach. After stating that "it's starting to rain," the magician disappears behind a cloth on the sand and reappears on a small platform back in the theater auditorium. To further illustrate the magical interpenetration of the two distant places, Copperfield brings

5 Even when not recording the performance for television, Copperfield would hire a satellite truck that would stay outside the theater, to reinforce the idea that there was a live broadcast from Hawaii (Young, Britten, and Copperfield 2020).

the Hawaiian weather with him: A tilt of the camera reveals water drops on its lens and his reappearance is accompanied by thunder, lightning, and rain pouring down on astonished spectators. For the finale, Copperfield returns to the stage, pours out a handful of sand, and strikes a pose.

"Portal" exemplifies the trouble taken to establish credibility in magic performances on television. While these practices are also found in stage performances, in mediatized ones, their importance increases dramatically. For television, Copperfield has to disprove the assumptions that the illusion might be achieved by "a camera trick," that the show, and the beach sequence in particular, may have been pre-recorded and edited. To do this, great care is taken to establish continuity between the two places. For instance, Michael's getting his feet wet in the ocean serves to prove that "Hawaii" is neither a painted background in an adjacent studio nor a greenscreen effect. Disproving video manipulation is a challenge that has accompanied mediated magic performances ever since the possibility has been understood by audiences. A televised magic show has to please two possible audiences: the one in the theater from which volunteers are pulled and the one watching from home via their televisions. The first audience establishes credibility for the second one, vouching for an authenticity that is otherwise inaccessible. The televised Hawaii also stands in for the televised space watched by the television audience—this continuity therefore also serves to connect the space of the television show with their physical environment.

Performing Magic via Video Conferencing

This example shows how magic has previously been shaped by its migrations across media and by finding ways to compensate for the difference to in-person experiences. While the same arguments concerning performance magic's unsuitability for transmission are brought up against online shows, some practitioners have nonetheless pioneered performances via the internet. Demonstrations of magic tricks, debunking videos, and tutorials on YouTube and similar platforms have been attracting unprecedented numbers of viewers in recent decades, but these formats, though somewhat different under the conditions of the internet, mostly follow the one-to-many mass-media model of television shows. This changed in 2020, when the Covid-19 pandemic forced magicians, like many other artists, to find solutions for interactive online performances. This is a challenge for two reasons that have already played in relation to television: performance magic as an art form relies heavily on (1) making the apparently supernatural phenomena happen in the same physical space occupied by the spectators and (2) interacting with audience members who take on the role of assistants or of authenticating observers. Adapting their performances to virtual spaces and to software initially developed for other purposes, some magicians successfully transitioned to shows via video conferencing software. Working against the long-standing conviction that magic only "works" in person, approaches emerged to productively use the new virtual environment for illusionistic purposes and to exploit its possibilities.

As already discussed, television was initially devised as a tool for two-way communication, "a form of visual telephony" (Roberts 2019, 21). While its successful practical implementation required a shift to one-way transmission, video conferencing technologies that first became reality in the 1930s enabled a return to the two-way model envisioned by Victorian electrical engineers. The American Telephone and Telegraph Company (formerly the Bell Telephone Company) presented an improved version of their "iconophone" to a selected public in 1930. A device that looks similar to this one (see *Popular Mechanics* 1930, 892), while also referencing later developments, is depicted in Stanley Kubrick's science fiction classic *2001: A Space Odyssey* (1968), which features a video phone call from a space station to earth. Outside of futuristic cinematic magic, AT&T's first commercial videotelephone service was a failure: When Kubrick's futuristic vision was released, videophone booths were being closed after an unsuccessful four-year run in three major US cities (Mäkinen 2007, 37). Succeeding models, too, failed commercially. Videotelephony only became part of our everyday reality with the advent of the internet, smartphones, and, particularly, during the remote-working peak caused by the Covid-19 pandemic.

As a performance space, the virtual environment offers a unique combination of in-person and television performance. On the one hand, performances are mediated visually and acoustically via camera, microphone, and screen, as with television. On the other hand, they maintain a liveness that is close to in-person shows, with the possibility of real-time two-way interaction between performer and audience, both as a whole and as individuals. In addition, the wholesale adoption of video conferencing in 2020, across a broad demographic and for events ranging from business meetings to education to social interaction, engendered a kind of trust in the medium. People experienced video conferencing (often engaging deeply with the technology for the first time) as a stand-in for the in-person events that were suspended due to the pandemic. Most users of video conferencing software do not perceive it as a technology prone to audiovisual manipulation via special effects and editing. While video conferencing makes the same tools available that are at hand in any performance mediated via a camera, spectators are not necessarily aware of this, tending to believe that what they see via the camera is precisely what they would see if they were in the room with the performer, without the suspicion that would accompany an identical illusion if it were viewed on a television program like Copperfield's. Thus, because of the context of pandemic-era video conferencing, the technology itself creates the impression of being a neutral transmission device that imitates an in-person situation as closely as possible. Because video conference calls are typically perceived in this way, seemingly without offering much possibility of manipulation, it is the technology itself that does the authenticating work that magicians took pains to accomplish in television shows. This, too, is rooted in today's use as much as in the history of two-way video communication, which, since its earliest conceptions, has been regarded as an electric "window through which one could see a distant place," including "[w]itnesses [who] sometimes responded to the interaction with a television screen as if there was no mediation involved" (Roberts 2019, 213–14).

As large numbers of magicians shifted their performances to online video conferencing platforms, they broadly utilized one of two different approaches: either attempting to adapt the tricks they had performed in person to the new setting or trying to develop new material for this novel performance space. The former approach had the advantage of being a quicker route to getting performances going but also featured an inherent issue: in adapting material designed for a different setting to work online compromises invariably had to be made, and the best-case outcome was a performance that would have been better in an in-person setting. The latter approach, while requiring more work and creativity initially, produced a remarkably broad selection of work specifically designed to create magical experiences that could only ever happen via video conferencing. An example of a trick developed using the latter approach is another form of teleportation.

Example: NFW (2020)

On a video call the magician invites two spectators to help with an experiment in teleportation: Anne, in America, and her sister, Sam, in Spain. Anne and Sam's video feeds are added to the performer's, beside one another, and they are asked to get the pack of cards and envelope that they were told to bring to the show. Anne shows everyone on the call that the envelope she has is empty, seals it up, and then holds it in between her hands, making sure it stays in view of the camera. Sam then names a random playing card, the Queen of Hearts, and holds her sealed pack of cards up in one hand. The magician then reaches over to where Sam's video is showing and mimes grabbing something invisible from it. They look at it, smile, and then throw it toward Anne's video feed.

A puzzled Sam is invited to break the seal on her pack of cards and take the pack out of the box. She confirms that she picked the Queen of Hearts, and that she could have named any card. She then shows the cards, one by one to the camera. The entire pack is in order, until she gets to the hearts. The Ten of Hearts is there, as is the Jack, but then the next card is the King of Hearts. Sam goes through the entire pack and confirms that the Queen has completely vanished.

Anne confirms that the envelope has been between her hands the entire time, that it was empty, and that nobody has gone anywhere near it. She then tears it open, looks inside, and starts laughing. She tells everyone that there is now one card inside. On removing the card, she reveals the Queen of Hearts. Sam and Anne's videos expand to fill the screen, as the magician fades away into the background, and the audience offer the astonished sisters a round of applause.

While a chosen card appearing in a spectator's hands has been a standard effect for centuries (see, for example, Breslaw 1795) the fact that this performance is taking place via video conferencing offers a range of novel possibilities.

Perhaps the most obvious of these is distance. Whilst Copperfield required extensive technological apparatus to frame his broadcast from Hawaii, the fact that the card transposition takes place on a video conferencing platform makes it entirely uncontrived that the card travels from Spain to America. It does not, however, diminish the drama in the effect. Making a spectator's chosen playing card appear in their hands is always a good trick, but making the card travel from one continent to another adds a whole new degree of impossibility. While distance is key in the narrative of the trick, the idea of making it disappear, as with Copperfield's magical journey to see an estranged father, is also present. Sam and Anne would only be able to share this magical experience of an in-person show at huge expense and inconvenience, but via video conferencing, they can share it conveniently from the comfort of their homes.

The fact that Anne and Sam participate in the show from a domestic setting is, in itself, important. In-person magic shows tend to happen in public spaces, whether theaters, bars, or restaurants, rather than in homes, and because of this magicians go to great lengths to prove that the items they are using are every-day objects rather than special apparatus (see Maskelyne and Devant 1946 [1910], 119–20). Because all the spectators are in a domestic setting for this show, and they have brought the props themselves, that suspicion is eliminated. The setting also removes the need for the elaborate proof that Copperfield required to demonstrate that he and Michael have actually traveled to Hawaii. Anne and Sam are in their homes, locations that may be recognized by other family members on the call, so there is no question that the card has been teleported from one place to the other. Thus, the invasion of private spaces by videotelephony that has called for associations with surveillance and caused concern throughout the technology's history since the nineteenth century (see Roberts 2019, 42, 79, 138, 212–14), becomes an advantage for performance magic because it effortlessly adds authentication.

Even the mechanics of Anne and Sam's participation in the show is facilitated by the use of video telephony. Getting audience members on stage is always a challenge at in-person shows, creating dead time and logistical challenges. Via video conferencing anyone in an audience can instantaneously be brought "on stage" without delay or considerations about seating positions, etc. The technology also allows the wider audience to see the on-stage helpers more clearly than in any other setting, enjoying every nuance of their reaction to the impossibility in which they play a part.

This example highlights just a few of the ways in which video conferencing has shaped the performances that magicians are now giving, and as more conjurers spend time exploring the possibilities, many new opportunities will appear. Video conferencing also offers advantages of television, like control over the studio setting for the performer (lighting, sound, multiple cameras, etc.) and the facility to do physically small magic in a way that large audiences can see. In addition, other forms of audience interaction that would traditionally be the preserve of in-person performance are facilitated, and even expanded upon, in the video conferencing setting, thanks to functionality like breakout rooms, voting, reactions, and text chat.[6]

Conclusion

At first glance, the integration of performance magic in a video conferencing environment seems problematic. The same has been said about television, and while magic is still often regarded as unsuitable for television, it was hugely successful in this medium during the twentieth century. In 1895, magic's first transition to an audiovisual medium was a smooth one, when magicians became pioneers of early cinema. At this point, filming the tricks was enough, before magicians like Georges Méliès turned to the invention and implementation of cinematic illusions, i.e., special effects, which, for a period of time, worked as an attraction in their own right (see Gunning 1986). Six decades later, when it came to television, the presentation of special effects turned into the thing that had to be believably ruled out in order for magic performances to be effective. Our exemplary analysis of performance magic on television showed how it was adapted to work with and despite this medium's specificities, primarily establishing credibility by disproving the use of image manipulation. Moreover, magicians were also working against the characteristics of media in general, by making television appear as a neutral window onto the show.

Crucially, in the next shift of audiovisual channels, to the context of video conferencing, magicians encountered the same problem in principle but, for the time being, without the need to disprove image manipulation. Unlike television, video conferencing performances do not require the same amount of authentication because the medium is connoted differently, seemingly promising a higher degree of "unmanipulated liveness" to the spectators. The perception of video conferencing as a "window" has accompanied it since the technology's early days, and it proved useful

6 For an in-depth exploration of the practical, methodological possibilities video conferencing affords the magician, see Houstoun and Thompson (2021).

to magicians when they shifted to video conferencing applications. This is paradoxical because, for instance, Zoom very overtly offers the possibility of image manipulation to its users through the use of video filters and virtual backgrounds—a technique going back to the illusionistic practice of the black screen that originated in performance magic. And yet, it is not typically thought of as prone to image manipulation outside of these obvious effects. Rather, in the tradition of videotelephony, it is perceived as a neutral transmitter, a window to another space.

In the context of illusionistic performance practice, we have shown, video conferencing platforms allow for a combination of some of the best aspects of in-person and mediated performances. While the field is a relatively new one, there are advantages to be found in the realm of online shows for performing artists, with further development ongoing. The space combines the convenience of a global reach without the necessity to travel with the possibilities of digital effects and audience interaction.

Far from having failed, and despite having had a hard time asserting itself in everyday life for a number of reasons, videotelephony has been on the minds of inventors, engineers, businesspeople, authors, and filmmakers for about 150 years. The countless fantasies and debates it sparked, and the numerous inventions it stimulated attest to the fact that videotelephony responds to a persistent desire, while further, its long-time image as a future or futuristic technology evidences its "magical" qualities.

References

Barnouw, Erik. 1981. *The Magician and the Cinema*. New York: Oxford University Press.

Breslaw, Philip. 1997 [1795]. "For a Person to chuse a Card, you not supposed to know what it is, and then for the person to hold the cards between his Finger and Thumb, to strike them all out of his Hand, except the very Card he had taken." In *Breslaw's Last Legacy*, 92–93. Wichita: Stevens Publishing (Facsimile).

Burns, R. W. 1998. *Television: An International History of the Formative Years*. London: Institution of Electrical Engineers.

Caveney, Mike. 2009. "The Masters of the Golden Age." In *Magic, 1400s–1950s*, edited by Noel C. Daniel, 338–397. Cologne : Taschen.

Chaplin, Charles, dir. *Modern Times*. 1936. USA : Charles Chaplin Productions.

Davis, Erik. 2004. *TechGnosis: Myth, Magic, and Mysticism in the Age of Information*. Updated ed. London: Serpent's Tail.

Debruge, Peter. 2020. "'The Present': L.A. Theater Review." *Variety*, May 15, 2020. https://variety.com/2020/legit/reviews/the-present-review-helder-guimaraes-geffen-stayhouse-1234608463.

During, Simon. 2002. *Modern Enchantments: The Cultural Power of Secular Magic*. Cambridge: Harvard University Press.

Enns, Anthony. 2015. "Spiritualist Writing Machines: Telegraphy, Typtology, Typewriting." *Communication +1* 4 (1), art. 11.

"Far-Off Speakers Seen as Well as Heard Here in a Test of Television." *New York Times*, April 8, 1927: 20.

Fraps, Thomas. 2021. "Illusions, Magic and the Aesthetics of the Impossible." In *Illusions in Cultural Practice: Productive Deceptions*, edited by Katharina Rein, 53–70. London: Routledge.

Gernsback, Hugo. 1922. *Radio for All*. Philadelphia: J. B. Lippincott.

Gunning, Tom. 1986. "The Cinema of Attraction: Early Film, Its Spectator and the Avant-Garde." *Wide Angle* 8 (3 & 4 "Narrative/Non-Narrative"): 63–70.

Gunning, Tom. 1995. "An Aesthetic of Astonishment. Early Film and the (In)Credulous Spectator." In *Viewing Positions: Ways of Seeing Film*, edited by Linda Williams, 114–33. New Brunswick: Rutgers University Press.

Gunning, Tom. 2008. "Uncanny Reflections, Modern Illusions. Sighting the Modern Optical Uncanny." In *Uncanny Modernity. Cultural Theories, Modern Anxieties*, edited by Jo Collins and John Jervis, 68–90. Houndmills, Basingstoke, Hampshire: Palgrave Macmillan.

Hagen, Wolfgang. 2002. "Die entwendete Elektrizität. Zur medialen Genealogie des 'modernen Spiritismus.'" In *Grenzgänge zwischen Wahn und Wissen: Zur Koevolution von Experiment und Paranoia 1850–1910*, edited by Torsten Hahn, Jutta Person, and Nicolas Pethes, 215–39. Frankfurt on the Main: Campus-Verlag.

Hanna, William, and Joseph Barbera, dirs. 1962. "The Space Car." *The Jetsons*, season 1, episode 4. USA: Hanna-Barbera Productions et al.

Held, Tobias. 2020. *Face-to-Interface: Eine Kultur- und Technikgeschichte der Videotelefonie*. Marburg: Büchner.

Houghton, Natasha et al. 2021. "Cross-disciplinary Perspectives on the Transition to Remote Education." *BMJ Simulation & Technology Enhanced Learning* 7: 586–89.

Houstoun, Will, and Steve Thompson. 2021. *Video Chat Magic*. s.l.: Vanishing Inc. Magic.

Kneebone, Roger, Will Houstoun, and Natasha Houghton. 2021. "Medicine, Magic, and Online Performance." *The Lancet* 398 (10314), November 20, 2021. https://doi.org/10.1016/S0140-6736(21)02485-5.

Kubrick, Stanley, dir. *2001: A Space Odyssey*. 1968. UK/USA: MGM & Stanley Kubrick Productions.

Lamont, Peter. 2006. "Magician as Conjurer: A Frame Analysis of Victorian Mediums." *Early Popular Visual Culture* 4 (1): 21–33.

Lang, Fritz, dir. *Metropolis*. 1927. German Reich: UFA.

Mäkinen, Lauri. 2007. "Mobile Videophone." In *Topical Evolution Paths of Mobile Multimedia Services: Proceedings of the Research Seminar on Telecommunications Business,* edited by Sakari Luukkainen, 37–41. Helsinki University of Technology Publications in Telecommunications Software and Multimedia. http://www.tml.hut.fi/Opinnot/T-109.7510/2007/Proceedings_2007.pdf.

Maskelyne, Nevil, and David Devant. 1946 [1910]. *Our Magic: The Art in Magic, the Theory of Magic, the Practice of Magic*, edited by Paul Flaming. Berkeley Heights: Fleming Book.

McLuhan, Marshall. 1994 [1964]. *Understanding Media. The Extensions of Man.* Cambridge: The MIT Press.

M'Govern, Chauncy Montgomery. 1899. "The New Wizard of the West: An interview with Tesla, the modern miracle-worker, who is harnessing the rays of the sun; has discovered ways of transmitting power without wires and of seeing by telephone; has invented a means of employing electricity as a fertiliser; and, finally, is able to manufacture artificial daylight." *Pearson's Magazine* VII (41) (May): 291–97.

Natale, Simone. 2016. *Supernatural Entertainments: Victorian Spiritualism and the Rise of Modern Media Culture.* University Park: The Pennsylvania State University Press.

Rein, Katharina. 2015. "Mind Reading in Stage Magic: The 'Second Sight'" *communication +1* 4 (1): article 8. https://doi.org/10.7275/R50C4SPB.

Rein, Katharina. 2017. "Magicians and Early Cinema." In *A History of Cinema Without Names/2*, edited by Diego Cavallotti et al., 169–78. Milan: Mimesis.

Rein, Katharina. 2019. "Vanishing Technology: Transparency of Media in Stage Magic." In *Media Archaeology and Intermedial Performance: Deep Time of the Theatre*, edited by Nele Wynants, 99–114. Cham: Palgrave Macmillan.

Rein, Katharina. 2023. *Techniques of Illusion: A Cultural and Media History of Stage Magic in the Late Nineteenth Century*. London: Routledge.

Roberts, Ivy. 2019. *Vision of Electric Media: Television in the Victorian and Machine Ages.* Amsterdam: Amsterdam University Press.

Sconce, Jeffrey. 2000. *Haunted Media: Electronic Presence from Telegraphy to Television.* Durham, NC: Duke University Press.

Scott, Ridley, dir. 1982. *Blade Runner*. USA: Warner Bros. et al.

Silverman, Kenneth. 1996. *Houdini!!! The Career of Ehrich Weiss; American Self-Liberator, Europe's Eclipsing Sensation, World's Handcuff King & Prison Breaker.* New York: HarperCollins.

Solomon, Matthew. 2010. *Disappearing Tricks: Silent Film, Houdini, and the New Magic of the Twentieth Century*. Urbana: University of Illinois Press.

Steinmeyer, Jim. 2005. *Hiding the Elephant: How Magicians Invented the Impossible.* London: Arrow.

Swiss, Jamy Ian. 2007. "In Search of Street Magic." *ANTINOMY* 9. https://jamyianswiss.com/blog/in-search-of-street-magic.

Swiss, Jamy Ian. 2022. "Conjuring on the Small Screen." In *Magic: A Companion*, edited by Katharina Rein. Oxford: Peter Lang, 175–183.

"Telephone Users See Each Other as They Talk." 1930. *Popular Mechanics* 53, no. 6 (June): 892.

Verne, Jules (and Michel). 1889. "In the Year 2889." In *The Forum* VI (February), edited by Lorettus S. Metcalf: 662–77. Hathi Trust. https ://hdl.handle.net/2027/mdp.39015008192885.

Verne, Jules. 1892. *Le Château des Carpathes*. Paris : J. Hetzel et Compagnie. https ://fr.wikisource.org/wiki/Le_Château_des_Carpathes.

Young, Richard, Britten, Noel and David Copperfield. 2020. "Portal," edited by Will Houstoun. *The Magic Circular* 114 (1249): 238–39.

Zemeckis, Robert, dir. *Back to the Future II*. 1989. USA: Universal Pictures, Amblin Entertainment, and U-Drive Productions.

Dis/Abling Video Conferences
A Video- and Auto-Ethnographic Exploration of Remote Collaboration Situations

Tom Bieling, Beate Ochsner, Siegfried Saerberg, Robert Stock, and Frithjof Esch

This article addresses the question of access to video conferencing from the different perspectives of crip techno-science (Hamraie and Fritsch 2019), media ethnography, design theory, and the politics of inclusion during the SARS-CoV-2 pandemic. First, we describe a situation where a blind user and his sighted work assistant create access to a basically inaccessible online conferencing system via several workarounds. This example confronts the two different sensory-technical perspectives of both involved and diverse actors. In the second example, which is even more complex than the first one, a team of variously disabled employees who are blind, deaf or hard of hearing, and non-disabled colleagues and work assistants work out a branched path through a partially inaccessible video conferencing software. One of the team's work sessions we recorded serves as empirical material for the subsequent media and autoethnographic analyses. The focus is on the intertwined loops of interaction and communication between diverse sensory repertoires, technical tools, and social negotiation processes, which all together create a highly complex network of sensory-socio-technical dimensions.

The paper analyses the two examples of circumventing "restricted access" (Ellcessor 2016) and shows how social negotiation processes attempt to construct hacks and workarounds that undermine or circumvent technologically inaccessible solutions for those that do not correspond to the so-called preferred user (Ellcessor 2016, 77; Ellis, Kao and Bitman 2020, 17). While those workarounds, detours, or shortcuts represent a temporary, socially and technologically conceivable solution, they nevertheless demand a great deal of time and organizational work (Schabacher 2017). A screen reader–ready website, revised screen reader logic, or a screen reader–accessible sharing function (example 1), as well as the efforts trying to synchronize spoken and written as well as heard and read language in distributed remote communication (example 2), are just two concrete examples that, on the one hand, demand improved design. Thus, another perspective, that of design theory, comes into play. Participatory, cocreative, and codesigned processes become meaningful on their own. On the other hand, these examples reveal the underlying political concept of in- and exclusion. However, one may doubt that a fully accessible technical solution for all diverse groups can or should be possible. Though, from a political point of view, this demand is justified, since otherwise the dynamics of political processes would lead to a chronic undersupply of accessible technological solutions. Dealing with such solutions will always involve socio-technical tinkering or everyday design in the making. Much more important, however, is the degree to which this will take place as an open, ongoing process which leads from temporary workarounds and hacking to design improvement and back again.

Workarounds on an Online Conference Platform: Collaboration between a Sighted and a Non-Sighted Person

In November 2021, the authors of this article held a workshop on the topic of this present paper at the "University:Future Festival" (U:FF). There, we discussed digital barriers and workarounds in the context of video conferences and accessibility during the SARS-CoV-2 pandemic from an interdisciplinary perspective and by drawing on approaches from disability studies, media studies, and design studies. We joined the conference remotely from five different localities. The digital UFF was hosted by the "Higher Education Forum on Digitalization" in partnership with the "Foundation for Innovation in Higher Education in Germany." The invitation to this virtual event stated that they "focus on interaction and encounter like never before" and on "exchange at eye level and accessibility."[1] In our case, unfortunately, accessibility turned out to be a complicated issue as the conference platform hampered the participation of people with visual disabilities. Thus, while problematizing the issue of access work and digital frictions with our workshop, we were once again confronted with the ambivalent promises—and failures—of current innovative technologies aiming to "fix," among other things, the societal problems caused by the global pandemic. In the following section, Frithjof Esch and Siegfried Saerberg refer to their experiences as a work assistant and a blind user concerning registration and sign-up to the UFF's Online Conference Platform "Let's Get Digital"—which differs significantly from applications like Zoom or BigBlueButton—from two different points of views: the sighted and the blind.

Sighted Point of View

Some days before the Festival, Siegfried Saerberg and I have registered and signed up on the UFF platform and done some pretesting to get comfortable with it. How-

---

1 https://festival.hfd.digital/en/archive/universityfuture-festival-2021/festival.

ever, it has taken us more than two hours and even then, full access for the blind user has not been accomplished. The drama has contained three acts of inaccessibility, as is retrospectively described below:

The Eternal Back and Forth between Key Combinations

As we are located in different places, we have connected via the "Quick Assist" feature in Microsoft Windows so that I can see Siegfried's screen. First, Siegfried clicks on the corresponding link in the invitation email. Subsequently, the landing page for the registration opens in his browser and the structure of the website is easily recognizable for me as a sighted person. There is a registration window with the log-in data and a button that says "check in now," which is not apparent to Siegfried. Laboriously, he uses the tab key to find the button. But as the button is not labeled, the screen reader could not read its description. So far it has been an eternal back and forth between key combinations. Siegfried is annoyed and stressed by the simultaneous and contradictory voices from the screen reader and me. We cannot find any workarounds, which is why I take over the remote control and confirm the button to log in.

When the End Is the Starting Point

After confirming, the lobby page of the event platform opens. Automatically, a pop-up window appears over the page in which the privacy policy of the platform has to be confirmed. I am confronted with a visual barrier to confirm the privacy policy. For Siegfried, however, this window and thus the policy is not perceptible but he can already navigate in the lobby with his screen reader, which again is not apparent to me. Moreover, he has to move through the entire page to reach the pop-up window. "You may have to go all the way to the end to get to the beginning," I say. It works. But then it turns out that the private policy buttons are not labeled correctly. Finally, he chooses the path of least resistance. He confirms all preset privacy settings.

When Moving Backward Brings You Forward

We are exploring the lobby page of the platform now. We are pleased to realize that it can be navigated via screen reader. But access to the virtual lecture room for pre-testing purposes is not possible. At this point, Siegfried is using the provided direct room link from the invitation email. However, before accessing the lecture room, a pop-up window with a "continue" and "cancel" button appears to confirm entry. I try to follow the red screen reader frame to visually orient myself and see where Siegfried is. Everything runs simultaneously. A huge effort for me and Siegfried. I have no idea who is interacting with what. After about thirty clicks with the tab key, the red screen reader frame shows me that Siegfried is on the right button. However, the labeling is wrong. I read, "continue" or "cancel." Siegfried hears, "Get your ticket here button." But then, by coincidence, we come to a new workaround. Siegfried tabs forward. I ask him to go back one step. And behold: backward, the screen reader reads the labeling of the buttons correctly. "When moving backwards brings you forward," says Siegfried. Finally, he is entering the lecture room.

Blind Point of View

I, Siegfried, as the second part of the work assistance relationship in this socio-technical ensemble of humans, software, and materiality, can say that synchronization is indeed a problem to be overcome. First of all, the acoustics of the spoken words must be put into a communicational order via negotiation. If this is not done, the machine voice of the screen reader and the human voices of the two persons in the work-assistance-relationship often speak at the same time, which makes communication very arduous for me. Since these agreements on the order of words are always made in concrete situations and also acoustically, there is a high potential for mutual interference.

A second need of synchronization is to be found in the spatio-temporal order in which a web page is perceived by the sighted reader and the reader hearing via screen reader. Frithjof often describes his reading practices on a homepage as being based on an economy of visual attention due to graphic design elements which give him a flexible movement in space and time. In contrast, the culture of blind web page reading as designed today, prescribed by screen reader logic, is characterized by only following a very strict one-dimensional sequence. Here, there is no spatial juxtaposition, only a temporal sequence. In spatial view, reading would have to choose between jumping down on the page, moving further down the line, or skipping diagonally. The screen reader culture knows only forward or backward, which varies only in its speed. Therefore, in both personal and professional contexts I have often observed a sighted assistant say, "Right next to it is the button you need to click on."

Disability and Remote Collaboration during the Pandemic

The story of our online conference experience related above is not an isolated episode—neither in academia nor in everyday life. It is an example of how society "consciously and unconsciously, has built in disability into digital technologies" (Goggin and Newell 2003, 147) and how disability is "constructed in and through technology" (12). This experience was embedded in the situation of a global and ongoing health crisis where, since its beginning, many measures were taken to contain the pandemic. Social distancing and prolonged periods of lockdowns brought about radical changes in everyday life, work, and education. Thereby, the pandemic has disproportionately impacted people with disabilities from heterogeneous backgrounds (Spinney 2022; Garland-Thomson 2020). At the same time, many people with disabilities engaged in mutual aid campaigns and advocacy or invented novel forms of convivial and creative practices (Ginsburg, Mills, and Rapp 2020). These initiatives demonstrate the manifold ways in which people with disabilities used their expertise to critically and practically engage with the pandemic and its spatio-temporal constraints (Shields et al. 2020).

Technology is often given a positive connotation in this context for it provides the necessary conditions for creating a sense of conviviality and perhaps even supporting well-being. One striking example is the massive proliferation of remote collaboration tools during the course of the pandemic. Video conferencing platforms were implemented in many professional and educational settings as a workaround—they allowed the continuation of work in times of social distancing and other governmental measures. However, standardized operating procedures inscribed via platforms such as Zoom would have to be adapted through improvisation and workarounds to make them more accessible for disabled people (Ginsburg, Mills, and Rapp 2020).

With regard to the academy and higher education that are often criticized for not facilitating proper participation of students or scholars with disabilities (Dolmage 2017), there are several aspects to consider regarding issues of access in terms of the built environment and digital learning architectures. For instance, during the pandemic teaching was mostly transferred to video conference and learning platforms with varying degrees of success and access options (Ellcessor 2021; Ellis, Kao, and Bitman 2020) radically impacting universities' network traffic and information infrastructure. Research, in many cases, was reorganized as well. Ashley Shew from Virginia Tech reported in an article published by *Nature* that a research project addressing technology development mainly by non-disabled developers and daily practices with assistive devices by people with disabilities was easily done remotely:

> I planned my work intending to recruit disabled students among my researchers. Most of the studies can be done remotely … And because of its disability-led design, my team's project is pandemic-proof. … we conduct our interviews by text, e-mail, Zoom and other means. Because we planned for disabled people to lead and participate in the research, we're well prepared for the current situation—or for any other. (Shew 2020)

Shew highlights that many not-yet-disabled colleagues would have adopted their work, research, and teaching according to disability-led hacks during the pandemic. At the same time, the STS scholar laments that disabled people were most affected by the pandemic, yet their merits regarding remote collaborations and doing work from home were not recognized as they should be. While stressing the benefits of remote research, Shew does not directly address the question of how remote collaboration—be it research or teaching—might be complicated by the respective technological frameworks and data infrastructures. Yet elsewhere she emphasizes the bias and problematic impact of technologies that disfavor certain corporealities or neurodiverse ways of being by coining the concept of "technoableism" (2020).

Digital media such as video conferencing platforms are ambivalent in the ways they aim to foster participation through complex infrastructures (Star and Bowker 2002, 242). These technologies can be considered as frictional, political, and partially open infrastructures. They form contradictory "emancipatory devices" that transform into "uneven ensembles for articulating political engagements" (Houston, Gabrys, and Pritchard 2019, 845). Hence, the workaround implemented through video conferencing systems might not be appropriate for all people. Against this background and associated with the complex socio-technological issues, it is necessary to reconceptualize the concept of accessibility and the ways in which remote collaboration systems "provide" access for their users. We hence propose that access is not a fixed condition or stable configuration. The abovementioned experience with the UFF platform and many other examples demonstrate this poignantly. Thus, access can rather be understood as "a relational, unstable phenomenon that both grants benefits and interpellates individuals into larger social systems that may be empowering, exploitative, or both" (Ellcessor 2016, 7).

As we will see in more detail below, remote video conferencing gets more complex when adding the need for (automatic) closed captioning, audio description, sign language interpretation, or CART (Hickman 2019). One finds some cues for the challenges of accessible video conferencing in academia when looking at guidelines for teaching involving students with disabilities from the University of Cologne (Melzer et al. 2020) or the "Mapping Access" project information about "Teaching in time of Covid-19." Note that projects like "Mapping Access" (Critical Design Lab, Hamraie 2018) and theoretical approaches like the "Crip Technoscience Manifesto" (Hamraie and Fritsch 2019) stress that people with disabilities are makers and actively participating through their expertise in hacking, tinkering with, and working around non-accessible media environments building non-innocent, situated, interdependent, and sometimes messy human-technology arrangements. Accessing video conferences as we describe it in this chapter might be considered part of "the crip politics of interdependence," that is, "a technoscientific phenomenon, the weaving of relational circuits between bodies, environments, and tools to create non-innocent, frictional access" (Hamraie and Fritsch 2019).

While acknowledging the productive and positive impact of video conference systems during the pandemic for the "pandemic preferred user" (Ellis, Kao, and Pitman 2020) and partly also for disabled people, we argue that such platforms often incorporate a "potential for inaccessible elements" (Kent 2020, 269). Therefore, the following sections explore the material and performative doing of frictional access by focusing on remote collaboration settings in German higher education: we analyze a second example of a video conference situation during the pandemic at the Centre for Disability Studies ZeDiSplus in Hamburg. There we focus on the diverse sensory practices within groups of persons with varying disabilities and abilities using digital communications like Zoom Technologies and other assistive applications. By drawing on conversational analysis and auto-ethnographic reflections (Ellis, Adams, and Bochner 2010) we map socio-technical assemblages of people, diverse sensory practices, and (individual) media devices (like screen readers) that became problematic during the pandemic which fosters the remote working practices. Tracing the communicative processes, individual bodily techniques, and different sensory enactments during video conferences provides us with the possibility of approaching the various translational processes that are at stake in these media settings involving eye-camera contacts, bodily techniques, spoken words that are being transcribed and appear on screens, shared documents or chat messages read (or not) by screen readers, lip reading complicated by low video quality, and other factors.

Video Conferencing at ZeDiSplus

In times of the global SARS-CoV-2 pandemic, working remotely from home has become the new normal: while before the pandemic, many employers considered it an unnecessary effort to make accommodations supporting remote working for people with disabilities, the world has changed with the impact of the SARS-CoV-2 pandemic. Many organizations had to send their employees—whether disabled or not—home to work remotely, enabled by digital assistive technologies. Assessing the effects of the pandemic for workers with disabilities, some researchers see a "silver lining" (Schur, Ameri, and Kruse 2020) in the evolving teleworking options. A recent post on Microsoft Accessibility Blog says it best: "The biggest source of knowledge right now are your employees, especially those in your disability employee communities" (Lay-Flurrie 2020). Meanwhile, others draw attention to the growing risks and disadvantages for workers with disabilities within the labor market (Morgan 2021). However, they all point out that we still know little about how people with disabilities engage in remote work or act in videoconference meetings during and (hopefully) after SARS-CoV-2.

The Centre for Disability Studies ZeDiSplus is a striking example of inclusive workplaces assembling socio-technical arrangements of people, diverse sensory practices, and digital media devices. Compared to our first example ("Sighted Point of View"), all members of the working meeting know each other and have a longer experience in organizing their work environment together. Consequently, they are used to coping with various technical devices such as laptops, additional beamers for speech-to-text-translations, personal assistants, and screen readers. Since SARS-CoV-2, however, they—like all of us—could no longer meet in person, which made further additional technical systems like the video conferencing system Zoom necessary. This, in turn, requires further workarounds like prior arrangements as to who speaks or writes for whom in which channel and in which function. We videotaped a selection of video conferences at the Centre to examine the experiences with and manifold effects of videoconferencing technologies and practices in interaction with previously established human and/or technical assistance.

The following analysis is based on the video footage. The researcher was not present at the meeting and did not claim any other footage or further information except for the responsibilities and functions of the participants. The analysis focuses on processes of synchronization between spoken and written as well as heard and read language in distributed remote communication. These "normally" not visible or audible translational practices and agreements become visible and audible in the opening agenda process where temporal overlapping and misunderstandings are produced by the pre-arranged settings of different socio-technical translations like speech to text, text to speech, or speech to speech, which prove to be highly complex in execution. It thus became evident that it is not a matter of getting things done but of doing itself. We will see that this complex sociotechnical network of things, differently abled persons, as well as technologies requires constant adaptation work like, for example, previously agreed workarounds, which on the basis of a reorganization of both the standard software Zoom and specific assistive technologies, determines which person or software (e.g., screen reader) is responsible for which translation (speech-to-text, text-to-speech) in which media (live transcription, spoken language, text chat).

Some facts in advance: The chosen video conferencing system is Zoom; seven participants are involved in the selected scene. Their functions vary from passive or active participant, personal assistant, and translator from speech to text and from speech to speech. Different persons take on several functions:

- Alex[2], researcher at the ZeDiS and chair of this meeting, is blind. Alex uses the computer audio for speech contributions in the meeting and headphones for the text-to-speech translation via screen reader. Alex, whose video is turned on, is furthermore supported by a personal assistant, Kim.
- Kim, personal assistant to Alex, uses the computer audio. Kim's video is turned off in the situation in question. Kim is additionally tasked with writing the speech-to-speech translation of Leslie by Kai in the chat. This is important for Alex to be able to distinguish who Kai is speaking for.

2 In order to preserve the anonymity of the persons involved, we work with gender-neutral pseudonyms.

- Leslie, researcher at the ZeDiS, uses the computer audio and video. Leslie is deaf and uses speech-to-text-translation to follow the meeting. Since Leslie's pronunciation is not always understandable for the other participants, Kai translates the inputs (speech-to-speech) if necessary.
- Kai uses the computer audio; the video is switched off. As a project coordinator at the ZeDiS, Kai participates in the meeting, and, if necessary, functions as a speech-to-speech interpreter for Leslie.
- Luca functions as a speech-to-text interpreter for the deaf participants. The computer audio is muted; the video is turned off; we only see the shared screen on which the text appears.
- Noa, researcher at the ZeDiS, is deaf and uses speech-to-text-translation. The video is turned on, but Noa does not actively participate in the selected video clip.
- Mika works as an editor on the book projects. The video is turned on, but, like Noa, Mika is not actively involved in the clip.

Transcription of Zoom Meeting, ZeDiSplus (June 14, 2020)

Table 1

TC	Persons/Functions	Display	Sound
00:00	Leslie takes his/her/their turn, and Luca, the speech to text interpreter, starts to translate.	I would like to add ...	*I would like to add ...*
00:04	The translation stops and a question mark appears.	I would like to add?	
00:04–00:10	Leslie continues without translation.		
00:11	Luca deletes "to add" and replaces it by "our" and adds another question mark.	I would like to ... our?	*I would like to... our?*
00:12–00:15	Kai takes his/her/their turn, Luca immediately starts to caption.	I would like to add under the category of miscellaneous.	*I would like to add under the category of miscellaneous.*

00:16	**Alex interrupts.**		**Stop, just one moment, before you continue, I'm missing the ...**
00:19	Leslie (re)takes his/her/their turn without being immediately translated.		
00:20	With Leslie speaking, Kai is barely understandable but is translated into text.	I'm trying to repeat what Leslie said.	*I'm trying to repeat what Leslie said.*
00:21	Alex interrupts; Luca notes what Alex says.	Sorry, I see, but we agreed that Kim should write this in the chat, but I don't have anything from Kim.	*Sorry, I see, but we agreed that Kim should write this in the chat, but I don't have anything from Kim.*
00:23	Kim intervenes.		Yes, but Kai has already translated what Leslie said.
00:29	**Confusion of voices, interruption of the agenda**		
00:31	Alex takes his/her/their turn.		Okay, what? I didn't get that now ... ?
00:33	Kai explains.		When Luca doesn't caption, I'll see that, and then I try to repeat what I understood.
00:40	**Situation is cleared up, Alex sums up.**		**Sorry, now I see, I didn't get that.**

Analysis

The meeting essentially serves the purpose of bringing participants to a common level of information regarding past as well as future events or projects. The chair has just opened the meeting and the participants are asked for further suggestions for the meeting agenda. After discussing the order of the topics, Alex, the chair, suggests starting with the general question about how each participant, in particular Noa, is doing. However, Alex continues, they should move this topic to the end of the agenda. While Alex is still speaking, Leslie takes his/her/their turn, but the captioning does not start immediately. Alex responds to Leslie and adds to also be interested in knowing how Leslie is doing. At this point the recording begins.

Leslie continues, this time the captioning works: "I would like to add." After the first words, however, the translation stops and a question mark appears. Leslie continues while Luca deletes the words "to add" and replaces them with "our," followed by another question mark. Leslie keeps on talking. After some seconds, Luca deletes the "our" and the question mark. Kai takes his/her/their turn as a translator for Leslie. Luca immediately starts captioning Kai without signaling that Kai translates what Leslie said: "I would like to add under the category of miscellaneous ..."

At this point, the confusion starts: Alex, who only receives what the screen reader translates, apparently thinks that Kai speaks for her-/himself/them and interrupts: "Stop, just one moment, before you continue, I'm missing the ..." While Alex speaks without being captioned, Leslie speaks up again. Kai tries to "repair" the situation by explaining: "I'm trying to repeat what Leslie said." Since neither Kai nor Luca clarify on whose behalf they speak or write, Alex thinks that Kai, the coordinator of the research group, wants to bring in her/his/their own agenda. Only after the explicit clarification of what role Kai currently plays, Alex understands that Kai translated what Leslie said. However, the problem is not completely cleared: previously, the participants of the meeting had agreed on a different system of distributed remote speaking. Alex is referring to that and says: "Sorry, I see, but we agreed that Kim should write this [Kai functioning as a translator to Leslie, B.O.] in the chat; but I don't have anything from Kim." While Alex talks, Luca captions. Kim intervenes to clarify: "Yes, but Kai has already translated what Leslie said." For Kim, it seems unnecessary to write down Leslie's contribution in the chat, since Kai has already translated it into speech and Alex has heard this. This is also true, but Alex cannot distinguish in which role Kai is speaking, therefore differentiation is needed. Since everybody now tries to clear up the situation, the scene culminates in a renewed confusion of voices. Alex can't understand Kim and continues: "Okay, what? I didn't get that now." Kai finally clears up the situation: "When Luca doesn't caption, I'll see that, and then I try to repeat what I understood." It is interesting that this meta-explanatory utterance by Kai isn't captioned. Alex concludes: "Sorry, now I see, I didn't get that." The meeting continues.

In situations like this, it is almost impossible to decide from the outside, whether the interruption is due to individual dis/abilities, technical delays, or misunderstood meta-information. These situations need both explanations and synchronization to allow continuation. If it can be assumed that workarounds like the one described above enable comparatively complex situations. Nevertheless, frictions and specific modifications of dis/abled temporalities prevent the sociotechnical zigzag from running (Schabacher 2017, 13–14; Löffler 2017, 137). The combination of text and voice-based translation services as well as a prearranged distribution of written translations between subtitles and chat, only work if everyone strictly adheres to the workaround. In the present situation, the interplay between the different actors ultimately fails to synchronize Luca's captioning of Leslie's statement via Kai's translation as well as Kim's transfer of the same into the chat. Thus, Luca needs time to understand what is being said before typing it, and we furthermore note a considerable delay before the live translation appears or disappears (is deleted). So, it takes time for Kai to realize that Luca hasn't understood, etc. Thus, Luca's captioning fails, Kai starts a bit late, Luca deletes the text written before in favor of what Kai is now saying, and Kim does not feel the need to pick it up again. Alex interrupts, thinking that Kai wants to introduce a separate agenda item. However, Alex would like to postpone this. Kai starts again and realizes only belatedly that Alex does not understand that it is "only" a translation. When Kai and Kim try to clarify the situation, the buzz of voices is already so great that Alex cannot understand them acoustically.

As the example showed, meta-practices of communication that deal with the different abilities and temporalities draw our attention to the relationality and precarity of social, cultural, and technical infrastructures of communication that necessitates constant repair work (Star and Ruhleder 1996). And this does not mean to restore a presupposed continuity or common abledness. According to Schabacher, workarounds are specific embedded practices which involve a labor-intensive detour. In a certain sense, they are robust, and, at the same time, since they represent a specific solution to a problem which occurs in a concrete situation at a particular time, they have to be highly flexible and cannot be directly addressed in advance (Schabacher 2017, 13). So although it was agreed beforehand that certain forms of speech would be written in the chat to create distinctiveness, this did not work. So the situation has to be interrupted and resolved on very short notice to clarify what has happened. After the communicative order has been restored, the conversation continues. Against this backdrop, interferences or discontinuities between the social and the technical, between materiality and meaning-making, shouldn't only be understood as a problem to be solved; they should rather be regarded as a chance to enable situative "crip" reorderings or productive deviations from the norm with a nonstandardized character and relevant political and ethical implications (Star and Bowker 2002, 242). As such, workarounds are continuous processes which "operate ... with what is currently available, be it people, things or information, in order to informally establish solutions" (Schabacher 2017, 23)—even if previously agreed workarounds seem to fail, as Siegfried Saerberg from an auto-ethnographic point of view will now elaborate on.

Autoethnographic Exploration of Video Conferencing at ZeDiSplus

In this section, I Siegfried Saerberg—and Alex in the little piece that has just been analysed and interpreted—reflect on the logic of doing auto-ethnography (Ellis, Adams, and Bochner 2010). A reflection of this kind is essential for any auto-ethnography especially in a context in which the perspectives involved are so heterogeneous as they are in situations of digital access connected to the issue of disability. Therefore, I will turn to an interpretation of the situation at stake. The section follows the written manuscript for a short presentation I held at the Siegen conference in autumn of 2020. There I used a special technique: Because I did not have a braille-manuscript at hand, because I wanted to read it out exactly, and because there was no time to learn the text by heart, I let my screen reader read the text via headphone to me before I repeated it loudly for my audience. Therefore, I always structure the sentences into little pieces as shown beneath.

Meta-Auto-Ethnographic Observation

Yesterday
when I listened
to the little piece
that had just been analyzed and interpreted above,
I did not really understand anymore
what I was doing in that situation
Why?
Experience builds up moment by moment.
Every moment that once took place
is important for this process
Auto-ethnography tries to record
all inner movements
and outer observations
that a subject of experience has.
But it is always behind
the flow of events
especially if this flow is fast.
If the movement of feeling
is changing a lot
it is difficult to remember
when exactly
joy turned into anger,
relaxation transformed into stress
and so on.

You only remember
a general, synoptic pattern,
that relates events to feelings.
Phenomenological description of events,
cognitions, perceptions, and interactions
tries to reconstruct
as exact as possible
what and how it happened.
Succession is important as well.
It tries to find little bits of
interpretive action
in interaction and communication.
In a polythetic
step by step way,
to use Husserl's term (Husserl 1954),
But often it finds big pieces
of clustered meaning.
Thus,
you can only find an action
in a monothetical piece of working practice.
For lack of those kinds
of polythetic description
of the process of becoming,
I now turn
to the monothetic built up experience
in the form of receipt knowledge ("Rezeptwissen")
in Schütz's term (Schütz and Luckmann 1979/1984).

Captioning

Captioning has its own speed
which is always slower than spoken words
When I (remember that I am Alex in the situation at stake) speak,
I always try to accommodate
to its speed.
I feel uncomfortable
to hear my colleagues, usually Kai or Luca, say:
"slower please."
I divide a sentence
into little pieces
like an actor.

Sometimes I can hear
the sound of the keyboard
which Luca uses.
In this case
I can synchronize
my speaking to the sounds
s/he makes with the keyboard.
In other cases
when I don't hear the keyboard
I rely on my inner sense
of the flow of time.
It is difficult for me
to understand one colleague (Leslie).
Generally, Kai understands those utterances quite well,
Luca nearly the same.
The others rely on the captioning
but I cannot hear the captioning,
because the screenreader does not read it.
Often in consequence
there was a response to Leslie's utterance
before I knew
what Leslie has said.
A first solution to this problem was
that Kai voiced this utterance for me.
But that created a lot more of speaking.
To solve this problem
we decided in June
that Kim
should type this utterance
in a summarized form
via the chat.

Interpretation of the Situation

In the situation at stake
two interpretations of my actions
are possible:
I misunderstood Kai's utterance
as an "own utterance"

- *Speaking on one's own behalf,*
  - *Where the word "I" in this example signifies Kai –*
  - *and not as voicing*
- *Reading out loud the utterance of another person,*
  - *Where "I" in this example signifies Leslie.*

But could I really have misunderstood
the different sound of voice
that exists
between "own utterance" and voicing?
Having listened to it again yesterday,
I understood it at once as voicing.
Second interpretation is:
I wanted to reaffirm
this newly found solution (chat-typing)
against the old solution
which contained in voicing.
But I misinterpreted
the aim of the voicing
which for all others
was to clear the meaning
of the utterance from Leslie
for the captioning.
This
because I did not monitor
the question mark in the captioning.
Now you may ask: "Why all these complicated communicative interactions?"
To answer this, first of all, I have to say that captioning is not accessible for me, the blind user, via the shared screen in Zoom. And we have not found any technical hack as a solution to this problem.
Second, and as a consequence of this, we had to work around this lack of accessibility in various communicative ways.

Workarounds as Design Resources

These observations demonstrate the potential of digital technologies to both enable and prevent access and participation in society. We argue that inclusion/exclusion are deeply interrelated and connected to the ways technologies are being designed. Consequently, "access is a variable relationship between numerous material, and cultural, social factors" (Ellcessor 2016, 12). This analysis raises at least two important issues: On the one hand, it sets a focus on the concept of design as empowerment. On the other, it sheds light on the aspect of participation in the design process itself. Constantly driven by the question of how conclusions can be drawn, like for instance in this case, conclusions from an everyday video conference situation, which may lead to a set of design guidelines, in which the full range of possible or impossible types of use can be addressed. The question is how to move from the observation of such improvised "error handling" (Schabacher 2017, 14) and workarounds to the mode of solution-oriented development that does justice to the diversity of the different modes and contexts of use and thus also to the diversity of non-/users such as "non-preferred or disabled user" (Ellis, Kao, and Bitman 2020). Understanding workarounds might be a first step.

Questioning the access work related to digital technologies is closely linked to the concept of Non-Intentional Design (NID), which was introduced by Uta Brandes and Michael Erlhoff into design theory and research (Brandes and Erlhoff 2006) and refers to the conversion or repurposing (Schüttpelz 2006) of things ("*Zweckentfremdung der Dinge*") contrary to their original design intention. Function or meaning of designed objects and software architectures such as video conferencing platforms can therefore shift, a process that Brandes describes as the "production of things through use" (Brandes, Stich, and Wender 2009, 10). Take, for example, the party project "Remote Access" of the Critical Design Lab. It drew on the platform Zoom and aimed to translate disability party culture to the virtual realm. Standard operations were shifted by using "remote access as a method for organizing pleasure and kinship" in March 2020. The initiative produced Zoom in a situated setting that strived for inclusion of users with bodily and cognitive variability: "The participation guide for the party took as given that Zoom is not an ideal platform ... So some partygoers volunteered to be access doulas—to troubleshoot and accompany—alongside captioners and sound and video describers. People dropped in and out of the party, danced, tried on clothes, lay in bed, kept their video off, chatted, were alone on camera or not" (Ginsburg, Mills, and Rapp 2020). Our two examples mentioned above point in a similar direction: they also reframe standard communicative settings enacted through video conferencing while not transforming free-time online activities, but professional practices relevant for the daily processes of higher education.

Eventually, design is realized in the process of use meaning that the user has a far less passive role than is often assumed. From a design (research) perspective, practices of tinkering (see Hamraie 2018), DIY practices, everyday hacks, and workarounds (see Schabacher 2017) are important resources of information. The ability to find a "knack" (German: *Kniff*) in dealing with technical objects or infrastructures that is able to "grasp the zigzag of technical beings" (Latour 2014, 312, cited in Löffler 2017, 139) allows us to understand what is really missing. At this point, we should once again refer to the "Crip Technoscience Manifesto," in which

Hamraie and Fritsch put maker and design practice at the center and emphasize how much disabled people are experts and designers of everyday life (2019). Design principles extend across at least two dimensions: on the one hand, to the inclusive process itself, and on the other, to resulting process outcomes that are inclusive in their design or manageability (Bieling 2019). However, in many research and development processes, a strong underrepresentation of disabled people can be observed (Hendren 2020). In the course of what the sociologist Madeleine Akrich has described as the "I-methodology," researchers tend to use their own expectations of certain products or circumstances as a frame of reference for design decisions or research tendencies (Akrich 1995). There is a risk of not considering, penetrating, or even registering the claims and needs of "other" groups of people (Bessing and Lukoschat 2013, 29).

Considering the diverse situations of video conferencing analyzed above, we aim to further problematize design issues with regard to such platforms (Ang et al. 2022) that enable but also disable remote communicative practices.[3] Our observations and discussions around the relations of sensory and bodily differences and digital technologies led us to think about, for example, how screen readers read web pages and that their path to the appropriate button is often very long in time sequence. With regard to blind or visually impaired people, there is also often no highlighting of single buttons or elements via color or size of letters, only a strict rule of succession. In the case of video conferencing, the technical construction of screen sharing often turns out to be inaccessible for a blind user. Workarounds are essential for all these synchronization problems in socio-technological settings. However, it is quite conceivable that a technical improvement drawing on processes of codesign in screen reader logic, web page design, and the video conferencing platforms could optimize the initial conditions for workarounds here. Screen readers could have additional functional logic in spatial terms or in highlighting certain elements. Web pages could be structured more intuitively for people with visual disabilities. Both optimizations would need to be explored in a participatory, contextualized, codesigned setting.

Outlook

Understanding media and disability as mutually constitutive and as being enacted in sensory media practices is necessary to ethnographically explore and conceptualize new applications such as video conference systems. These platforms had a massive impact during the SARS-CoV-2 pandemic, which also extends to higher edu-

---

3 We strive to further develop our interdisciplinary approach and construct a full research project to thoroughly tackle these questions.

cation and its possibilities to continue teaching and research under difficult and constrained conditions. Yet their implications for heterogeneous populations and user groups with disabilities remain to be fully explored. The analysis of our empirical material demonstrates some of the enabling potentials but mainly points to the uneven character of such "emancipatory devices" (Houston, Gabrys, and Pritchard 2019, 845) that promise a technological "fix" to complex social and political issues. Creating more inclusive environments for remote collaboration in the future depends on how cocreation and remote digital prototyping will involve programmers, designers, and humanities scholars with varying bodily and sensory abilities.

By assembling people with varying abilities and disabilities, their social relations, senses, and (assistive) devices around a data stream, the remote collaboration situations under analysis transgress a notion of accessibility limited to a technological dimension. Rather, by drawing on recent scholarship about crip technoscience and tinkering practices, we argue that access is an ongoing collaborative effort involving access knowledge and access work. Hence, access work is to be understood as a situated assemblage engaging various human and nonhuman actors. These collaborative processes are composed of situated workarounds, individual hacks, and improvisations that lay bare the creative potential of bringing together people with varying dis/abilities, sensory practices, experiences, and digital media. Whether and how such platforms and their users might enact inclusive collaboration processes in just ways remains to be explored in future research.

Acknowledgments

We would like to thank the research students Edda Payar (Konstanz) and Celine Keuer (Berlin) who supported us in finalizing the manuscript of this contribution.

References

Akrich, Madelaine. 1995. "User Representations. Practices, Methods and Sociology." In *Managing Technology in Society: The Approach of Constructive Technology Assessment*, edited by Arie Rip, Thomas J. Misa, and Johan Schot, 167–84. London.

Ang, Jazz R. X., Ping Liu, Emma McDonnell, and Sarah Coppola. 2022. "'In This Online Environment, We're Limited': Exploring Inclusive Video Conferencing Design for Signers." *CHI Conference on Human Factors in Computing Systems*, edited by Simone Barbosa, 1–16. ACM Digital Library. New York: Association for Computing Machinery.

Bessing, Nina, and Helga Lukoschat, eds. 2013. *Innovation durch Perspektivenvielfalt: Impulse für die industrielle Praxis aus der Gender- und Diversity-Forschung*. Leverkusen: Budrich.

Bieling, Tom. 2019. *Inklusion als Entwurf. Teilhabeorientierte Forschung über, für und durch Design*. Board of International Research in Design. Basel: Birkhäuser; De-Gruyter.

Brandes, Uta, and Michael Erlhoff. 2006. *Non Intentional Design*. Köln: DAAB.

Brandes, Uta, Sonja Stich, and Miriam Wender. 2009. *Design by Use: The Everyday Metamorphosis of Things*. Board of International Research in Design. Basel: Birkhäuser.

Dolmage, Jay. 2017. *Academic Ableism: Disability and Higher Education*. Ann Arbor: University of Michigan Press.

Ellcessor, Elizabeth. 2016. *Restricted Access: Media, Disability, and the Politics of Participation*. New York: New York University Press.

Ellcessor, Elizabeth. 2021. "Three Vignettes in Pursuit of Accessible Pandemic Teaching." *Communication, Culture and Critique* 14 (2): 324–27. https://doi.org/10.1093/ccc/tcab010.

Ellis, Carolyn, Tony E. Adams, and Arthur P. Bochner. 2010. "Autoethnography: An Overview." *Forum Qualitative Sozialforschung* 12 (1). https://doi.org/10.17169/fqs-12.1.1589.

Ellis, Katie, Kai-Ti Kao, and Tim Pitman. 2020. "The Pandemic Preferred User." *FC* 17 (2). https://fastcapitalism.journal.library.uta.edu/index.php/fastcapitalism/article/view/390.

Garland-Thomson, Rosemarie. 2020. "Ensuring Health Equity for Persons with Disabilities." *Radcliffe Institute for Advanced Studies. Harvard University*. https://www.radcliffe.harvard.edu/video/ensuring-health-equity-persons-disabilities.

Ginsburg Faye, Mara Mills, and Rayna Rapp. 2020. "Lessons from the Disability Front: Remote Access, Conviviality, Justice." In *ConTactos*, edited by Diana Taylor and Marcial Godoy-Anativia. Hemipress. https://contactos.tome.press/lessons-from-the-disability-front.

Goggin, Gerard, and Christopher Newell. 2003. *Digital Disability the Social Construction of Disability in New Media*. Lanham, MD: Rowman & Littlefield.

Hamraie, Aimi. 2018. "Mapping Access: Digital Humanities, Disability Justice, and Sociospatial Practice." *American Quarterly* 70 (3): 455–82. https://doi.org/10.1353/aq.2018.0031.

Hamraie, Aimi, and Kelly Fritsch. 2019. "Crip Technoscience Manifesto." *Catalyst* 5 (1) : 1–33. https ://doi.org/10.28968/cftt.v5i1.29607.

Hendren, Sara. 2020. *What Can a Body Do? How We Meet the Built World*. New York: Riverhead Books.

Hickman, Louise. 2019. "Transcription Work and the Practices of Crip Technoscience." *Catalyst* 5 (1): 1–10. https://doi.org/10.28968/cftt.v5i1.32081.

Houston, Lara, Jennifer Gabrys, and Helen Pritchard. 2019. "Breakdown in the Smart City: Exploring Workarounds with Urban-Sensing Practices and Technologies." *Science, Technology, & Human Values* 44 (5): 843–70. https://doi.org/10.1177/0162243919852677.

Husserl, Edmund. 1954. *Die Krisis der europäischen Wissenschaften und die transzendentale Phänomenologie. Eine Einleitung in die phänomenologische Philosophie*. Den Haag: Nijhoff.

Kent, Mike. 2020. "Social Media and Disability: It's Complicated." In *The Routledge Companion to Disability and Media*, edited by Katie Ellis, Gerard Goggin, Beth Haller, and Rosemary Curtis, 264–74. London: Routledge.

Larkin, Brian. 2004. "Degraded Images, Distorted Sounds: Nigerian Video and the Infrastructure of Piracy." *Public Culture* 16 (2): 289–314. https://doi.org/10.1215/08992363-16-2-289.

Latour, Bruno. 2014. *Existenzweisen. Eine Anthropologie der Modernen*. trans. Gustav Roßler. Berlin: Suhrkamp.

Lay-Flurrie, Jenny. 2020. "What the Disability Community Can Teach Us about Working Remotely." *Microsoft Accessibility Blog*, April 1, 2020. https://blogs.microsoft.com/accessibility/inclusive-remote-working.

Löffler, Petra. 2017. "Zick-Zack. Bruno Latours Umwege." *ilinx—Berliner Beiträge zur Kulturwissenschaft* 4: 137–43.

Melzer, Conny, Karolin Schäfer, Thomas Kaul, and Pamela Perniss. 2020. "Hinweise zum Nutzen von Zoom in Seminaren: Zusatzinformationen bei Teilnehmenden mit Sinnesbeeinträchtigung." https://www.hf.uni-koeln.de/data/main/File/2020-04-21-Nutzungshinweise-Zoom-Zusatzinfo-bei_Hoerschaedigung.pdf.

Mills, William R., S. Sender, J. Lichtefeld, N. Romano, K. Reynolds, M. Price et al. 2020. "Supporting Individuals with Intellectual and Developmental Disability during the First 100 Days of the COVID-19 Outbreak in the USA." *Journal of Intellectual Disability Research* 64 (7): 489–96. https://doi.org/10.1111/jir.12740.

Morgan, Paula. 2021. "How the Labor Market has Changed for People with Disabilities." *Forbes*, 27 December. https://www.forbes.com/sites/paulamorgan/2021/12/27/how-the-labor-market-has-changed-for-people-with-disabilities.

Raymond, Eric S., ed. 1991. *The New Hacker's Dictionary*. Cambridge, MA: MIT Press.

Schabacher, Gabriele. 2017. "Im Zwischenraum der Lösungen. Reparaturarbeit und Workarounds." *ilinx—Berliner Beiträge zur Kulturwissenschaft* (4): xiii–3. https://doi.org/10.18452/18683.

Schüttpelz, Erhard. 2006. "Die medienanthropologische Kehre der Kulturtechniken." In *Archiv für Mediengeschichte*. Kulturgeschichte als Mediengeschichte (oder vice versa?), 87–110.

Schütz, Alfred, and Thomas Luckmann. 1979/1984. *Strukturen der Lebenswelt*. Vol. 1 (1979); Vol. 2 (1984). Frankfurt a. M.: Suhrkamp.

Schur, Lisa A., Mason Ameri, and Douglas Kruse. 2020. "Telework after COVID: A 'Silver Lining' for Workers with Disabilities?" *Journal of Occupational Rehabilitation* 30 (4): 521–36. https://doi.org//10.1007/s10926-020-09936-5.

Shew, Ashley. 2020. "Let COVID-19 Expand Awareness of Disability Tech." *Nature* 581: 9. https://doi.org/10.1038/d41586-020-01312-w.

Shew, Ashley. 2020 "Ableism, Technoableism, and Future AI." IEEE *Technology and Society Magazine*, 39 (1): 40–85. https://doi.org/10.1109/MTS.2020.2967492.

Shields, Rob, Michael Schillmeier, Justine Lloyd, and Joost van Loon. 2020. "6 Feet Apart: Spaces and Cultures of Quarantine." *Space and Culture* 23 (3): 216–20. https://doi.org/10.1177/1206331220938622.

Spinney, Laura. 2022. "Pandemics Disable People—The History Lesson That Policymakers Ignore." *Nature* 602 (7897): 383–85. https://doi.org/10.1038/d41586-022-00414-x.

Star, Susan Leigh, and Karen Ruhleder. 1996. "Steps toward an Ecology of Infrastructure: Design and Access for Large Information Spaces." *Information Systems Research* 7 (1): 111–34.

Star, Susan Leigh, and Geoffrey Bowker. 2009. "How to Infrastructure." In *The Handbook of New Media: Social Shaping and Social Consequences of ICTs*, edited by Leah A. Lievrouw and Sonia Livingstone, 230–45. London: Sage.

Authors

Kim Albrecht visualizes cultural, technological, and scientific forms of knowledge. His diagrams unfold and question the structures of representation and explore the aesthetics of technology and society. Kim is a professor at the Film University Babelsberg Konrad Wolf, principal at metaLAB (at) Harvard, director of metaLAB (at) FU Berlin, and an affiliate of the Berkman Klein Center for Internet & Society at Harvard. Kim holds a PhD from the University of Potsdam in media theory and has exhibited at Harvard Art Museums, Four Domes Pavilion Wrocław, Ars Electronica Center, Cooper Hewitt, Cube design museum, ZKM Center for Art and Media Karlsruhe, Kaestner Gesellschaft, The Wrong Biennial, Istanbul Contemporary Art Museum, and Kunsthaus Graz, among other venues.

Christian Ulrik Andersen publishes on digital aesthetics and interface criticism and is a regular speaker and panel organizer at various media art festivals, events, and conferences. He organizes yearly research workshops in collaboration with transmediale festival, Berlin, and is coeditor of *A Peer-Reviewed Journal About*, an open-access journal that addresses the ever-shifting themes of art and digital culture. Currently, he is research associate at Centre for the Study of the Networked Image, London South Bank University, associate professor at the Department of Digital Design and Information Studies, Aarhus University, and director of Digital Aesthetics Research Center (DARC).

Alexandra Anikina is a media theorist and visual artist working imaginaries of technology and AI, feminist STS, affective infrastructures and technological conditions of knowledge production, governance, labor, and affect. She is a senior lecturer in media practices at Winchester School of Arts (University of Southampton) and codirector of Critical Infrastructure and Image Politics research group, and cocurator of media art festival IMPAKT 2018 Algorithmic Superstructures. Currently she is working on a monograph on procedural images as well as on the themes of techno-animism and postsocialist decolonising.

Maha Bali is associate professor of practice at the Center for Learning and Teaching at the American University in Cairo. She is cofounder of virtuallyconnecting.org (a grassroots movement that challenges academic gatekeeping at conferences) and cofacilitator of Equity Unbound (an equity-focused, open, connected intercultural learning curriculum, which has also branched into academic community activities *Continuity with Care*, *Socially Just Academia*, and a collaboration with OneHE: *Community-Building Resources*). She writes and speaks frequently about social justice, critical pedagogy, and open and online education. She blogs regularly at http://blog.mahabali.me and tweets @bali_maha.

Tilman Baumgärtel is a writer. He lives in Berlin and teaches media studies at Hochschule Mainz. Previously he was a professor at the University of the Philippines in Manila (2005–2009) and in the Department of Media and Communication at the Royal University of Phnom Penh (2009–2012). He has written and edited more than a dozen books on various aspect of media culture, including net art and net culture, Asian cinema, and the director Harun Farocki. He has served on the juries of the Warsaw Film Festival, the Pacific Meridian International Film Festival in Vladivostok, Transmediale in Berlin, among others.

Tom Bieling is professor for design theory at HfG Offenbach. Previously professor for design research and theory at Zentrum für Designforschung (HAW Hamburg), visiting professor in Trento, Cairo, and Hildesheim, research associate at Berlin University of the Arts (2010–19), and TU Berlin (2007–10). Chief editor at Designforschung.org, coeditor of the book series Design Meanings (Mimesis), and the BIRD Series (Birkhäuser, Board of International Research in Design). Cofounder of the Design Research Network. Falling Walls Young Innovator of the Year. Numerous awards and exhibitions worldwide. Recent books: *Inklusion als Entwurf* (2019), *Design (&) Activism* (2019), and *Gender (&) Design* (2020). Further information: www.tombieling.com.

Donatella Della Ratta is a writer, performer, and curator specializing in digital media and networked technologies, with a focus on the Arab world. She is associate professor of communications and media studies at John Cabot University, Rome. Donatella holds a PhD from the University of Copenhagen and is former affiliate of the Berkman Klein Center for Internet & Society at Harvard University. She managed the Arabic-speaking community for the international organization Creative Commons from 2007 until 2013. *Shooting a Revolution: Visual Media and Warfare in Syria* (Pluto Press, 2018) is her latest monograph. With Geert Lovink, Teresa Numerico, and Peter Sarram, she is coeditor of *The Aesthetics and Politics of the Online Self: A Savage Journey into the Heart of Digital Cultures* (Palgrave, 2022), for which she has authored the chapter "Reflecting on the Online Self through the Looking-Glass: From Autoethnography to Empathic Criticism."

Philipp Deny is a PhD candidate in the Media | Transformation Department at the Leibniz Institute for Educational Media and a member of the junior research group in the Leibniz-Science Campus—Postdigital Participation—Braunschweig. He studied media studies at the Braunschweig University of Art, where he received his master's degree with a thesis on performances of authenticity within contemporary club cultures. His dissertation is concerned with media practices and methods of media reflection in the context of school education.

Jan Distelmeyer is professor of media history and media theory in the European media studies program of the Potsdam University of Applied Sciences and the University of Potsdam. His current research focuses on the relationship between mediality and digitality with a special interest in interface processes as well as questions of automation and autonomy. Recent publications include: *Kritik der Digitalität* (Springer VS, 2021; English translation: *Critique of Digitality*, Palgrave Macmillan, 2022) and *Algorithmen & Zeichen: Beiträge von Frieder Nake zur Gegenwart des Computers* (Kadmos, 2022, together with Sophie Ehrmanntraut & Boris Müller). Further information: http://distelmeyer.emw-potsdam.de.

Frithjof Esch is a neurodivergent staff member of the endowed chair for disability studies and participation research (Prof. Dr. Siegfried Saerberg) at the Evangelische Hochschule für Soziale Arbeit & Diakonie, Stiftung Das Rauhe Haus in Hamburg. He studied social economics and sociology at the University of Hamburg. He is particularly interested in researching social atmospheres and the phenomenology of disability in the context of digitality.

Winfried Gerling is professor of concept and aesthetics of new media in the European media studies program of the Potsdam University of Applied Sciences and the University of Potsdam. His research focuses on the practical and theoretical reflection of photographic media, digital aesthetics, media environments, and media art. His recent books include *Versatile Camcorder: Looking at the GoPro Movement* (Berlin: Kadmos, 2021; edited with Florian Krautkrämer) and *Bilder Verteilen: Fotografische Praktiken in der digitalen Kultur* (Bielefeld: transcript, 2018; coauthored with Susanne Holschbach and Petra Löffler). Further information: http://gerling.emw-potsdam.de.

Will Houstoun is performer in residence at the Imperial College/Royal College of Music Center for Performance Science (London), holds a literary fellowship from the Academy of Magic Arts (Los Angeles), and is a past winner of the European

Magic Championships. His PhD explored the history of magical education, and his work as a consultant includes contributions to productions such as Martin Scorsese's *Hugo*, the BBC's *Wolf Hall*, and the West End's *The Twilight Zone* and *The Prince of Egypt*. Will has also worked with organizations like the United Nations Development Programme, KPMG, and the World Economic Forum, and he edits the Magic Circle's 115-year-old periodical, *The Magic Circular*.

Martina Leeker teaches in aesthetic theory and practice at the Department of Art and Music, University of Cologne. She had been a guest professor at the Digital Cultures Research Lab (DCRL), Leuphana University Lüneburg. Her research interests include digital cultures, art and technology, critique in digital cultures, theater and digitality, artistic research. She undertakes art-based research in performing lectures with artificial figures. She leads the project The Respectful Nettheatrechannel on theater and digitality.

Geert Lovink is a Dutch media theorist, internet critic, and author of *Uncanny Networks* (2002), *Dark Fiber* (2002), *My First Recession* (2003), *Zero Comments* (2007), *Networks Without a Cause* (2012), *Social Media Abyss* (2016), *Organisation after Social Media* (with Ned Rossiter, 2018), *Sad by Design* (2019) and *Stuck on the Platform* (2022). He is professor of art and network cultures at the Amsterdam University of Applied Science (HvA), where he founded the Institute of Network Cultures.

Irina Kaldrack is lecturer in society and digitalization at Bauhaus University Weimar. She was full professor in media studies at Hunan Normal University, Changsha, China (2022) and visting professor of cultures of knowledge in the digital age at the Braunschweig University of Fine Arts (2015–2021). Her research interests include the theory and history of digital media cultures in Europe and China, the design and designability of digital cultures in the context of social transformations and sustainability, and methods of media studies at the intersection of design processes, artistic practice, and media studies research. She is the author of *Imaginierte Wirksamkeit: Zwischen Performance und Bewegungserkennung* (2011). She has coedited eight books, including *Preferable Futures: Trans-formation Design in Digital Cultures* (2022) and *Throwing Gestures* (2021), as well as issue 21 of *Zeitschrift für Medienwisseschaft*, on artificial intelligence (2019).

Kalani Michell is assistant professor of European languages and transcultural studies at the University of California, Los Angeles. She has published on a variety of film and media topics, such as experiments with ink in 16 mm film (in *Re-Animationen*), the circulation of the *Mona Lisa* in pornographic set designs (in *CineAction*), and a computer game that restages waiting for a performance by Marina Abramović (in *kultuRRevolution*). Her recent publications are about comics, sound studies, and image captions: the comics storyboard in Christian Petzold's filmic shot composition (in *Storyboarding*), the emergence of academic podcasts in film and media studies (in *Format Matters*), and scholarly writing that performs a critique of seemingly neutral text-image relationships (in *New Perspectives on Academic Writing*).

Olga Moskatova is professor for media theory at the University of Art and Design Offenbach am Main (Germany). Her main research fields include theory and aesthetics of visual media, materiality of media, immunitary media dispositifs, and networked images. Selected publications: *Images on the Move: Materiality—Networks—Formats* (2021); *Networked Images in Surveillance Capitalism* (special issue of *Digital Culture & Society* 2021, co-edited with Anna Polze and Ramon Reichert); *Male am Zelluloid. Zum relationalen Materialismus im kameralosen Film* (2019).

Laura Katharina Mücke has been a research and teaching associate at the University of Mainz (Germany) since 2022 and a doctoral researcher in film theory in the Department of Theatre, Film and Media Studies at the University of Vienna (Austria) since 2019. Her PhD project is entitled "Anti | Immersion? Toward Empowering/Enfeebling Structures of an All-Inclusive Concept." Mücke's articles have been published in, for instance, *Ambiances: International Journal of Sensory Environment, Architecture and Urban Space*; she also coedited the 2022 special issue on *Messy Images* in the German film journal *montage AV* and cotranslated Roger Odin's book *Les espaces de communication: Introduction à la semiopragmatique* from French.

Beate Ochsner, professor of media studies (University of Konstanz), spokeswoman of the research unit Media and Participation, where she heads the project "Technosensory Processes of Participation." Research interest: audiovisual productions of disability, participation/Teilhabe, gaming and (dis)ability. Coeditor of *Affizierungs- und Teilhabeprozesse zwischen Organismen und Maschinen* (Springer, 2020, with S. Nikolow and R. Stock). Recent publications: "Kulturwissenschaftliche Disability Studies," *Handbuch Disability Studies* (2022, 201–219); "Die Zukunft des smarten Hörens hat begonnen," *Physiognomien des Lebens* (2020); and "Documenting Neuropolitics," *Documentary and Disability* (2017). OrcidID: 0000-0002-6041-9968.

Søren Bro Pold is an associate professor at Aarhus University. He has published on digital media aesthetics and the interface in its different forms, such as electronic literature, net art, software art, creative software, urban and mobile interfaces, activism, critical design, and digital culture. His main research field is interface criticism, which discusses the role and the development of the interface for art, literature, aesthetics, culture, and IT. Together with Christian Ulrik Andersen, he authored *The Metainterface: The Art of Platforms, Cities and Clouds* (2018). He was cochair of the ELO 2021 conference Platform (Post?) Pandemic (2021) and chair of the Dariah

EU project Electronic Literature (e-lit) and Covid-19. Further information: http://pure.au.dk/portal/en/pold@cavi.au.dk.

Robert Rapoport's work focuses on the confluence of image sequencing and automation. He received his PhD (DPhil) from the University of Oxford in 2016 with a dissertation titled "The Iterative Frame: Algorithmic Video Editing, Participant Observation, and the Black Box," which brought visual anthropology into dialogue with procedural approaches to video editing. His postdoctoral work was at the Centre for Digital Cultures at Leuphana University Lüneburg. Since 2017 he has lectured at Leuphana in the Digital Media BA, developing a curriculum to teach theory alongside critical media production. He has been a visiting scientist at the Institute of Culture and Aesthetics of Digital Media (ICAM) at Leuphana University and a guest lecturer at the University of St Gallen. Recent talks include "End User Narrative," a keynote at Stiftung Niedersachen's workshop KI als Werkzeug in Film und Medien. Further information: www.iterativeframe.com.

Katharina Rein was awarded a PhD from the Humboldt University in Berlin for her prize-winning dissertation on the media and cultural history of stage magic in the late nineteenth century. Currently a lecturer at the University of Potsdam, she previously worked at the Bauhaus University Weimar and lectured at the Free University Berlin. She was a member of the international research project Les Arts Trompeurs: Machines, Magie, Médias (2015–2018). Rein is the author of *Techniques of Illusion* (2023) *Gothic Cinema* (2023), and the editor of *Illusions in Cultural Practice* (2021) and *Magic: A Companion* (2022). Her academic essays have appeared in four languages.

Robert Stock is assistant professor for cultures of knowledge at the Institute for Cultural Theory and History at Humboldt-Universität zu Berlin. His main research interests are digital media and dis/ability, cultural animal studies, and postcolonial memory politics. He is a cofounder of the research network Dis-/Abilities and Digital Media (2020–2023). A recent publication is "Broken Elevators, Temporalities of Breakdown and Open Data: How Wheelchair Mobility, Social Media Activism and Situated Knowledge Negotiate Public Transport Systems," *Mobilities* (2022). OrcidID: https://orcid.org/0000-0002-2256-0928.

Siegfried Saerberg is professor for disability Studies and Teilhabeforschung at Evangelische Hochschule für Soziale Arbeit & Diakonie, Stiftung Das Rauhe Haus and director of ZeDiSplus (Zentrum für Disability Studies und Teilhabeforschung) in Hamburg, Germany. Siegfried specializes in sensory ethnography, auto-ethnographic approaches, phenomenology, and disability studies. He is author of "The Sensorification of the Invisible Science, Blindness and the Life-World," *Science,*

Technology & Innovation Studies (2011); and *"Geradeaus ist einfach immer geradeaus": Eine lebensweltliche Ethnographie blinder Raumorientierung* (UVK 2006).

Vera Tollmann is a visiting researcher and lecturer at the Institute for Culture and Aesthetics of Digital Media (ICAM) at Leuphana University Lüneburg. She completed her doctorate at the Hochschule für Bildende Künste Hamburg in 2020. A revised version titled *Sicht von oben: "Powers of Ten" und Bildpolitiken der Vertikalität* was published by Spector Books in 2023. She was a research associate at Universität der Künste Berlin from 2015 to 2017 and at the Institute for Media, Theatre and Popular Culture, University of Hildesheim from 2019 to 2021. She is coeditor of *Proxy Politics* (Archive Books 2017), and her writing includes an essay on the history of video telephony (*Cabinet*, September 2020) and "Proxies" (*Uncertain Archives*, MIT Press 2021, with Wendy Hui Kyong Chun and Boaz Levin). She was cocurator of the exhibition *Sensing Scale* (Kunsthalle Münster, 2021). Her research interests include the politics of visibility, feminist critique of technology, remote sensing practices, media art, and AI in digital cultures.

Axel Volmar is a currently a guest professor at the Institute of Music and Media at the Humboldt-University in Berlin. Previously, he was a research associate in the research initiative Transforming Infrastructure" at the University of Konstanz (2022–2023) and in the Collaborative Research Center "Media of Cooperation" at the University of Siegen (2016–2022). From 2014 to 2016, he was a postdoctoral fellow of the Andrew W. Mellon Foundation in the Department of Art History and Communication Studies at McGill University. His research is in the areas of media history, media theory, and the praxeology of media, intersecting with the history of science, infrastructure studies, format studies, and disability studies. Volmar is author of *Klang-Experimente: Die Auditive Kultur der Naturwissenschaften 1761–1961* (2015) and coeditor of various edited volumes, including *Format Matters: Standards, Practices, and Politics in Media Cultures* (2020), *Media Infrastructures and the Politics of Digital Time: Essays on Hardwired Temporalities* (2021), *Interrogating Datafication: Towards a Praxeology of Data* (2022), and *Rethinking Infrastructure Across the Humanities* (2023).

Andreas Weich is a postdoctoral research fellow in the Media | Transformation Department at the Leibniz Institute for Educational Media where he is the head of the junior research group in the Leibniz-Science Campus—Postdigital Participation—Braunschweig. He previously worked at TU Braunschweig in the field of media education and directed the coordination office for media studies at the Braunschweig University of Art. He received his PhD in media studies from Paderborn University with a thesis on the genealogy and mediality of profiling technologies and practices. His research focusses on media constellations, media theory, educational theory, (post)digital media cultures, media education, and media literacy.

Axel Volmar is currently a guest professor at the Institute for Music and Media at Humboldt-Universität zu Berlin. His research is on media history, media theory, and the praxeology of media, intersecting with the history of science, infrastructure studies, and disability studies.

Olga Moskatova is a professor for media theory at University of Art and Design Offenbach am Main. Her fields of research include theory and aesthetics of visual media, materiality of media, networked images and media of immunization.

Jan Distelmeyer is a professor of media history and media theory in the European Media Studies program of Fachhochschule Potsdam and Universität Potsdam. His current research focuses on the relationship between mediality and digitality with a special interest in interface processes as well as questions of automation and autonomy.